# Modeling Magnetospheric Plasma Processes

## Geophysical Monograph Series

Including

**IUGG Volumes**
**Maurice Ewing Volumes**
**Mineral Physics Volumes**

# GEOPHYSICAL MONOGRAPH SERIES

**Geophysical Monograph Volumes**

1 **Antarctica in the International Geophysical Year** *A. P. Crary, L. M. Gould, E. O. Hulburt, Hugh Odishaw, and Waldo E. Smith (Eds.)*
2 **Geophysics and the IGY** *Hugh Odishaw and Stanley Ruttenberg (Eds.)*
3 **Atmospheric Chemistry of Chlorine and Sulfur Compounds** *James P. Lodge, Jr. (Ed.)*
4 **Contemporary Geodesy** *Charles A. Whitten and Kenneth H. Drummond (Eds.)*
5 **Physics of Precipitation** *Helmut Weickmann (Ed.)*
6 **The Crust of the Pacific Basin** *Gordon A. Macdonald and Hisahi Kuno (Eds.)*
7 **Antarctica Research: The Matthew Fontaine Maury Memorial Symposium** *H. Wexler, M. J. Rubin, and J. E. Caskey, Jr. (Eds.)*
8 **Terrestrial Heat Flow** *William H. K. Lee (Ed.)*
9 **Gravity Anomalies: Unsurveyed Areas** *Hyman Orlin (Ed.)*
10 **The Earth Beneath the Continents: A Volume of Geophysical Studies in Honor of Merle A. Tuve** *John S. Steinhart and T. Jefferson Smith (Eds.)*
11 **Isotope Techniques in the Hydrologic Cycle** *Glenn E. Stout (Ed.)*
12 **The Crust and Upper Mantle of the Pacific Area** *Leon Knopoff, Charles L. Drake, and Pembroke J. Hart (Eds.)*
13 **The Earth's Crust and Upper Mantle** *Pembroke J. Hart (Ed.)*
14 **The Structure and Physical Properties of the Earth's Crust** *John G. Heacock (Ed.)*
15 **The Use of Artificial Satellites for Geodesy** *Soren W. Henricksen, Armando Mancini, and Bernard H. Chovitz (Eds.)*
16 **Flow and Fracture of Rocks** *H. C. Heard, I. Y. Borg, N. L. Carter, and C. B. Raleigh (Eds.)*
17 **Man-Made Lakes: Their Problems and Environmental Effects** *William C. Ackermann, Gilbert F. White, and E. B. Worthington (Eds.)*
18 **The Upper Atmosphere in Motion: A Selection of Papers With Annotation** *C. O. Hines and Colleagues*
19 **The Geophysics of the Pacific Ocean Basin and Its Margin: A Volume in Honor of George P. Woollard** *George H. Sutton, Murli H. Manghnani, and Ralph Moberly (Eds.)*
20 **The Earth's Crust: Its Nature and Physical Properties** *John G. Heacock (Ed.)*
21 **Quantitative Modeling of Magnetospheric Processes** *W. P. Olson (Ed.)*
22 **Derivation, Meaning, and Use of Geomagnetic Indices** *P. N. Mayaud*
23 **The Tectonic and Geologic Evolution of Southeast Asian Seas and Islands** *Dennis E. Hayes (Ed.)*
24 **Mechanical Behavior of Crustal Rocks: The Handin Volume** *N. L. Carter, M. Friedman, J. M. Logan, and D. W. Stearns (Eds.)*
25 **Physics of Auroral Arc Formation** *S.-I. Akasofu and J. R. Kan (Eds.)*
26 **Heterogeneous Atmospheric Chemistry** *David R. Schryer (Ed.)*
27 **The Tectonic and Geologic Evolution of Southeast Asian Seas and Islands: Part 2** *Dennis E. Hayes (Ed.)*
28 **Magnetospheric Currents** *Thomas A. Potemra (Ed.)*
29 **Climate Processes and Climate Sensitivity (Maurice Ewing Volume 5)** *James E. Hansen and Taro Takahashi (Eds.)*
30 **Magnetic Reconnection in Space and Laboratory Plasmas** *Edward W. Hones, Jr. (Ed.)*
31 **Point Defects in Minerals (Mineral Physics Volume 1)** *Robert N. Schock (Ed.)*
32 **The Carbon Cycle and Atmospheric $CO_2$: Natural Variations Archean to Present** *E. T. Sundquist and W. S. Broecker (Eds.)*
33 **Greenland Ice Core: Geophysics, Geochemistry, and the Environment** *C. C. Langway, Jr., H. Oeschger, and W. Dansgaard (Eds.).*
34 **Collisionless Shocks in the Heliosphere: A Tutorial Review** *Robert G. Stone and Bruce T. Tsurutani (Eds.)*
35 **Collisionless Shocks in the Heliosphere: Reviews of Current Research** *Bruce T. Tsurutani and Robert G. Stone (Eds.)*
36 **Mineral and Rock Deformation: Laboratory Studies—The Paterson Volume** *B. E. Hobbs and H. C. Heard (Eds.)*
37 **Earthquake Source Mechanics (Maurice Ewing Volume 6)** *Shamita Das, John Boatwright, and Christopher H. Scholz (Eds.)*

38 **Ion Acceleration in the Magnetosphere and Ionosphere** *Tom Chang (Ed.)*

39 **High Pressure Research in Mineral Physics (Mineral Physics Volume 2)** *Murli H. Manghnani and Yasuhiko Syono (Eds.)*

40 **Gondwana Six: Structure, Tectonics, and Geophysics** *Garry D. McKenzie (Ed.)*

41 **Gondwana Six: Stratigraphy, Sedimentology, and Paleontology** *Garry D. McKenzie (Ed.)*

42 **Flow and Transport Through Unsaturated Fractured Rock** *Daniel D. Evans and Thomas J. Nicholson (Eds.)*

43 **Seamounts, Islands, and Atolls** *Barbara H. Keating, Patricia Fryer, Rodey Batiza, and George W. Boehlert (Eds.)*

44 **Modeling Magnetospheric Plasma** *T. E. Moore and J. H. Waite, Jr. (Eds.)*

45 **Perovskite: A Structure of Great Interest to Geophysics and Materials Science** *Alexandra Navrotsky and Donald J. Weidner (Eds.)*

46 **Structure and Dynamics of Earth's Deep Interior (IUGG Volume 1)** *D. E. Smylie and Raymond Hide (Eds.)*

47 **Hydrological Regimes and Their Subsurface Thermal Effects (IUGG Volume 2)** *Alan E. Beck, Grant Garven, and Lajos Stegena (Eds.)*

48 **Origin and Evolution of Sedimentary Basins and Their Energy and Mineral Resources (IUGG Volume 3)** *Raymond A. Price (Ed.)*

49 **Slow Deformation and Transmission of Stress in the Earth (IUGG Volume 4)** *Steven C. Cohen and Petr Vaníček (Eds.)*

50 **Deep Structure and Past Kinematics of Accreted Terranes (IUGG Volume 5)** *John W. Hillhouse (Ed.)*

51 **Properties and Processes of Earth's Lower Crust (IUGG Volume 6)** *Robert F. Mereu, Stephan Mueller, and David M. Fountain (Eds.)*

52 **Understanding Climate Change (IUGG Volume 7)** *Andre L. Berger, Robert E. Dickinson, and J. Kidson (Eds.)*

53 **Plasma Waves and Istabilities at Comets and in Magnetospheres** *Bruce T. Tsurutani and Hiroshi Oya (Eds.)*

54 **Solar System Plasma Physics** *J. H. Waite, Jr., J. L. Burch, and R. L. Moore (Eds.)*

55 **Aspects of Climate Variability in the Pacific and Western Americas** *David H. Peterson (Ed.)*

56 **The Brittle-Ductile Transition in Rocks** *A. G. Duba, W. B. Durham, J. W. Handin, and H. F. Wang (Eds.)*

57 **Evolution of Mid Ocean Ridges (IUGG Volume 8)** *John M. Sinton (Ed.)*

58 **Physics of Magnetic Flux Ropes** *C. T. Russell, E. R. Priest, and L. C. Lee (Eds.)*

59 **Variations in Earth Rotation (IUGG Volume 6)** *Dennis D. McCarthy and William E. Carter (Eds.)*

60 **Quo Vadimus *Geophysics for the Next Generation* (IUGG Volume 10)** *George D. Garland and John R. Apel (Eds.)*

61 **Cometary Plasma Processes** *Alan D. Johnstone (Ed.)*

## Maurice Ewing Volumes

1 **Island Arcs, Deep Sea Trenches, and Back-Arc Basins** *Manik Talwani and Walter C. Pitman III (Eds.)*

2 **Deep Drilling Results in the Atlantic Ocean: Ocean Crust** *Manik Talwani, Christopher G. Harrison, and Dennis E. Hayes (Eds.)*

3 **Deep Drilling Results in the Atlantic Ocean: Continental Margins and Paleoenvironment** *Manik Talwani, William Hay, and William B. F. Ryan (Eds.)*

4 **Earthquake Prediction—An International Review** *David W. Simpson and Paul G. Richards (Eds.)*

5 **Climate Processes and Climate Sensitivity** *James E. Hansen and Taro Takahashi (Eds.)*

6 **Earthquake Source Mechanics** *Shamita Das, John Boatwright, and Christopher H. Scholz (Eds.)*

## IUGG Volumes

1 **Structure and Dynamics of Earth's Deep Interior** *D. E. Smylie and Raymond Hide (Eds.)*

2 **Hydrological Regimes and Their Subsurface Thermal Effects** *Alan E. Beck, Grant Garven, and Lajos Stegena (Eds.)*

3 **Origin and Evolution of Sedimentary Basins and Their Energy and Mineral Resources** *Raymond A. Price (Ed.)*

4 **Slow Deformation and Transmission of Stress in the Earth** *Steven C. Cohen and Petr Vaníček (Eds.)*

5 **Deep Structure and Past Kinematics of Accreted Terranes** *John W. Hillhouse (Ed.)*

6 **Properties and Processes of Earth's Lower Crust** *Robert F. Mereu, Stephan Mueller, and David M. Fountain (Eds.)*

7 **Understanding Climate Change** *Andre L. Berger, Robert E. Dickinson, and J. Kidson (Eds.)*

8 **Evolution of Mid Ocean Ridges** *John M. Sinton (Ed.)*

9 **Variations in Earth Rotation** *Dennis D. McCarthy and William E. Carter (Eds.)*

10 **Quo Vadimus *Geophysics for the Next Generation*** *George D. Garland and John R. Apel (Eds.)*

## Mineral Physics Volumes

1 **Point Defects in Minerals** *Robert N. Schock (Ed.)*

2 **High Pressure Research in Mineral Physics** *Murli H. Manghnani and Yasuhiko Syono (Eds.)*

Geophysical Monograph 62

# Modeling Magnetospheric Plasma Processes

**Gordon R. Wilson**
*Editor*

American Geophysical Union

Published under the aegis of the AGU Geophysical Monograph Board.

**Library of Congress Cataloging-in-Publication Data**

Modeling magnetospheric plasma processes / Gordon R. Wilson, editor.
p. cm. — (Geophysical monograph : 62)
Papers from the Second Huntsville Workshop on Magnetosphere/Ionosphere Plasma Models, held Oct. 11–13, 1989, at the Tom Bevill Center on the campus of the University of Alabama in Huntsville; sponsored by NASA/Marshall Space Flight Center, the Center for Space Plasma and Aeronomic Research (CSPAR) at the University of Alabama in Huntsville, and the UAH Division of Continuing Education.
ISBN 0-87590-761-X
1. Space plasmas—Mathematical models—Congresses.
2. Magnetosphere—Mathematical models—Congresses. I. Wilson, Gordon R. II. Huntsville Workshop on Manetosphere/Ionosphere Plasma Models (2nd : 1989 : University of Alabama in Huntsville) III. George Marshall Space Flight Center. IV. University of Alabama in Huntsville. Center for Space Plasma and Aeronomic Research. V. University of Alabama in Huntsville. Division of Continuing Education. VI. Series.
QC809.P5M64 1991 91-10779
CIP



Printed in the United States of America.

# CONTENTS

# CONTENTS

## THE INNER MAGNETOSPHERE/PLASMASPHERE

# PREFACE

The ultimate goal of modeling of the plasma in Earth's environment is an understanding of the magnetosphere and ionosphere as a coupled global system. To achieve this goal requires a coordinated effort between models applied to different spatial scales. The desire to model this system on a global scale is leading to models which encompass larger and larger regions. The ever-increasing availability of computing resources has allowed models to expand to 2 and 3 dimensions. At the other extreme are the micro-scale processes which transfer energy to individual particles within the global system. As more detailed observations become available the necessity for accurately including such processes in the global models becomes more apparent. Then it becomes a question of how to incorporate the necessary physical processes from all scale sizes into a model of a global system. It now seems clear that such multi-scale scenarios exist where micro-scale processes provide energy to the plasma which flows outward from Earth into the distant magnetotail before returning to the near-Earth regions. The challenge of incorporating all relevant processes into a model of this entire plasma path is a formidable one. The existence of separate models of the separate steps along this pathway leads directly to efforts to fuse models with different scales into a single, self-consistent treatment.

In an effort to address the current state of this fusion (or confusion, as the case may be) the Second Huntsville Workshop on Magnetosphere/Ionosphere Plasma Models was held October 11-13, 1989 at the Tom Bevill Center on the campus of The University of Alabama in Huntsville. The purpose of the workshop was to bring together researchers to discuss the important processes, occurring at various spatial scales, which must be included in a realistic global magnetosphere/ionosphere model. While the workshop was billed as a "plasma" model affair, during the organization it became clear that at the current level of sophistication of these models there was considerable interest in the latest models of magnetic field configurations and current systems. Thus there was significant attention given to these topics. The agenda was originally structured along boundaries (perhaps arbitrary ones) of spatial scale: micro-scale, meso-scale, and global-scale. This organized the presented material well. However, after reviewing the contributions to this manuscript we reverted to the standard division along regional boundaries.

The meeting was organized into three days with each morning and early afternoon devoted to invited presentations and the late afternoons reserved for contributed posters. The invited papers were selected by the program committee which was divided into three groups covering the three different spatial scales to be discussed. The committee members were:

J. J. Sojka, Utah State University
L. R. Lyons, The Aerospace Corporation
R. J. Walker, UCLA
T. I. Gombosi, The University of Michigan
R. J. Moffett, University of Sheffield
R. L. Lysak, The University of Minnesota
B. H. Mauk, Applied Physics Laboratory, JHU
H. Okuda, Princeton University
P. J. Palmadesso, Naval Research Laboratory
M. K. Hudson, Dartmouth College
P. M. Kinter, Cornell University

We are indebted to these individuals for their efforts in defining the program and securing the participation of the invited speakers. We are also grateful to M. K. Hudson, R. L. Lysak, and L. R. Lyons for chairing the sessions and moderating the discussions.

This monograph exists because of the help of many individuals. We wish to thank 42 individuals who consented to acting as referees for these papers. In keeping with the policies of the AGU, the papers in this monograph underwent peer review prior to being accepted for publication. We wish to thank those associated with the AGU, particularly Judy Holoviak, Maury Cain, and Michael Schultz for their help in securing approval for this monograph and guiding it to its final form. Our secretarial staff at the Center of Space Plasma and Aeronomic Research at UAH, Jo Peddycoart and Kathy Wagner, deserves special thanks for their assistance in keeping this project on track through numerous phone calls and letters, letters and more letters.

This workshop was sponsored by the Space Science Laboratory at the Marshall Space Flight Center, the Center for Space Plasma and Aeronomic Research (CSPAR) at The University of Alabama in Huntsville and the UAH Division of Continuing Education. We further acknowledge financial support from the Air Force Geophysics Laboratory and the National Science Foundation, as well as, the sponsoring agencies of all the participants. The assistance from the National Science Foundation and the Air Force Geophysical laboratory allowed us to provide travel assistance for ten graduate students who attended the workshop.

G. R. Wilson
The University of Alabama in Huntsville

M. O. Chandler
NASA/Marshall Space Flight Center

October, 1990

# INTRODUCTION

Most of us have imagined a future age when spacecraft routinely travel the region of space between the Earth and Moon, and perhaps less routinely, between our planet and others in our solar system. For the most part, we can readily extrapolate from the exploration and settlement of North America to the exploration and settlement of moons and other planets. When such an age arrives, the environmental conditions in space will be followed with the same level of interest and immediacy with which we currently track the motions of weather systems within our thin terrestrial atmosphere. In addition to the dangers posed by energetic solar particles, the general state of the atmosphere, ionosphere, and magnetosphere will be of great interest with regard to drag in low Earth orbit and geomagnetic activity affecting electromagnetic wave propagation. A market will thereby be created for "solar-terrestrial weather" reporting and prediction.

In the present we pursue space physics research mainly for the intellectual challenge of understanding the basic plasma processes upon which the behavior of space plasma systems depends, though the knowledge of its impact on space systems, communications, and the ground power distribution system provides a more practical motivation as well. A very large literature of theoretical concepts has been put to the test of explaining over 30 years of space observations, which have become ever more detailed and comprehensive with our increasing experience in space. While evolution of the fundamental concepts we use to account for our observations has by no means ceased, we are in a phase in which the complexities of magnetospheric geometry, across a wide range of length and time scales, must be dealt with in order to quantitatively test our understanding.

In order to test concepts initially expressed in terms of analytic theory, under conditions of complex geometry and boundary conditions, we are forced to resort to the use of numerical techniques for modeling system behavior. Similarly, in order to digest and assimilate the vast store of data which we have collected and will continue to collect, we are forced to resort to the development of empirical models expressing the mean state of the magnetosphere in terms of three dimensional distributions of numerous scalar and vector quantities. The latter technique, well-established for the geomagnetic field, has proved in that case to be extremely valuable as a basis for numerous derivative modeling efforts which are thereby spared the necessity to compute a self-consistent magnetic field distribution. The establishment of an empirical model creates a natural interface between efforts which seek to account for that distribution, and others which seek to identify the consequences of its existence. They relieve us of attempting to solve for everything at once, while injecting physical reality into the discussion of model results.

The First Huntsville Workshop on Plasma Models (1986) was conceived to provide a forum for discussion of physical and empirical models and the feasibility of their synthesis into a unified view of magnetospheric plasmas leading toward a predictive capability for "Solar Terrestrial Weather". The Second Huntsville Workshop on Plasma Models (1989) generally has continued this theme, but with an increasing focus on the three-dimensionality of the interaction of the solar and terrestrial magnetic fields and atmospheres. This requires descriptions of physical processes which are both micro- and macro-scale in nature and which act over localized as well as global regions of space. The papers in this monograph represent some of the most recent efforts to bridge the gap between these scale sizes and to advance towards a unified magnetosphere model.

The magnetosphere is, in essence, a system of reservoirs (of either energy or mass) coupled by a network of conduits (again, of either energy or mass); conduits which represent not only transport paths but also one or more physical processes which provide the actual transport mechanism. Figure 1 represents one view of these reservoirs and the conduits (and associated processes) which link them. The shading highlights areas and/or processes which are discussed in this monograph. On a global scale the primary interaction between Sun and Earth is the balance of magnetic and plasma pressures which forms the magnetopause. Changes in the streaming solar wind produce variations in the nature of this interaction. Information regarding these variations is transmitted to the ionosphere, in the region of the cusp, by intense currents which are associated with wave activity, aurorae, and particle precipitation. Sibeck discusses the response of the magnetospheric and ionospheric systems to such changes in the dynamic pressure of the solar wind. The interaction of the solar wind with the magnetopause also results in a large-scale current layer formed at the magnetopause. The stability of this layer is addressed, through numerical simulations, in a paper by Okuda. While these interactions are considered global in nature many of the physical processes act on a much smaller scale. The effect of micro-scale instabilities on large-scale plasma dynamics is examined by Ganguli. He discusses the large-scale manifestations of these processes such as anomalous viscosity. Investigation of such anomalous effects provides a means of incorporating micro-scale physics into macro-scale models.

One of the largest manifestations of the solar-terrestrial interaction is the magnetotail which includes the boundary layer and plasma sheet. This region is dominated by the "stretched" magnetic field but is heavily influenced by the loading of plasma into the tail by transport from other regions of the magnetosphere. Significant processes include merging, neutral sheet acceleration, and plasmoid formation. Birn and Schindler discuss the quasi-steady nature of the tail field under a variety of conditions. In a separate paper Hesse and Birn investigate the three-dimensional magnetotail with emphasis on field-line reconnection and the evolution of plasmoids. Lui discusses the transport of plasma into, and its effect on, the tail region. While plasma transported into the tail has a direct effect on the tail itself the plasma also undergoes significant changes as a result of acceleration

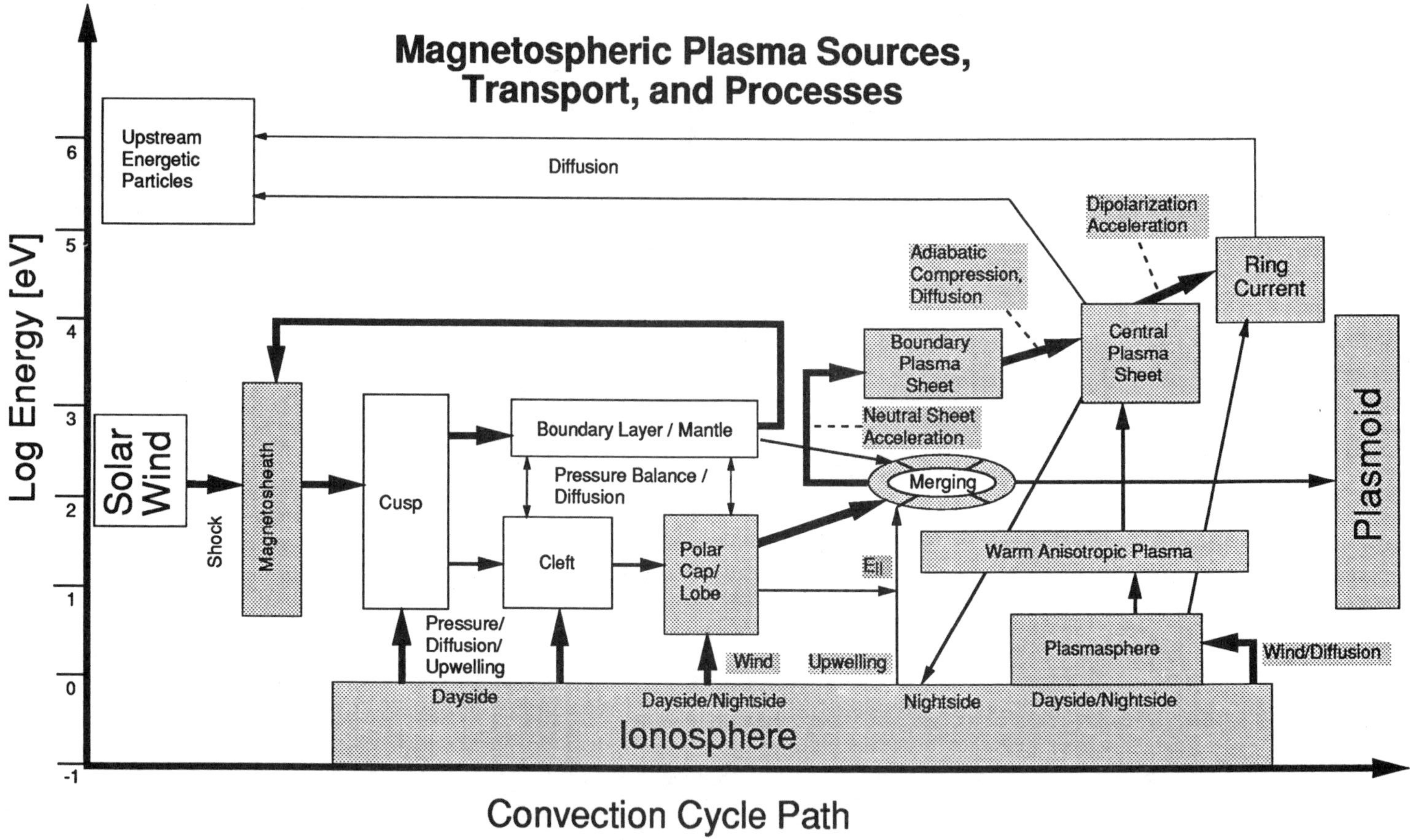


Fig. 1. Schematic representation of magnetospheric particle populations, energy and mass transport paths, and associated processes. The vertical axis represents particle energy and the horizontal axis is a rough representation of the convection paths of particles within the magnetosphere. Shading highlights regions and processes of the magnetosphere which are addressed in this monograph.

and loss processes. The papers by Speiser et al. and Moore and Delcourt present models of these processes and discuss the resulting nature of the "transformed" plasma.

In order to understand the role of plasma of terrestrial origin within the magnetosphere one must understand the sources of this plasma as well as the nature of the interactions between this plasma and the various magnetic and electric field conditions which exist throughout the magnetospheric "cavity". The source of plasma of terrestrial origin within the magnetosphere is the ionosphere. Ho et al. present a model of ion outflows from the polar cap ionosphere which provides a means of investigating the nature of the outflow of ions from this region under a variety of conditions such as in the presence of wave heating or parallel electric fields. Papers by Tetreault and Otani et al. further address the effects of waves and double layers within acceleration regions.

While the plasmasphere is perhaps the most stable reservoir of plasma within the magnetosphere this dense region of the inner magnetosphere provides a sharp boundary within which waves and particles can readily interact. Separate papers by Takahashi, Huang, and Qian et al. address the subject of wave generation within the inner magnetosphere and the effects of these waves on the resident plasma within this boundary. Wave-particle and particle-particle interactions in this region provide the means for transferring energy and momentum from the outer magnetosphere to the plasmasphere. Papers by Guiter and Gombosi and Lin et al. cover the area of plasma flow within closed magnetic flux tubes and the effects of localized heating and scattering of the plasma due to these interactions at the boundary.

Several papers dealing with observational results are included in this monograph and provide a means of accessing the validity of the various models. Validation of an ionospheric model is the subject addressed by Richards et al. through comparisons to radar observations at low latitudes. Pollock et al. present observations from two satellites which are used to infer the presence of potential drops along polar cap field lines. Providing the basic plasma observations necessary to allow models to be tested, Decreau et al. present observations of the character of electrons in the polar cap while Craven et al. show results from an empirical model of ions temperatures within the plasmasphere. These papers provide experimental basis for inputs to models as well as provide interesting topics for future research.

While the synthesis of all of these efforts into a complete model of the entire magnetosphere is still a goal to be reached for, compilations, such as this manuscript, of research efforts covering all size scales and varied techniques represent the budding stages of that synthesis.

T. E. Moore and M. O. Chandler
Magnetospheric Physics Branch
Space Science Laboratory
NASA Marshall Space Flight Center
Huntsville, Alabama

# Solar Wind/Magnetopause

# THE MAGNETOSPHERIC AND IONOSPHERIC RESPONSE TO SOLAR WIND DYNAMIC PRESSURE VARIATIONS

D. G. Sibeck

The Johns Hopkins University Applied Physics Laboratory
Laurel, Maryland 20723

*Abstract.* We discuss the characteristics of solar wind dynamic pressure variations associated with solar wind tangential discontinuities and transient processes at the Earth's bow shock. We quantify existing models for the transient magnetospheric and ionospheric response to the solar wind dynamic pressure variations. The largest variations in solar wind dynamic pressure drive large amplitude (~1 $R_E$) magnetopause motion with radial velocities of ~60 km $s^{-1}$ and produce transient dayside ionospheric flows of 2 km $s^{-1}$ which are organized into double convection vortices. Ground magnetometer signatures are most pronounced under the auroral ionosphere, where they reach 60–300 nT, and under the equatorial electrojet.

## Introduction

This paper represents an effort to survey, update, and quantify current theoretical and observational work concerning the effects which small scale features in the solar wind have upon the magnetopause, magnetosphere, and ionosphere. It has recently become increasingly clear that sudden, brief, large amplitude variations in the solar wind dynamic pressure are common and that they drive significant magnetopause motion. Within the magnetosphere, this motion is associated with transient electric fields, field-aligned currents, and bursts of precipitating particles, all of which transmit information about the magnetopause motion to the dayside auroral ionosphere. The electric field drives transient ionospheric flows and the precipitating particles enhance ionospheric conductivities. The enhanced conductivities allow the electric fields to drive large ionospheric currents which produce correspondingly large ground magnetic field perturbations under the auroral ionosphere. Additionally, the magnetospheric motion launches compressional waves into the magnetosphere. These couple to resonant azimuthal oscillations in the outer magnetosphere. They also reach the low-latitude ionosphere, where they produce the strongest ground magnetometer perturbations under the equatorial electrojet.

## Solar Wind Features

The solar wind can be highly inhomogeneous, even on brief time scales. Although we are most interested in solar wind dynamic pressure variations on short time scales, the dynamic pressure has not been studied systematically. Indeed, studies devoted to solar wind plasma parameters on short time scales are rare. Burlaga [1968] presented examples of plasma density, velocity, and temperature variations attending interplanetary magnetic field (IMF) tangential discontinuities. These examples indicate that the solar wind density can briefly increase and decrease by factors of 2–3 for 5–10 min periods. However, a statistical study reported by Solodyna et al. [1977] indicates that few tangential discontinuities bring density variations greater than 35%.

None of these studies made use of high time resolution solar wind plasma data. Consequently, it is important to consider what evidence can be gleaned from the many studies which only considered the magnetic field characteristics of IMF tangential discontinuities. According to Burlaga [1969], tangential discontinuities typically arrive at a rate of about 1 $hr^{-1}$ at a distance of 1 AU from the Sun. Siscoe [1968] showed that the surfaces of these discontinuities tend to lie along the IMF garden hose spiral, i.e. parallel to the IMF orientation. The extent of discontinuities transverse to the Earth-Sun line is not well known. Several examples of discontinuities extending some 100 $R_E$ transverse to the Earth-Sun line have been presented [Burlaga and Ness, 1969], but at other times strikingly different features are seen over distances as small as 30 $R_E$ [Crooker et al., 1982; Sibeck et al., 1989c]. The results presented by Siscoe et al. [1968] indicate that the magnetic field strength varies by 10% or more across over half the observed tangential discontinuities. However, differences of a factor of 2 are rare. For typical IMF strengths of 6 nT, 10% field strength variations correspond to 0.6 nT.

As our main interest is in determining the change in solar wind dynamic pressure across tangential discontinuities, we seek some way to estimate this parameter from IMF measurements. Burlaga [1968] and subsequent authors have suggested both that the solar wind temperature and the sum of the solar wind thermal and magnetic pressures remain constant during the passage of solar wind tangential discontinuities. Then we estimate the variation in the solar wind density across a discontinuity as

$$\delta n = \delta p_B / k(T_i + T_e), \qquad (1)$$

where $\delta p_B$ $(=(B_2^1 - B_2^2)/2\mu_o)$ is the change in the magnetic pressure and $T_i$ and $T_e$ are the ion and electron plasma temperatures. For typical values of $B_1$ = 6 nT and $B_2$ = 6.6 nT, $T_i = 2 \times 10^4$ °K, and $T_e = 1 \times 10^5$ °K, we expect typical density variations on the order of 1.8 $cm^{-3}$, although much larger variations are possible. Thus relatively weak, small scale, IMF features may be accompanied by variations of about 30% about the typical solar wind density of 6 $cm^{-3}$.

Similarly, the passage of such discontinuities indicates substantial variations in the solar wind dynamic pressure ($nMV^2$). With the above assumptions (and one more: that the radial velocity remains constant), the change in solar wind dynamic pressure associated with a tangential discontinuity is given by

$$\delta(nMV^2) = MV^2 \delta n = MV^2 \delta p_B / k(T_i + T_e). \qquad (2)$$

Modeling Magnetospheric Plasma Processes
Geophysical Monograph 62

Assuming a typical solar wind velocity (V) of 400 km $s^{-1}$ and the same parameters as above, we find that a change in the solar wind dynamic pressure of about 0.5 nPa would be common. Since the typical solar wind dynamic pressure lies between 1 and 2 nPa, the variations associated with tangential discontinuities are quite substantial.

Besides intrinsic solar wind tangential discontinuities, there are other sources for solar wind dynamic pressure variations. Spangler et al. [1988] suggested a modulational instability of waves generated at the Earth's bow shock as the source of wave packets with periods of 4–5 minutes. Sibeck et al. [1989] and Fairfield et al. [1990] presented numerous examples of 1–2 nPa solar wind dynamic pressure variations observed just upstream of the quasi-parallel bow shock. These quasi-periodic variations with periods of about 5–10 min can be distinguished from solar wind tangential discontinuities because the density and magnetic field strength vary in phase. Fairfield et al. [1990] presented evidence that the IMF orientation controlled the solar wind dynamic pressure during these events, in the sense that the pressure was reduced during periods when the IMF was nearly parallel to the Earth–Sun line. They argued that the reduction should be considered a phenomenon associated with the quasi-parallel bow shock. On the grounds of these case studies, we might anticipate that variations in the IMF associated with solar wind rotational discontinuities are transformed into variations in the solar wind dynamic pressure at the Earth's bow shock.

## Magnetopause Effects

Whether of intrinsic solar wind origin or generated at the bow shock, variations in the solar wind dynamic pressure drive significant magnetopause motion. The bow shock transition converts regions of enhanced dynamic pressure into regions of enhanced magnetosheath thermal and magnetic pressure. Because the magnetopause lies at the locus of points where magnetosheath and magnetospheric magnetic and thermal pressures balance, each region of enhanced pressure briefly compresses the magnetopause. Figure 1 shows that when the IMF has a garden hose spiral orientation, the compression first strikes the post-noon equatorial magnetopause (1), spreads to the pre-noon, dusk, and polar magnetopause (2), and then proceeds to move down the both flanks and the polar magnetotail in the anti-sunward direction (3). A series of brief solar wind dynamic pressure increases would produce a sequence of concentric ripples on the magnetopause surface, expanding outward from the point of first contact, rather like the concentric rings which a stone cast into a pool of water produces. It is worth emphasizing again that although the ripples typically move anti-sunward with the magnetosheath flow, we expect them to move sunward on the early post-noon magnetopause when the IMF has a spiral orientation.

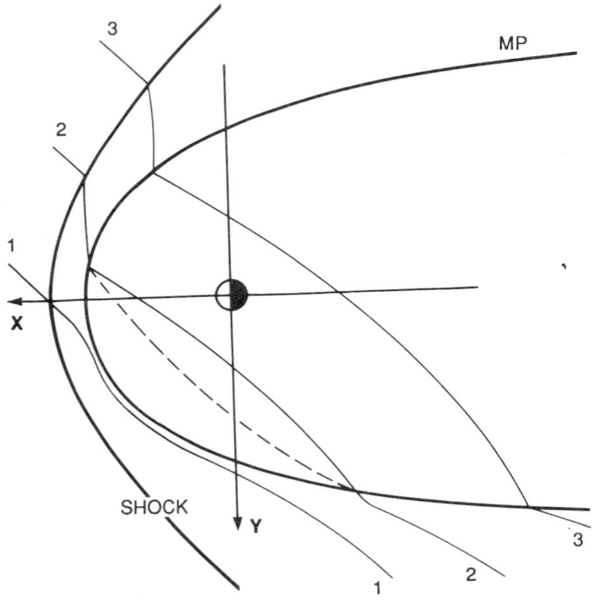


Fig. 1. Solar wind discontinuities are aligned with the garden hose spiral IMF. They first intercept the early post-noon magnetopause (1), then sweep across the pre-noon, dusk, and polar magnetopause (2), before moving antisunward down the magnetotail (3).

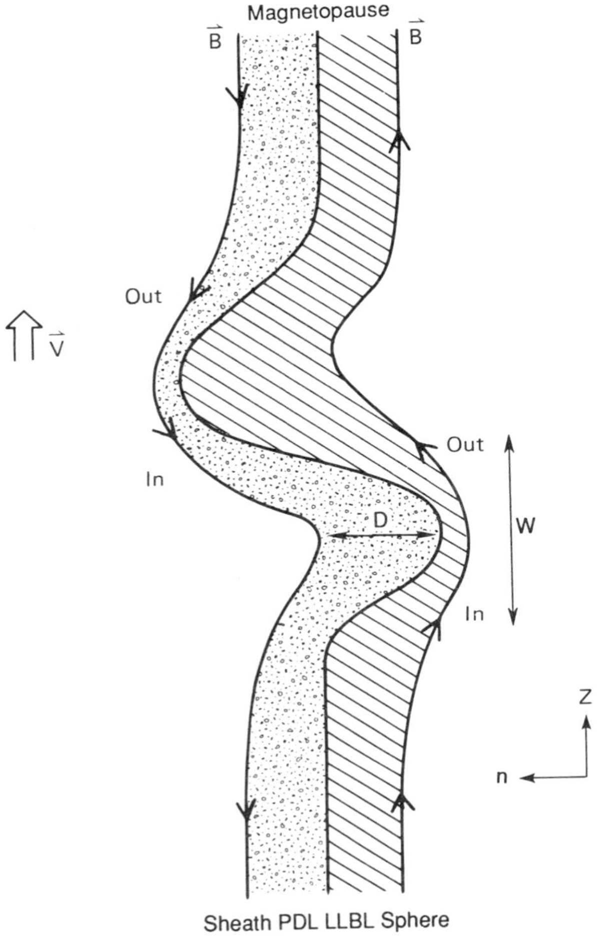


Fig. 2. The passage of northward moving ridge and trough on the magnetopause. The LLBL lies just inside the dayside magnetopause; the overlapping plasma depletion and energetic particles layers lie just outside. A spacecraft near the magnetopause observes oscillatory magnetic field signatures associated with the wavy magnetopause motion and may move from one region to another.

Following Sibeck [1990], let us now consider some of the detailed small scale features of the magnetopause motion. Figure 2 shows the passage of a northward (+z) moving trough in the magnetopause position. The trough has depth D and width W. The low-latitude boundary layer (LLBL), a region of magnetosheath-like plasma on northward pointing magnetospheric magnetic field lines, lies just inside the magnetopause.

The plasma in this region is flowing in the same direction as that in the magnetosheath, but at reduced velocities. The plasma depletion and energetic particles layers lie just outside the magnetopause on southward pointing magnetosheath magnetic field lines. The plasma depletion layer is a region of magnetosheath plasma with depressed densities and enhanced magnetic field strengths, whereas the energetic particle layer is a region of streaming particles which have escaped from the magnetosphere. These two layers are not necessarily coextensive.

The magnitude of the solar wind dynamic pressure variation determines the amplitude of the magnetopause motion. The subsolar magnetopause lies at the point where solar wind dynamic pressure (converted into magnetosheath magnetic and thermal pressures) balances magnetospheric magnetic and thermal pressures. For simplicity, it is common to suppose that the pressure of the Earth's dipole magnetic field dominates the thermal pressure within the magnetosphere. Then pressure balance indicates that the position of the subsolar magnetopause varies as the sixth root of the solar wind dynamic pressure. From the review of solar wind dynamic pressure variations above, we see that factor of 2 variations in the solar wind dynamic pressure are not uncommon, although they probably result primarily from processes occurring at the Earth's bow shock rather than represent intrinsic solar wind features. A factor of 2 increase in the solar wind dynamic pressure, say from 1 to 2 nPa, results in a 12% decrease in the position of the subsolar magnetopause. Because the subsolar magnetopause typically lies near 10 $R_E$ from Earth, such a variation in magnetopause position corresponds to a ~1 $R_E$ decrease in the magnetopause position.

The duration of the solar wind/magnetosheath pressure enhancement determines the width of the trough in the magnetopause. A duration of 2 min and a magnetosheath velocity of 200 km $s^{-1}$ correspond to widths of 4 $R_E$. Assuming that the trough has a sinusoidal shape, taking the amplitude of the wave as the depth of the trough, and the wavelength as twice its width, one expects slopes in the magnetopause position of no more than ~38° from the y-z plane.

We now consider the radial magnetopause velocity. According to Baumjohann et al. [1983], the magnetopause initially moves at a velocity of

$$v_o = \{1 - (n/n')^{1/2}\} \, V\cos\Psi \tag{3}$$

in response to a solar wind density increase from n to n'. Here $\Psi$ is the angle between the solar wind bulk velocity and the magnetopause normal. Furthermore, Baumjohann et al. [1983] suggest that the magnetopause velocity decays exponentially according to the following equation

$$v(t) = v_o \, e^{-\gamma t}, \tag{4}$$

where $\gamma$ is given by

$$\gamma = (n/n')^{1/2} 3V\cos\Psi/R, \tag{5}$$

with R being the distance of the magnetopause from Earth. Now let us estimate the initial magnetopause velocity and the time required for the magnetopause to come to rest. Assume that the solar wind velocity is 400 km $s^{-1}$, the density increases from n = 5 $cm^{-3}$ to n' = 10 $cm^{-3}$, and the angle of incidence is $\Psi = 0°$. Then the initial magnetopause velocity is 117 km $s^{-1}$. If the magnetopause is initially at 10 $R_E$, the magnetopause velocity decays by a factor of e in about 75s. Due to the exponential decay, spacecraft typically observe magnetopause velocities in the lower half of the range from 0–120 km $s^{-1}$, consistent with results reported by Berchem and Russell [1982]. The magnetopause motion is associated with an electric field. Magnetopause velocities of 60 km $s^{-1}$ and outer magnetospheric magnetic field strengths of 60 nT indicate magnetospheric electric fields of 3.6 mV $m^{-1}$.

Consider the plasma signatures seen by satellites just inside and outside the magnetopause. Figure 2 shows that during the passage of a trough in the magnetopause position, a satellite just inside the magnetopause may enter the LLBL, where it will observe a flowing mixture of magnetosheath and magnetospheric plasmas on northward magnetic field lines. Similarly, during the passage of a ridge in the magnetopause position, a satellite just outside the magnetopause may briefly enter the plasma depletion and energetic particle layers, where it will observe a mixture of reduced-density magnetosheath plasma and streaming energetic magnetospheric particles on magnetosheath magnetic field lines of intensified strength.

Figure 2 shows that wavy magnetopause motion is associated with an oscillating magnetic field component normal to the nominal (unperturbed) magnetopause, i.e. $B_n$. We have already estimated that the slope of the magnetopause during the wavy motion does not exceed 38°. If we take a typical value for the magnetospheric magnetic field strength just inside the dayside magnetopause, say 60 nT, we estimate that $B_n = B \sin 38°$, or 37 nT.

Spacecraft in the magnetosheath and magnetosphere associate a specific bipolar signature with the passage of ridges and troughs in the magnetopause position. They observe −,+ $B_n$ signatures when ridges have a component of motion parallel to the ambient magnetic field, and +,− $B_n$ signatures when troughs have component of motion parallel to the ambient magnetic field. The sense of the $B_n$ signatures reverses for ridge and trough motion anti-parallel to the magnetic field. Farrugia et al. [1987] have demonstrated that the predicted magnetic field and plasma velocity perturbations attending the passage of wavy magnetopause motion are proportional and colinear. In general, the wavy magnetopause motion will move northward north of the equator and southward south of the equator.

It is important to note that spacecraft in the magnetosheath will observe stronger bipolar magnetic field signatures associated with the passage of ridges than with troughs, and that the opposite is true for spacecraft in the magnetosphere. The signatures would be similar were it not for the fact that the ridges and troughs propagate at transonic velocities. Studies of flow outside wavy boundaries indicate that at transonic velocities the disturbance field is stronger over crests in the boundary than valleys. Furthermore, we note that spacecraft in the magnetosheath associate the passage of ridges in the magnetopause position with enhanced magnetic field strengths because they enter the plasma depletion layer, and that spacecraft in the magnetosphere associate the passage of troughs with enhanced magnetic field strengths because the magnetosphere is compressed.

Thus criteria for event selection such as bipolar magnetic field signatures normal to the magnetopause, enhanced magnetic field strengths, and mixtures of flowing plasmas will cause the preferential selection of troughs in the magnetopause position for observing spacecraft in the magnetosphere and ridges in the magnetopause position for spacecraft in the magnetosheath. During periods of southward IMF, spacecraft in both the magnetosheath and the magnetosphere will associate wavy magnetopause motion with bipolar +,− $B_n$ signatures normal to the magnetopause, whereas spacecraft south of the equator will associate such motion with −,+ $B_n$ signatures.

The wavy magnetopause motion is associated with a field-aligned current in the magnetosphere [Southwood and Kivelson, 1990]. This current is determined by Ampere's Law

$$J_{parallel} = (B/\mu_o|B|) \cdot \nabla \times B. \tag{6}$$

For a background magnetospheric magnetic field (B) lying in the z direction and no gradients in the x direction, we find that it is a gradient of the perturbation magnetic field ($B_n$) normal to the nominal magnetopause which produces the field aligned current:

$$J_{parallel} = \delta B_n/\mu_o \delta y. \tag{7}$$

Such gradients are present whenever the troughs and ridges in the magnetopause position do not lie exactly parallel or perpendicular to the magnetospheric magnetic field lines, i.e. they are present for almost all magnetopause motion. If we take B as 60 nT, $\delta B_n$ as 37 nT, and $\delta y$ as 2 $R_E$ (i.e. half the width of the trough), we deduce peak field aligned currents of $2.3 \times 10^{-3}$ $\mu A/m^{-2}$. When mapped to the ionosphere, such current densities increase by a facter of $10^3$, and are therefore consistent with previously reported current densities in the low altitude cusp of J = 2.5 $\mu A\ m^{-2}$ [Iijima and Potemra, 1976].

Finally, we should consider the length of time a given magnetospheric magnetic field line is compressed. This time is much longer than the brief ~1–4 min required for a depression or ridge in the magnetopause surface to move past an observing satellite in the outer magnetosphere. After an increase in the solar wind dynamic pressure first strikes the subsolar magnetopause, it remains in contact with some portion of any given outer magnetospheric magnetic field line until it reaches the polar magnetopause. Previous timing studies [e.g., Figure 4 of Lockwood and Cowley, 1988] indicate that IMF (and solar wind features) first strike the post-noon magnetopause near 1400 LT and require some 10 minutes to reach the dusk terminator and 20 minutes to reach the dawn terminator. This result implies that a similar time is required for magnetosheath features to sweep north-south across the dayside magnetopause and reach the polar cap. Thus some portion of dayside magnetic field lines will be compressed for 10–20 min while a magnetosheath feature moves from the subsolar point to the polar region. The period of compression could be twice as long if the IMF had an unusual orientation and the magnetosheath feature first struck near one polar region and then moved across the entire dayside magnetopause to the other polar region.

## Magnetospheric Effects

A brief increase in the solar wind dynamic pressure disturbs the magnetopause and launches a compressional fast mode wave into the magnetosphere. This front advances into the magnetosphere at speeds which are generally considerably greater than the magnetosheath velocities themselves. As a consequence, Kaufmann and Konradi [1969] and Nishida [1978] suggested that the front outruns the magnetosheath discontinuity. Since the fast mode wave enhances magnetospheric magnetic field strengths, this implies that magnetospheric magnetic field strengths rise prior to the arrival of the magnetosheath discontinuity. The magnetopause moves outward ahead of the discontinuity to restore pressure balance. Thus, as shown in Figure 2, a ridge in the magnetopause position precedes each trough. Sibeck [1990] noted evidence suggesting that the LLBL is thicker during periods of northward IMF than southward IMF and suggested that it might also be colder and denser. The magnetospheric fast mode velocity would be less in such a colder, denser, LLBL. Consequently, Sibeck [1990] suggested that the magnetospheric fast mode wave would be less likely to outrun the magnetosheath discontinuity when the IMF was northward. Under these conditions, there will be no ridges in the magnetosheath position preceding the troughs, and the amplitude of magnetopause motion will be less for northward than southward IMF.

Southwood and Kivelson [1990] considered the propagation of the fast mode wave into the magnetosphere. They showed that an impulsive increase in the solar wind dynamic pressure launches a steep fast mode wave front which propagates inward from the magnetopause. The compression which this wave front induces is most significant near the magnetopause because the ambient magnetospheric magnetic field strength increases rapidly towards Earth. They also demonstrated that the impulsive fast mode wave front would couple to natural, azimuthal magnetospheric magnetic field resonances at specific L shells. Such localized resonances may endure long ( ~30 min) after the passage of the front.

The passage of the fast mode wave front enhances magnetospheric ELF-VLF wave activity which scatters magnetospheric particles [Perona, 1972]. Some scattered electrons enter the loss cone and reach the ionosphere. The scattering should be most pronounced in the outermost magnetosphere near local noon, where the increase in magnetospheric magnetic field strengths is greatest.

## Ionospheric Response

The pressure pulse driven magnetopause motion produces electric fields, field-aligned currents, ELF-VLF waves, and precipitating particles which communicate information on that motion to the base of the field lines in the ionosphere. Thus the ground and ionospheric signatures of transient motion at the magnetopause should be observed in the cusp, cleft, and mantle regions of the dayside auroral ionosphere. Here we summarize descriptions of the predicted signatures to be observed in those regions as discussed by Elphic [1988], Southwood and Kivelson [1990], Sibeck [1990], and Lui and Sibeck [1990].

Figure 3, adopted from Glassmeier et al. [1989] and modified by Sibeck [1990], depicts the magnetospheric plasma motion, field-aligned currents, and ionospheric plasma flows associated with wavy magnetopause motion. Inward magnetopause motion corresponds to an equatorward ionospheric flow and outward motion to a poleward flow. Thus the ridge-crest magnetopause system maps to an ionospheric double-vortex convection pattern. The double vortex pattern moves eastward or westward around the dayside auroral oval, just as the magnetopause motion moves eastward or westward around the dayside magnetopause. We recall that sunward motion is expected on the early post-noon magnetopause and suggest that the ionospheric convection vortices should move sunward at early post-noon local times. This east-west motion of the features should be consistent with the east-west motion at the magnetopause. A feature moving azimuthally around the magnetopause with a magnetosheath velocity of 100 km $s^{-1}$ would move through the auroral ionosphere with a velocity of 3.4 km $s^{-1}$.

The latitudinal extent of the convection vortices depends upon the amplitude of the solar wind dynamic pressure variations. The magnetospheric region over which the compression is felt exceeds the ~1 $R_E$ depth of the trough in the magnetopause position. Assuming that it is on the order of 3 $R_E$ and that one may use the ratio of magnetospheric to auroral zone ionospheric magnetic field strengths (0.001) to map scale sizes from

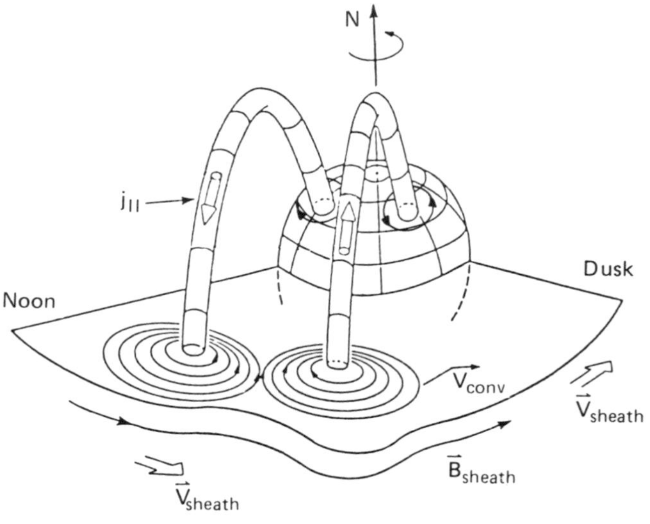


Fig. 3. Wavy magnetopause motion is associated with magnetospheric velocities, field-aligned currents, and ionospheric flows. This figure is adopted from Glassmeier et al. [1989] and Sibeck [1990].

the magnetopause to the auroral ionosphere, one predicts ionospheric scales sizes on the order of ~610 km. The longitudinal extent is comparable for solar wind/magnetosheath features which last 1–2 minutes, but greater for those lasting longer.

We can make rough estimates of the ionospheric electric fields, convection velocities, currents, and ground magnetic field perturbations associated with the convection vortices. According to Mozer [1970], one expects outer dayside azimuthal magnetospheric electric fields at L = 10 to map to ionospheric electric fields which are a factor of 30 greater. For the predicted 3.6 mV $m^{-1}$ outer magnetospheric electric fields, one expects 0.11 V $m^{-1}$ ionospheric electric fields. Following Glassmeier et al. [1989], we expect that such electric fields drive ionospheric flows of ~2 km $s^{-1}$ in the 60000 nT auroral zone magnetic field. The ground magnetic field perturbation due to a sheet Hall current driven by an ionospheric electric field is given approximately by $\delta b = \mu_o \Sigma_H E/2$. Typical values for the height-integrated Hall ionospheric conductivity in the cusp are on the order of 1 to 5 mhos [e.g., Vickrey et al., 1981], depending primarily on whether the cusp is sunlit. Thus the pressure pulse driven magnetopause motion should produce ground magnetic field perturbations on the order 60–300 nT. Note that the precipitating particles associated with the transient magnetospheric compression enhance conductivities, and therefore ground magnetic field perturbations, at the footprints of the outermost magnetospheric magnetic field lines.

It is important to recall that following the passage of the initial ridge and crest, poleward portions of the outermost magnetospheric magnetic field lines remains compressed for a period of 10–20 min, i.e. the time required for a magnetosheath feature to sweep over the face of the dayside magnetosphere and reach polar latitudes. As the discontinuity moves over the magnetopause surface, it progressively disturbs each portion of the field lines, producing ELF-VLF waves and energizing and scattering magnetospheric particles. The ELF-VLF waves can be observed at the base of the field lines and those particles which precipitate enhance ionospheric conductivities. Figure 4, adopted from Lui and Sibeck [1990], depicts the situation when a wide range of field lines at post-noon local-times is simultaneously compressed. The foot prints of these lines form an arc elongated east-west over a wide range of local times in the auroral ionosphere. The precipitating particles produce an elongated optical aurora in which ionospheric conductivities are enhanced. Thus ground magnetometers under the auroral arc observe enhanced perturbations. The compression of the LLBL which scatters particles and produces the arc also decreases the thickness of the LLBL and thereby enhances the electric field strength across the LLBL. These electric fields map to the region of the arc and drive enhanced ionospheric flows under the auroral arc in the same direction as those seen prior to the event. Finally, we note that when the LLBL is no longer compressed, particle precipitation will cease and the optical aurora will diminish. However, once the solar wind discontinuity reaches the polar magnetopause, it will compress the dense plasma in the mantle. An enhanced flux of particles will precipitate into the ionosphere at a latitude poleward of the preexisting LLBL arc. Thus the optical aurora will appear to jump poleward at the end of each event.

Finally, consider the situation under the enhanced conductivity equatorial electrojet. Electric fields associated with the fast mode compressional wave drive enhanced flows in the ionosphere at the equatorial electrojet. Thus ground magnetometers under the equatorial electrojet will observe enhanced H component increases associated with the magnetospheric compressions.

## Recent Statistical Results Concerning Impulsive Ground Events

Sufficient case studies of transient magnetospheric and ionospheric events have now been published to enable us to undertake a small statistical study. As an example of this approach, Figure 5 presents a

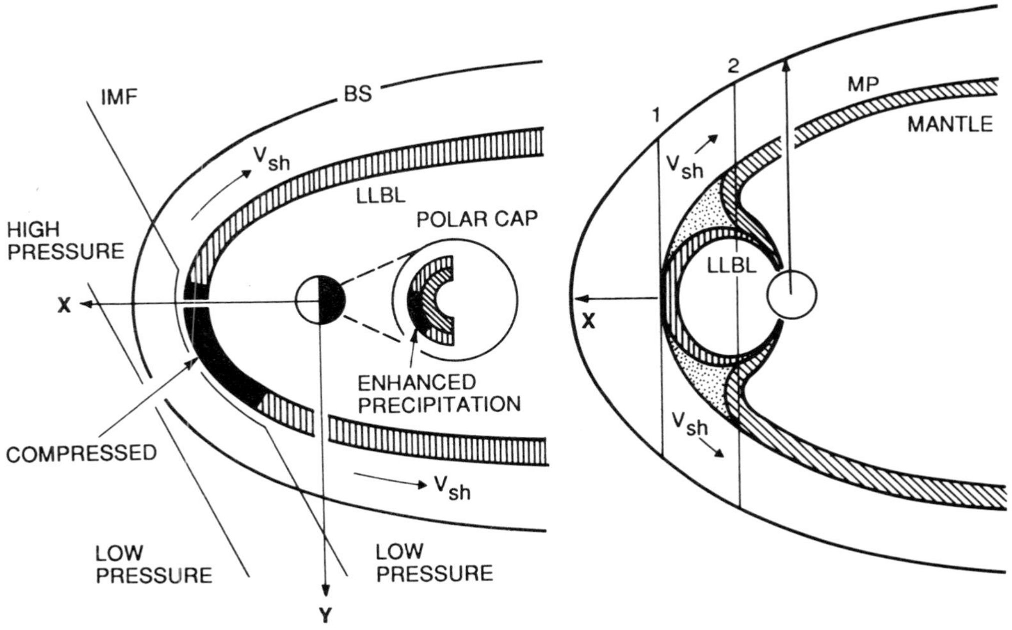


Fig. 4. Poleward portions of outer magnetospheric magnetic field lines remain compressed following the initial contact of the IMF/magnetosheath discontinuity with the magnetosphere. Particles in the low-latitude boundary layer are scattered and precipitate, producing an east-west arc in the auroral zone which endures for some 10–20 minutes. When the solar wind features no longer compress the dayside LLBL, they compress the polar mantle, which maps to the ionosphere poleward of the cusp. Thus, the initial arc in the auroral ionosphere fades and appears to jump poleward at the end of each magnetospheric compression.

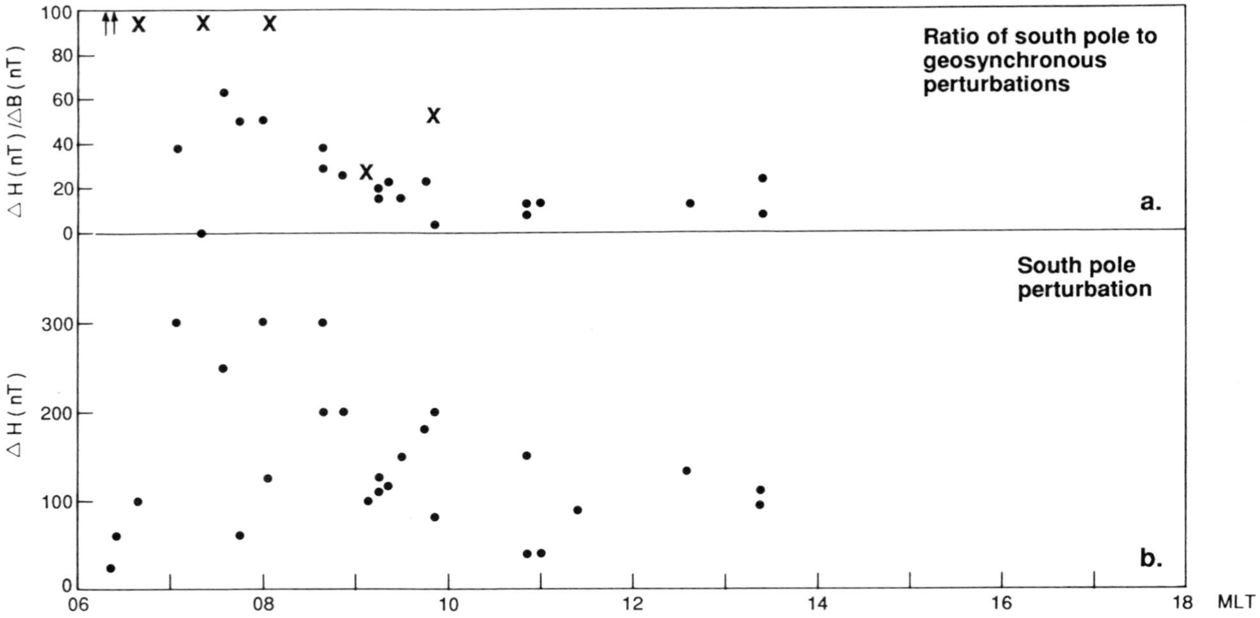


Fig. 5. The lower panel shows the peak-to-peak amplitude of reported transient ground magnetometer events at South Pole station as a function of local time. The upper panel shows the ratio of the amplitude of the ground signature to the strength of the corresponding compression in the geosynchronous magnetic field at a function of local time. Crosses mark ratios using GOES-2 observations; arrows and dots mark ratios using GOES-5 observations. Ground events with no corresponding compressional signature at geosynchronous orbit are plotted at ratios exceeding 90.

statistical comparison of impulsive high-latitude ground magnetometer events observed at South Pole station and compressions of the geosynchronous magnetospheric magnetic field observed by satellite GOES-2 or GOES-5. Figure 5b shows the distribution in local time of the 27 transient events presented by Lanzerotti et al. [1986; 1987; 1990], Bering et al. [1988; 1990], and Mende et al. [1990]. The reported events clearly occur predominantly at pre-noon local times, a point previously noted by Glassmeier et al. [1989] in their statistical study of similar events seen in Scandinavia. Figure 5b shows the amplitude of the events as a function of local time. The peak-to-peak variations in the H component are strongest near 07–08 LT and diminish towards local noon.

Figure 5a shows the ratio of the disturbance at the ground to the magnitude of the compression at geosynchronous orbit at the time of each impulsive ground event. Crosses mark events for which GOES-2 observations were used and arrows and dots mark events for which GOES-5 observations were used. Arrows and crosses plotted at ratios above 90 indicate events for which the ratio of the ground to geosynchronous disturbances was infinite, i.e. no measurable signatures was seen at geosynchronous orbit. The first point to note is that a compression was seen at geosynchronous orbit for all but 6 events, and that there is a reason why no signature was seen at geosynchronous orbit for each of these 6 events. For one of the ground events, that reported by Bering et al. [1990] as occurring at 1615 UT (1115 LT) on January 3, 1986, there were no geosynchronous data available. GOES-2 could have observed three of the remaining 5 events, but at the times of these events (marked by crosses), GOES-2 was located at 106.2° W, about 3 hours in local time earlier than the ground observations (and the times at which the events are plotted). The weak GOES-2 response during these events is consistent with the fact that geosynchronous SI signatures are weaker near dawn than near local noon [Kuwashima et al., 1985], presumably because geosynchronous satellites are further from the magnetopause at local dawn. The final two cases with no signature at geosynchronous orbit were observed by GOES-5 at the earliest local times (arrows), and had very weak ground signatures (Figure 5b). The signatures expected at geosynchronous orbit should have been weak for these events too.

Upon further inspection of Figure 5a, we note that the ratio of ground to geosynchronous signatures decreases towards local noon. This too is consistent with the results of Kuwashima and Fukunishi [1985] and Fairfield et al. [1990], and suggests that geosynchronous satellites are only useful in detecting the magnetospheric response to solar wind dynamic pressure variations when they are within several (~4) hours of local noon.

We need to consider the reasons why Bering et al. [1990] failed to find any signature in solar wind observations corresponding to the transient high-latitude auroral zone events which he examined. Possible explanations include the following: incorrect lag times, short scale lengths for solar wind features, low time resolution solar wind plasma measurements, and the generation of transient events at the Earth's bow shock. However, the single most important factor lies in the small amplitudes of the events which Bering et al. [1990] considered. The ground events which they studied had amplitudes of only several 10's of nT, e.g. ~20 nT in his Figure 4. We have earlier estimated that a 2 nPa solar wind dynamic pressure pulse might produce 60–300 nT transient ground magnetic field signatures. To produce a 20 nT transientground event in the dayside cusp, a solar wind dynamic pressure variation of only 0.06 to 0.33 nPa would be required. Variations this small are within the noise level of the solar wind dynamic pressure plots which Bering et al. [1990] present.

## Conclusions

In this paper, we described the characteristics of solar wind dynamic pressure variations. Most of the information we have about these features has been determined from low time resolution plasma measurements or inferred from high time resolution magnetic field measurements. Although high time resolution solar wind plasma observations are urgently needed, on the basis of what we now know, it seems likely that large amplitude, quasi-repetitive variations in the solar wind dynamic pressure are common.

These variations in the solar wind dynamic pressure have a dramatic, transient effect upon the magnetosphere. We quantified the characteristics of pressure pulse driven magnetopause motion and suggested that it may the most important contribution to magnetopause motion as a whole. Sibeck [1990] has successfully compared the predicted characteristics of pressure pulse driven magnetopause motion with statistical observations. It will, in future, be important to determine whether or not significant solar wind dynamic pressure variations were present during periods of transient magnetopause motion [e.g., Sibeck et al., 1989].

Electric fields, precipitating particles, and Birkeland currents transmit information upon the magnetopause motion to the auroral ionosphere. We quantified the ionospheric currents, velocities, and electric and magnetic fields associated with this motion. Lui and Sibeck [1990] have verified that observed characteristics of transient high-latitude auroral features are consistent with the model. In particular, they reexamined previously reported cases of transient optical aurora with greatly elongated east-west extents have been observed in the dayside cusp and cleft. At the close of each event, the aurorae appear to jump poleward. And the events tend to occur during intervals of highly variable solar wind dynamic pressure. It will be important to verify whether or not there were simultaneous observations of significant solar wind dynamic pressure or IMF variations at the times of other events [e.g., Sibeck et al., 1989b]. In some cases, this may have to be done by inference, i.e. by examination of geosynchronous and low-latitude ground magnetic field observations [Sibeck et al., 1989b; c; Lanzerotti et al., 1990].

*Acknowledgments.* The author thanks A. T. Y. Lui for helpful comments. This work was supported by NASA under Space and Naval Warfare Systems Command contract N00039-87-C-5301 of the Navy.

## References

Baumjohann, W., O. H. Bauer, G. Haerendel, H. Junginger, and E. Amata, Magnetospheric plasma drifts during a sudden impulse, *J. Geophys. Res.*, *88*, 9287–9289, 1983.

Berchem, J. and C. T. Russell, The thickness of the magnetopause current layer: ISEE 1 and 2 observations, *J. Geophys. Res.*, *87*, 2108–2114, 1982.

Bering, E. A., III, J. R. Benbrook, G. J. Byrne, B. Liao, J. R. Theall, L. J. Lanzerotti, C. G. Maclennan, A. Wolfe, and G. L. Siscoe, Impulsive electric and magnetic field perturbations observed over south pole: Flux transfer events?, *Geophys. Res. Lett.*, *15*, 1545–1548, 1988.

Bering, E. A., III, L. J. Lanzerotti, J. R. Benbrook, Z.-M. Lin, C. G. Maclennan, A. Wolfe, R. E. Lopez, and E. Friis-Christensen, Solar wind properties observed during high-latitude impulsive perturbation events, *Geophys. Res. Lett.*, *17*, 579–582, 1990.

Burlaga, L. F., Micro-scale structures in the interplanetary medium, *Sol. Phys.*, *4*, 67–92, 1968.

Burlaga, L. F., Directional discontinuities in the interplanetary magnetic field, *Sol. Phys.*, *7*, 54–71, 1969.

Burlaga, L. F. and N. F. Ness, Tangential discontinuities in the solar wind, *Sol. Phys.*, *9*, 467–477, 1969.

Crooker, N. U., G. L. Siscoe, C. T. Russell, and E. J. Smith, Factors controlling degree of correlation between ISEE 1 and ISEE 3 interplanetary magnetic field measurements, *J. Geophys. Res.*, *87*, 2224–2230, 1982.

Elphic, R. C., Multipoint observations of the magnetopause: Results from ISEE and AMPTE, *Adv. Space Res.*, *8*, 223–238, 1988.

Fairfield, D. H., W. Baumjohann, H. Luehr, and D. G. Sibeck, Upstream pressure variations associated with the bow shock and their effects on the magnetosphere, *J. Geophys. Res.*, *95*, 3773–3786, 1990.

Farrugia, C. J., R. C. Elphic, D. J. Southwood, and S. W. H. Cowley, Field and flow perturbations outside the reconnected field line region in flux transfer events: Theory, *Planet. Space Sci.*, *35*, 227–240, 1987.

Farrugia, C. J., M. P. Freeman, S. W. H. Cowley, D. J. Southwood, M. Lockwood, and A. Etamadi, Pressure-driven magnetopause motions and attendant response on the ground, *Planet. Space Sci.*, *37*, 589–607, 1989.

Friis-Christensen, E., M. A. McHenry, C. R. Clauer, and S. Vennerstrom, Ionospheric traveling convection vortices observed near the polar cleft: A triggered response to sudden changes in the solar wind, *Geophys. Res. Lett.*, *15*, 253–256, 1988.

Glassmeier, K.-H., M. Hoenisch, and J. Untiedt, Ground-based and satellite observations of travelling magnetospheric convection twin-vortices, *J. Geophys. Res*, *94*, 2520–2528, 1989.

Iijima, T. and T. A. Potemra, Field-aligned currents in the dayside cusp observed by Triad, *J. Geophys. Res.*, *81*, 5971–5979, 1976.

Kaufmann, R. L. and A. Konradi, Explorer 12 magnetopause observations: Large-scale nonuniform motion, *J. Geophys. Res.*, *74*, 3609–3627, 1969.

Kuwashima, M. and H. Fukunishi, Local time asymmetries of the SSC-associated hydromagnetic variations at the geosynchronous altitude, *Planet. Space Sci.*, *33*, 711–720, 1985.

Lanzerotti, L. J., L. C. Lee, C. G. Maclennan, A. Wolfe, and L. V. Medford, Possible evidence of flux transfer events in the polar ionosphere, *Geophys. Res. Lett.*, *13*, 1089–1092, 1986.

Lanzerotti, L. J., R. D. Hunsucker, D. Rice, L. C. Lee, A. Wolfe, C. G.Maclennan, and L. V. Medford, Ionospheric and ground-based response to field-aligned currents near the magnetospheric cusp regions, *J. Geophys. Res.*, *92*, 7739–7743, 1987.

Lanzerotti, L. J., A. Wolfe, N. Trivedi, C. G. Maclennan, and L. V. Medford, Magnetic impulse events at high latitudes: Magnetopause and boundary layer plasma processes, *J. Geophys. Res.*, *95*, 97–107, 1990.

Lockwood, M. and S. W. H. Cowley, Observations at the magnetopause and in the auroral ionosphere of momentum transfer from the solar wind, *Adv. Space Res.*, *8*, 281–299, 1988.

Lui, A. T. Y. and D. G. Sibeck, Dayside auroral activities and their implications for impulsive entry processes in the dayside magnetosphere, *J. Atmo. Terr. Phys.*, in press, 1990.

Mende, S. B., R. L. Rairden, L. J. Lanzerotti, and C. G. Maclennan, Magnetic impulses and associated optical signatures in the dayside aurora, *Geophys. Res. Lett.*, *17*, 131–134, 1990.

Mozer, F. S., Electric field mapping in the ionosphere at the equatorial plane, *Planet. Space Sci.*, *18*, 259–263, 1970.

Nishida, A., Geomagnetic Diagnosis of the Magnetosphere, Springer-Verlag, New York, pp. 1–37, 1978.

Perona, G. E., Theory of the precipitation of magnetospheric electrons at the time of a sudden commencement, *J. Geophys. Res.*, *77*, 101–111, 1972.

Potemra, T. A., H. Luehr, L. J. Zanetti, K. Takahashi, R. E. Erlandson, G. T. Marklund, L. P. Block, L. G. Blomberg, and R. P. Lepping, Multi-satellite and ground-based observations of transient ULF waves, *J. Geophys. Res.*, *94*, 2543–2554, 1989.

Sibeck, D. G., A model for the transient magnetospheric response to sudden solar wind dynamic pressure variations, *J. Geophys. Res.*, *95*, 3755–3772, 1990.

Sibeck, D. G., W. Baumjohann, R. C. Elphic, D. H. Fairfield, J. F. Fennell, W. B. Gail, L. J. Lanzerotti, R. E. Lopez, H. Luehr, A. T. Y. Lui, C. G. Maclennan, R. W. McEntire, T. A. Potemra, T. J. Rosenberg, and K. Takahashi, The magnetospheric response to 8 minute-period strong-amplitude upstream pressure variations, *J. Geophys. Res., 94*, 2505–2519, 1989a.

Sibeck, D. G., W. Baumjohann, and R. E. Lopez, Solar wind dynamic pressure variations and transient magnetospheric signatures, *Geophys. Res. Lett., 16*, 13–16, 1989b.

Sibeck, D. G., W. Baumjohann, and R. E. Lopez, Reply, *Geophys. Res. Lett., 16*, 1200–1202, 1989c.

Siscoe, G. L., L. Davis, Jr., P. J. Coleman, Jr., E. J. Smith, and D. E. Jones, Power spectra and discontinuities of the interplanetary magnetic field: Mariner 4, *J. Geophys. Res., 73*, 61–82, 1968.

Solodyna, C. V., J. W. Sari, and J. W. Belcher, Plasma field characteristics of directional discontinuities in the interplantary medium, *J. Geophys. Res., 82*, 10–14, 1977.

Southwood, D. J. and M. G. Kivelson, The magnetohydrodynamic response of the magnetospheric cavity to changes in solar wind pressure, *J. Geophys. Res., 95*, 2301–2310, 1990.

Spangler, S., S. Fuselier, A. Fey, and G. Anderson, An observational study of MHD wave-induced density fluctuations upstream of the Earth's bow shock, *J. Geophys. Res., 93*, 845–857, 1988.

Vickrey, J. F., R. R. Vondrak, and S. J. Matthews, The diurnal and latitudinal variation of auroral zone ionospheric conductivity, *J. Geophys. Res., 86*, 65–75, 1981.

NUMERICAL SIMULATIONS ON THE MAGNETOPAUSE CURRENT LAYER

H. Okuda

Princeton Plasma Physics Laboratory, Princeton University
Princeton, NJ 08543

Abstract. One–dimensional particle simulations are carried out in order to study the current layer between a plasma and a magnetic field such as seen at the magnetopause boundary layer. When a subsonic solar wind plasma flow impinges upon a vacuum dipole magnetic field, the width of the current layer is found much smaller than the ion gyroradius and is close to theoretically predicted geometric mean of the ion and electron gyroradii. The width remains essentially the same when the magnetic field is filled with a thermal plasma whose density is smaller than the incoming solar wind density. The width, therefore, remains much smaller than the ion gyroradius. It is found that a similar sharp current layer develops in a plasma confined in a magnetic field such as seen in laboratory and space plasmas.

## Introduction

Recent observations of the dayside magnetopause current layer allowed substantial progress in the understanding of the boundary layer between a plasma and the magnetic field [Haerendel and Paschmann, 1982; Paschmann, 1984; Elphic, 1987]. Magnetosheath plasma flow speeds vary from one to a few hundred km/s which are smaller than the sound speed, $v_s$ = 200–400 km/s, but remain larger than the Alfvén speed, $V_A$, which is of the order of 100 km/s for the average density, n = 10/cm$^3$, and magnetic field, B = 15 nT in the magnetosheath. The ion temperature is $T_i$ = 5–10 x 10$^6$ °K so that the magnetosheath ion gyroradius $\rho_i$ is about 100 km for a hydrogen atom.

More recently, the ISEE multispacecraft missions provided detailed information on the dayside magnetopause, such as its thickness is 400–1600 km, much larger than the corresponding ion gyroradius [Berchem and Russell, 1982]. Recent two–dimensional particle simulations suggest that the boundary layer thickness can be broadened beyond the ion gyroradius [Berchem and Okuda, 1990]. A high frequency drift instability, drift cyclotron and lower hybrid drift instabilities, with its frequency of the order of the harmonics of the ion gyrofrequency, $\omega \sim n\Omega_i$, is found to become unstable giving rise to anomalous diffusion of the current layer. The final thickness of the current layer found in the simulations, however, is a few ion gyroradii which is still much smaller than the space observations.

In this report we present results of one–dimensional numerical simulations on the formation of a current layer between a plasma and a magnetic field. In particular, the width of the current layer is studied in detail under various conditions. First a subsonic, unmagnetized plasma flow is injected into a dipole magnetic field. It is found that a sharp current layer is established whose width is much smaller than the ion gyroradius. The observed width is indeed close to the geometric means of the electron and ion gyroradii (hybrid gyroradius) and thus in good agreement with the classical theory [Longmire, 1963]. When a thermal plasma is present in the dipole magnetic field whose density is smaller than the incoming solar wind plasma, the width is found to remain much smaller than the ion gyroradius. Finally, a plasma confined in a magnetic field is considered by assuming that such a plasma is generated in a magnetic field. It is shown that a plasma expels the magnetic field generating a current layer at the boundary between a plasma and magnetic field. The width of the current layer is again much smaller than the ion gyroradius and is close to the hybrid gyroradius.

While these simulation results are consistent with the prediction from the classical theory of the boundary layer, they are in sharp contrast to the space observations at the dayside magnetopause. We discuss the importance of various plasma instabilities at the magnetopause current layer which may be responsible for the broadening of the current layer via anomalous plasma diffusion.

## Results From One–Dimensional Simulations

We shall first discuss results obtained from the one–dimensional simulation model which is essentially the same as the model developed earlier [Hasegawa and Okuda, 1967]. A sketch of the simulation model is shown in Fig. 1. A subsonic solar wind plasma flow located at x ≤ 0 is injected along the x–direction into the dipole magnetic field given by

$$B_z(x) = \hat{z}B_o\left[1/(1 - x/x_D)^3 - 1\right] \quad \text{for } x \geq 0 \qquad (1)$$
$$= 0 \quad \text{for } x \leq 0$$

where $\hat{z}$ is the unit vector in the z–direction and $x_D$ is the location of the dipole. Both the ions and electrons are assumed drifting Maxwellians located initially at x ≤ 0 given by

Modeling Magnetospheric Plasma Processes
Geophysical Monograph 62

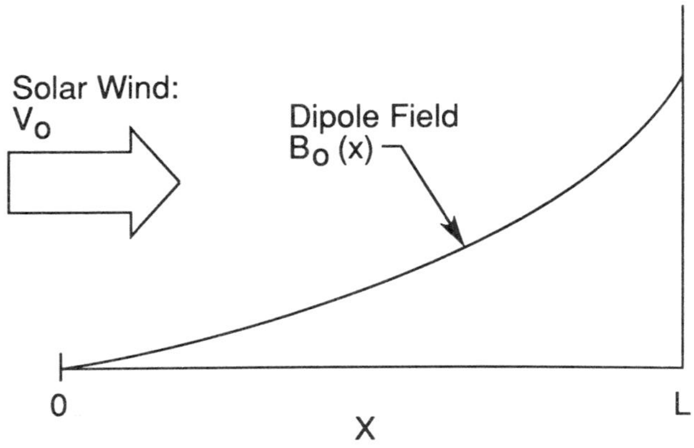


Fig. 1 A sketch of the simulation model. The unmagnetized subsonic solar wind initially located at $x \leq 0$ flows into the dipole magnetic field located at $0 \leq x \leq L_x$. A fully self–consistent interaction of the plasma with magnetic field is studied by means of particle simulation technique. Note only about one half of the solar wind particles enter the simulation system since $v_o << v_{te}$.

$$f_{e,i} = \frac{n_o}{2\pi v_{te,i}} \exp\left[-(v - v_o)^2/2v_{te,i}^2\right] \tag{2}$$

which are then allowed to stream toward the dipole field given by Eq. (1). A one–dimensional electromagnetic code was used in which Maxwell's equations are solved by using a finite difference method in x so that proper boundary conditions are easily imposed [Berchem and Okuda, 1990]. Another advantage of using a full electromagnetic code over the magnetostatic code previously used [Hasegawa and Okuda, 1967] is that the numerical algorithm in the electromagnetic code is much simpler and works well for arbitrary plasma beta in the presence of sharp discontinuities such as the magnetopause current layer. The price one must pay for employing an electromagnetic code is the use of a small time step determined from the speed of light [Lin et al., 1975]. This is, however, a small price to pay in one–dimension where the required computing time is small. It is essential here to follow the full dynamics of electrons in the code since we study a current layer which is much smaller than the ion gyroradius. More will be discussed later on this point. The boundary conditions are the following: the longitudinal electric field $E_x = 0$ at $x = L$ and for the transverse fields $E_y = 0$ at $x = 0$ and L are chosen where L is the system size of the simulation [Berchem and Okuda, 1990]. When the incoming solar wind particles are reflected back to $x = 0$, they are free to leave the system. $E_x = 0$ at $x = L$ assumes that the plasma remains charge neutral at a location away from the current layer. $E_y = 0$ at $x = 0$ assumes that the injected plasma is not magnetized so that no convection electric field exists there.

The system length is $L = 512\Delta$ where $\Delta$ is the grid length which is taken equal to the electron Debye length, $\lambda_e = (T_e/4\pi e^2 n_o)^{1/2}$ where $n_o$ is the average plasma density in the solar wind at $x \leq 0$. The number of particles per grid in the solar wind at $x \leq 0$ is $n_o = 100/\Delta$ and $T_i/T_e = 9$ in accordance with space observations [Berchem and Okuda, 1990]. The flow speed $v_o$ is chosen less than the ion thermal speed $v_i$ so that

$$v_o/v_i \approx 0.8 \tag{3}$$

for all the cases reported here. The location of the dipole $x_D$ is taken at $x_D = 7.5$ L and the skin depth is $c/\omega_{pe} = 5\Delta$ in this model where $\omega_{pe} = (4\pi e^2 n_o/m_e)^{1/2}$. The strength of the dipole field given by Eq. (1) is taken such that the electron gyrofrequency $\Omega_e = eB_o/m_e c$ is equal to $\Omega_e/\omega_{pe} = 2$ except for the last example of the confined plasma where $\Omega_e/\omega_{pe} = 3.5$ is assumed.

It is well known that the injected plasma is reflected back by the dipole magnetic field which is compressed by the plasma. The reflection takes place at a location where the plasma pressure balances with the magnetic field pressure,

$$p + n_o m_i v_o^2 \approx B^2/8\pi . \tag{4}$$

The compression of the magnetic field produces a sharp current layer at the reflection point where the width of the layer is predicted to be equal to the hybrid gyroradius [Longmire, 1963; Sestero, 1965],

$$\delta \approx (\rho_e \rho_i)^{1/2} = v_o/\Omega_H << \rho_i . \tag{5}$$

Here $\Omega_H = eB/(m_e m_i)^{1/2}c$ is the gyrofrequency of a particle whose mass is the geometric mean of the ion and electron masses. The width of the layer can be shown equal to, using Eq. (4),

$$\delta \approx c/\sqrt{2}\, \omega_{pe} \tag{6}$$

for $p \leq n_o m_i v_o^2$, where $\delta$ is independent of the electron–ion mass ratio and is essentially equal to the skin depth. Note $\omega_{pe}$ in Eq. (6) is the plasma frequency defined at the reflection point so that it is smaller than the plasma frequency in the solar wind at $x \leq 0$, since the density becomes small near the reflection point as we will find out.

Let us first study results of the simulations where the solar wind plasma is injected into a vacuum dipole magnetic field without any thermal plasma. In this case simulations using three different mass ratios, $m_i/m_e = 100$, 400 and 1600, are carried out. Shown in Figs. 2–4 are the results of the simulations using $m_i/m_e = 100$ at $\omega_{pe}t = 400$ in Figs. 2 and 3, and at $\omega_{pe}t = 1200$ in Fig. 4. The electron and ion densities shown in Fig. 2 (a) and (b) indicate the density decreases monotonically with x except for near $x \approx 200\ \Delta$ where there is a local density bump due to the reflection of the tip of the solar wind by the magnetic field. Note only the particles whose x–velocity is positive in the distribution given by Eq. (2) can enter the system which is about one–half of the total density for the electrons since $v_o << v_{te}$. Note not all the particles propagate into the magnetic field at the same velocity. In fact faster moving particles can penetrate into the magnetic field sooner since the distance of an injected particle can travel is directly proportional to the speed of that particle. This explains why the density decreases with x in Fig. 2 (a) and

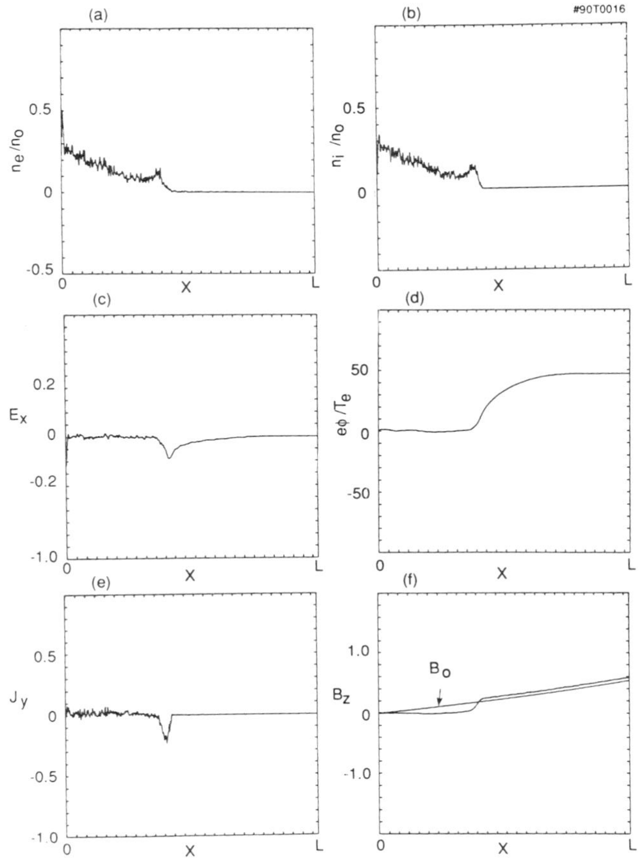


Fig. 2 Results of the one–dimensional simulation with $m_i/m_e = 100$ at $\omega_{pe}t = 400$. Shown are the (a) electron density, $n_e$, (b) ion density, $n_i$, (c) electrostatic field, $E_x$, (d) electrostatic potential, $\phi$, (e) current density, $J_y$, and (f) the total and initial magnetic field $B_z/B_o$.

(b). Therefore, particles propagating deeper into the magnetic field are primarily high energy particles, at least initially, which can be seen most clearly in the ion phase space plot shown in Fig. 3 (c) as we discuss shortly.

Shown in Fig. 2 (c) and (d) are the electrostatic field $E_x$ and the associated potential due to charge separation at the current layer. Note the electric field shown is normalized and is defined by $(e/m_e)E_x/\omega_{pe}{}^2\Delta$. Since the ions can penetrate deeper into the magnetic field while the electrons are reflected back sooner, a charge separation and hence the electric field must be set up. The magnitude of the potential must be equal to the ion energy since it is the electric field which stops ions. Therefore,

$$e\phi = \frac{1}{2} m_i v_i^2 \tag{7}$$

or

$$e\phi/T_e = \frac{1}{2}(m_i/m_e)(v_i/v_{te})^2$$

where $v_i$ is the reflected ion speed. If we had used $v_i = v_o \approx v_{ti}$, then the potential jump would be $e\phi/T_e = T_i/T_e = 9$ which is too small to explain the simulation result. As mentioned earlier, reflected ions are high energy particles whose speed is much larger than $v_o$. In fact the $x$–$v_x$ phase–space plot of the ions in Fig. 3 (c) indicates that the majority of the ions at this point have velocity $v_x \approx v_{te}$ which then can explain a large potential jump $e\phi/T_e \leq 50$. By this time some of the particles are already reflected, and therefore, the current layer and the corresponding compression of the magnetic field are already seen in Fig. 2 (e) and (f). Shown here are the normalized current density and magnetic field defined by $J_y/en_ov_o$ and $B_z/B_o$ as in Eq. (1).

The phase space of electrons and ions at $\omega_{pe}t = 400$ is shown in Fig. 3, where $x$–$v_x$ and $x$–$v_y$ plots are given for the electrons, (a) and (b), and ions, (c) and (d). The electron distribution in $x$–$v_x$, (a), is essentially the continuous drifting Maxwellian from the source located at $x \leq 0$ except for the tip near $x = 200\ \Delta$ where considerable acceleration in $v_x$ is seen due to the local electric field $E_x$. The $x$–$v_y$ phase–space, (b), indicates a sharp increase in $v_y$ near $x = 200\ \Delta$ due to the $c\,\underset{\sim}{E}_x \times \underset{\sim}{B}_z/B_o^2$ drift. The ion $x$–$v_x$ phase–space, (c), on the other hand, presents a distinct distribution where high energy ions propagate faster into the magnetic field forming a propagating beam even though the source distribution is a drifting Maxwellian [Okuda et al., 1987]. At a much later time, slower ions can also penetrate so that the beam nature of the propagating beam is less enhanced. The reason why the same beam is not seen in the electron $x$–$v_x$ phase–space is because the electrons are much lighter than the ions so that

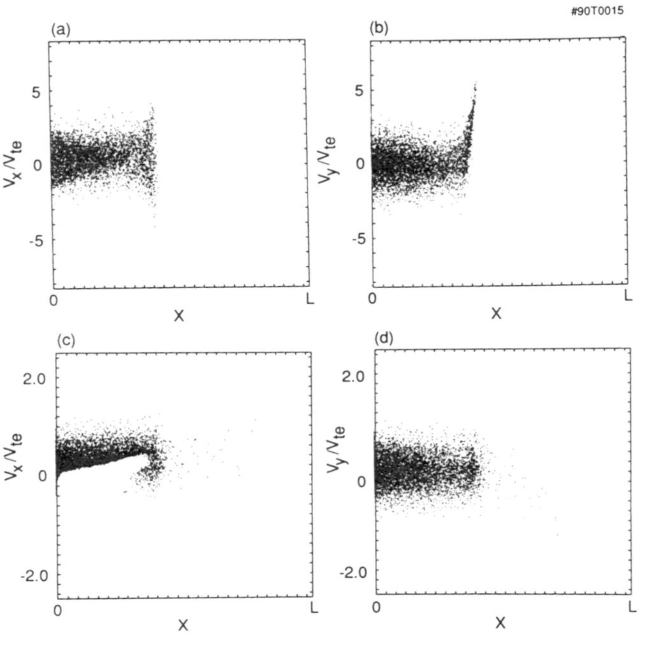


Fig. 3 Results of the one–dimensional simulation with $m_i/m_e = 100$ at $\omega_{pe}t = 400$. Shown are the electron phase space (a) $(x, v_x)$, (b) $(x, v_y)$, and ion phase space (c) $(x, v_x)$, and (d) $(x, v_y)$.

they are easily scattered by the fluctuating electric field associated with the charge separation due to propagation. There is no $v_y$ increase due to the $c\,\underset{\sim}{E}_x \times \underset{\sim}{B}_z/B_o^2$ drift of ions as shown in Fig. 3 (d) since the current layer is narrower than the ion gyroradius. Note a few ions can indeed penetrate deep into the magnetic field because of their large gyroradii.

Results of the simulations at a much later time, $\omega_{pe}t = 1200$, is given in Fig. 4 where the system reached more or less a steady state. Shown are the ion density, (a), current density (b), compressed dipole field (c), and the ion phase space $x$–$v_x$. The width of the current layer measured from Fig. 4 (b) is about

$$\delta \approx (20 \sim 30)\Delta \tag{8}$$

which compares with $\rho_e \approx 2.5\Delta$ and $\rho_i = 250\Delta$ so that $(\rho_e\rho_i)^{1/2} \approx 25\Delta$ in good agreement with Eq. (5).

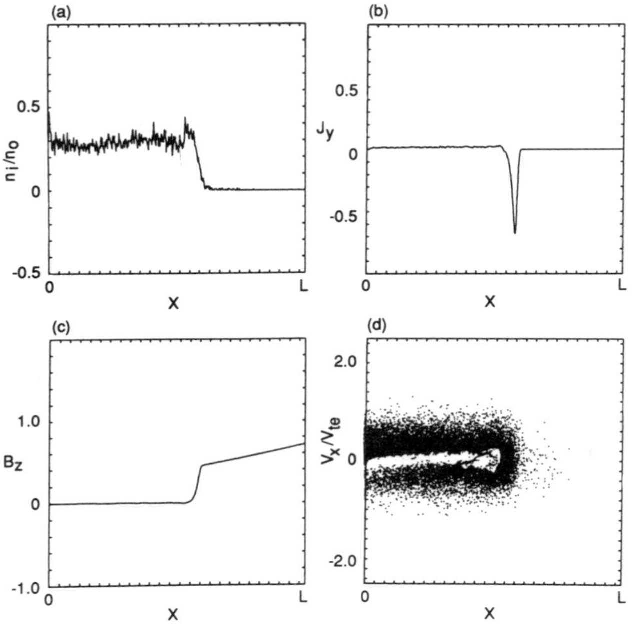


Fig. 4 Results of the one–dimensional simulation with $m_i/m_e = 100$ at $\omega_{pe}t = 1200$. Shown are the (a) ion density, $n_i$, (b) current density, $J_y$, (c) total magnetic field, $B_z/B_o$, and (d) ion phase space $(x, v_x)$.

Ion phase–space in Fig. 4 (d) shows a presence of counter–streaming ion beams which can become unstable with respect to ion–ion two–stream instability if $T_e \geq T_i$ is satisfied [Stringer, 1960]. For $x \leq 0$, the ion temperature is hotter than the electrons, however, as the high energy ions propagate into the magnetic field faster than the low energy ions, the effective ion temperature becomes progressively colder so that at some point, ion–ion instability may be triggered between the incoming ions and the reflected ones. A close examination of the ion phase–space in (d) reveals coherent modulations and trapped ions in the ion phase–space confirming a weak ion–ion two–stream instability is in operation. It is interesting to note, however, that the ion–ion two–stream instability has little effect on the width of the current layer and its main effect is the thermalization of the ion beams in velocity space.

In the following, we examine if the similar features hold for larger mass ratios. Figures 5 and 6 correspond to the simulations using $m_i/m_e = 400$ and 1600, respectively. In order to keep the flow subsonic and at the same time the reflection point at about the same location, the drift speed is reduced as the mass ratio is increased so that the relation given by Eq. (3) remains satisfied. The width of the current layer and the corresponding magnetic field profile shown in Fig. 5 (a) and (b) for $m_i/m_e = 400$ and in Fig. 6 (a) and (b) for $m_i/m_e = 1600$ remain almost the same as in Fig. 4 (b) and (c) despite the use of the larger mass ratios, confirming that the width of the current layer is much smaller than the ion grroradius.

The potential profile and the ion $x$–$v_x$ phase space shown in Fig. 5 (c) and (d) for $m_i/m_e = 400$ and in Fig. 6 (c) and (d) for $m_i/m_e = 1600$ are also very similar to Fig. 2 (d) and 4 (d). The ion phase space for the case of $m_i/m_e = 1600$, Fig. 6 (d), shows clearly the existence of holes and trapped ions which generate small potential structures in the upstream region of the boundary layer. These potential structures and the trapped ions do not modify the current layer and the shape of the magnetic field as shown in Fig. 6.

We would like to emphasize the importance of the electron inertia ($m_e$) in this problem in order to establish the current layer where the width is given by Eq. (5) or Eq. (6). As the electrons and ions compress the dipole magnetic field reaching the reflection point given by Eq. (4), electrons are first reflected which then hold the ions back preventing them from

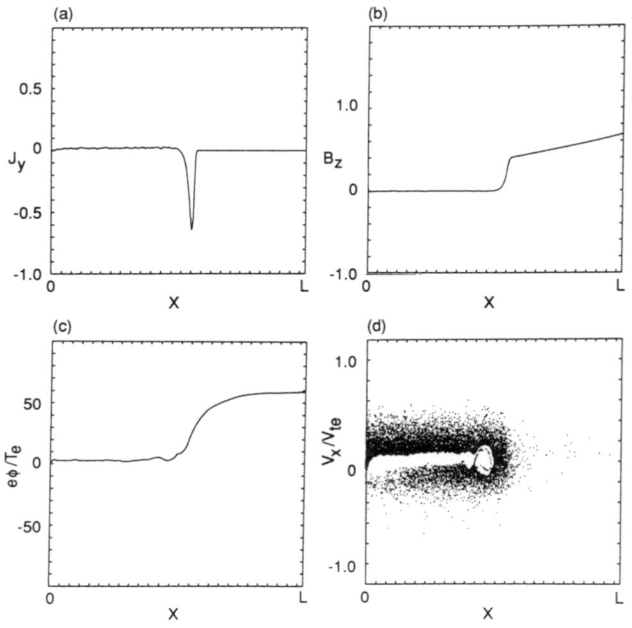


Fig. 5 Results of the one–dimensional simulation with $m_i/m_e = 400$ at $\omega_{pe}t = 2000$. Shown are the (a) current density, $J_y$, (b) total magnetic field, $B_z/B_o$, (c) electrostatic potential, $\phi$, and (d) ion phase space, $(x, v_x)$.

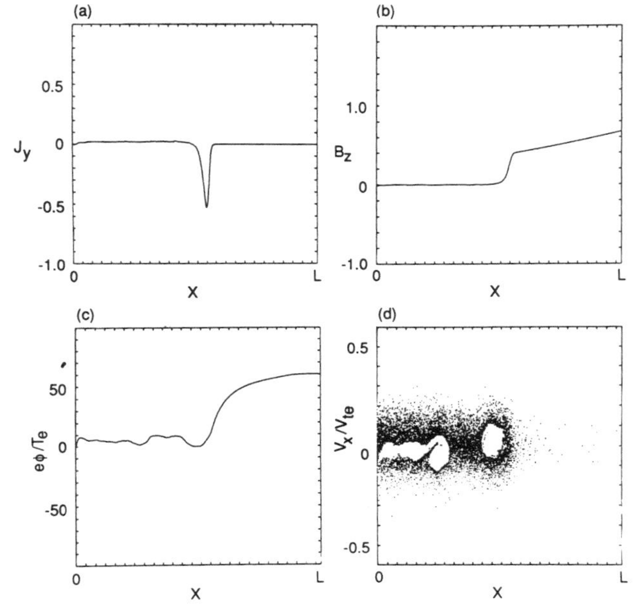


Fig. 6 Results of the one–dimensional simulation with $m_i/m_e$ = 1600 at $\omega_{pe}t$ = 5000. Shown are the (a) current density, $J_y$, (b) total magnetic field, $B_z/B_o$, (c) electrostatic potential, $\phi$, and (d) ion phase space, (x, $v_x$).

penetrating further into the magnetic field. This will then result in the width of the boundary layer essentially determined by the electron inertia, $\omega_{pe}$ given by the skin depth, $c/\omega_{pe}$. Therefore, any simulation models which neglect the electron inertia cannot be used to correctly produce the current layer of this type. One such example is the hybrid code where particle ions and massless electrons are usually followed with the assumption of charge neutrality. Such a code does not contain either the skin depth, $c/\omega_{pe}$, or the electron gyroradius and it will probably produce the width of the boundary layer of the order of the ion gyroradius or the inertial length, $c/\omega_{pi}$, both of which are much larger than the hybrid radius or the skin depth.

Next we consider what happens to the boundary layer when a thermal plasma is present in a dipole magnetic field. In order to simulate this, a low density thermal plasma is assumed trapped in a magnetic field at t = 0 whose density is uniform for $L/8 \leq x/\Delta \leq 7/8$ L. These particles may be considered magnetospheric or ionospheric origin at the magnetopause boundary layer. The density of the thermal plasma is initially taken to be 10% of the solar wind in the reservoir located at x $\leq 0$, however, the thermal plasma density can increase as the solar wind compresses the dipole magnetic field along with the thermal plasma. At the same time, the thermal plasma can also spread out by thermal motion at the same time. The temperature of the thermal plasma is taken equal to the solar wind electron temperature so that it is 1/9 of the solar wind ion temperature. Again the mass ratio of $m_i/m_e$ = 1600 is used in this example.

Shown in Fig. 7 are the instantaneous (a) thermal electron density, (b) thermal ion density, (c) beam electron density, and (d) beam ion density at $\omega_{pe}t$ = 2000. The thermal electrons are compressed by the solar wind together with the dipole magnetic field while the thermal ions can spread toward the solar wind. The thermal electrons are strongly magnetized so that they move with the magnetic field (frozen–in) while the ions are weakly magnetized so that they can move across the magnetic field because their gyroradius is very large. The beam electrons and ions behave similarly as shown in (c) and (d). While the beam electrons have a sharp boundary, the beam ions can spread out deep into the dipole magnetic field. Note the beam ions are much hotter than the electrons so that some of them have a large speed. These high energy ions can propagate further into the dipole magnetic field so long as the charge neutrality is not violated. In the presence of a thermal plasma, the beam ions need not move together with the beam electrons as they did in the absence of a thermal plasma. In the presence of beam and thermal ions, they can indeed mix with each other in contrast to beam and thermal electrons which cannot mix easily since both are frozen in their corresponding magnetic fields.

Shown in Fig. 8 are the instantaneous (a) Jy current density, (b) $B_z$ magnetic field and the initial dipole field, (c) longitudinal electric field, $E_x$, and (d) the electrostatic potential $\phi$ at $\omega_{pe}t$ = 2000. The current layer and the corresponding magnetic field ramp are clearly seen in (a) and (b) although the width of the layer is somewhat wider compared to the previous case where no thermal plasma was present. The width is about 30Δ which is still much narrower than the solar wind ion gyroradius which is about 320Δ at the layer. Both the electrostatic field, (c), and the potential, (d), at the current

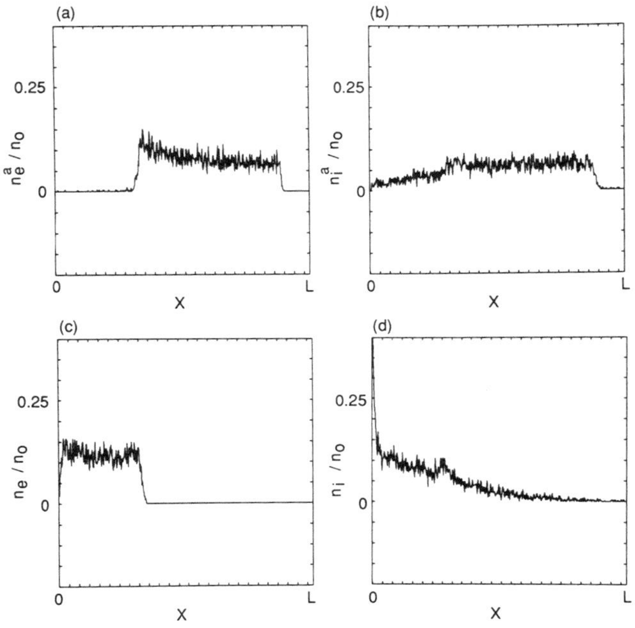


Fig. 7 Results of the one–dimensional simulation in the presence of an ambient plasma in the dipole field. $m_i/m_e$ = 1600 and $\omega_{pe}t$ = 2000. Shown are the (a) ambient electron density, $n_e^a$, (b) ambient ion density, $n_i^a$, (c) solar wind electron density, $n_e$, and (d) solar wind ion density, $n_i$.

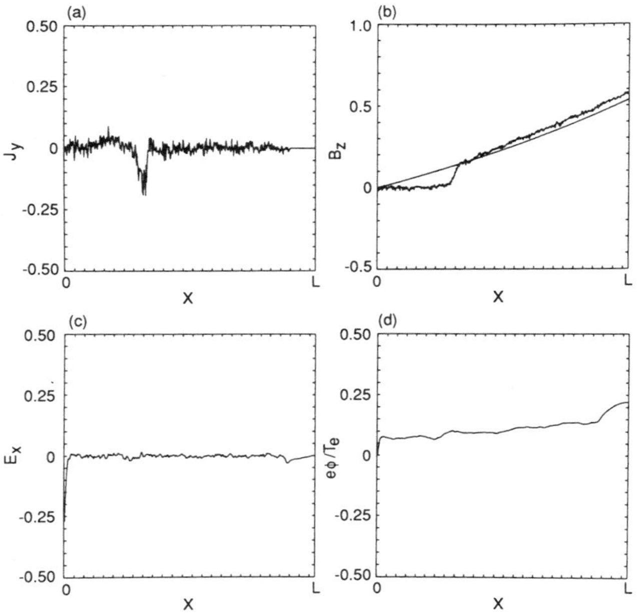


Fig. 8 Results of the one-dimensional simulation in the presence of an ambient plasma in the dipole field. $m_i/m_e$ = 1600 and $\omega_{pe}t$ = 2000. Shown are the (a) current density, Jy, (b) initial and total magnetic fields, $B_z/B_o$, (c) electrostatic field, $E_x$, and (d) electrostatic potential, $\phi$.

layer are much reduced in the presence of a thermal plasma since a charge separation can be neutralized by the thermal ions. One can still identify a dip in the electric field and a hump in the potential at the current layer near x = 140Δ although their amplitudes are much smaller than before.

In the absence of a thermal plasma, the current at the boundary was caused primarily by the electrons $c\,\underset{\sim}{E} \times \underset{\sim}{B}/B^2$ drift and the diamagnetic current given by

$$\underset{\sim}{J}_D = -\underset{\sim}{\nabla} p \times \underset{\sim}{B}/B^2 \tag{9}$$

[Berchem and Okuda, 1990]. In the presence of a thermal plasma, however, the $c\,\underset{\sim}{E} \times \underset{\sim}{B}/B^2$ current is much smaller and the diamagnetic current becomes very important. Since the diamagnetic current is proportional to ∇p, the beam and the thermal plasma produce currents which are oppositely flowing at the boundary layer. Since the solar wind pressure is much greater than the thermal plasma pressure, these two diamagnetic currents do not cancel out with each other. It is very interesting to learn then the current layer persists even in the presence of a thermal plasma which leads to charge neutralization. Even in the absence of an electric field, the current layer is produced by the diamagnetic currents generated by the solar wind electrons and the thermal electrons. The width of the layer in this case is also much smaller than the ion gyroradius since the sharp pressure profile is maintained by the electron gyration. Indeed the electron density profiles shown in Fig. 7 (a) and (c) reveal a presence of a very steep density gradient.

The solar wind and the thermal plasma reach an equilibrium with the dipole field at a later time. Shown in Fig. 9 are the (a) thermal electron density, (b) thermal ion density, (c) beam electron density, (d) beam ion density, (e) beam electron (x – $v_x$) phase space, and (f) beam ion (x – $v_x$) phase space at $\omega_{pe}t$ = 5500. The system is now in equilibrium with little changes in time. It is clearly seen that a sharp density discontinuity exists between the beam and thermal electrons while the ions are mixed up with each other regardless of their origins. The beam ion phase space indicates vortices inherent to ion–ion two–stream instability shown earlier in Fig. 6. The beam electrons phase–space, Fig. 9 (e), shows modulations associated in the instability.

Shown in Fig. 10 are the corresponding current profile Jy, (a), and the total magnetic field, (b). It is clear that both the current density and the corresponding magnetic field are similar to those without a thermal plasma verifying that a sharp

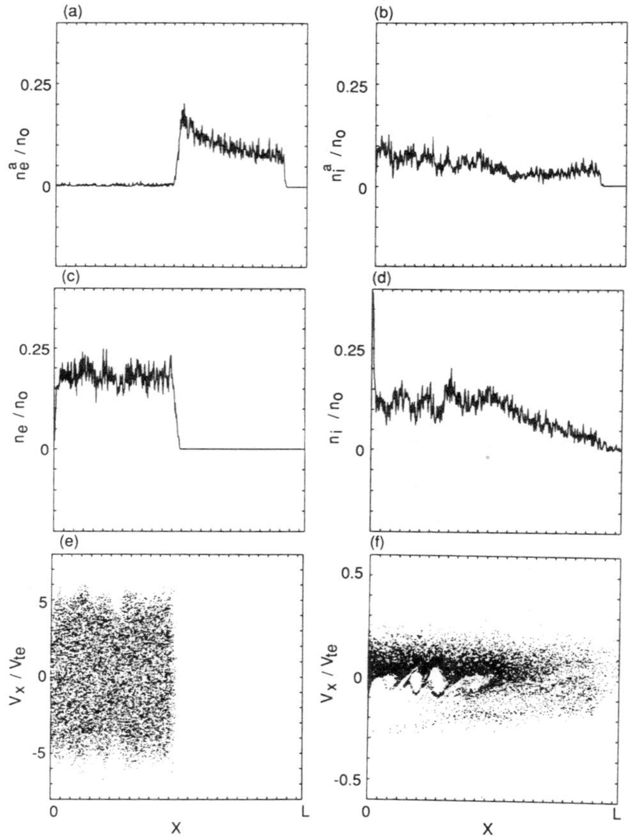


Fig. 9 Results of the one-dimensional simulation in the presence of an ambient plasma in the dipole field. $m_i/m_e$ = 1600 and $\omega_{pe}t$ = 5500. Shown are the (a) ambient electron density, $n_e^a$, (b) ambient ion density, $n_i^a$, (c) solar wind electron density, $n_e$, (d) solar wind ion density, $n_i$, (e) solar wind electron phase space, (x, $v_x$), and (f) solar wind ion phase space, (x, $v_x$).

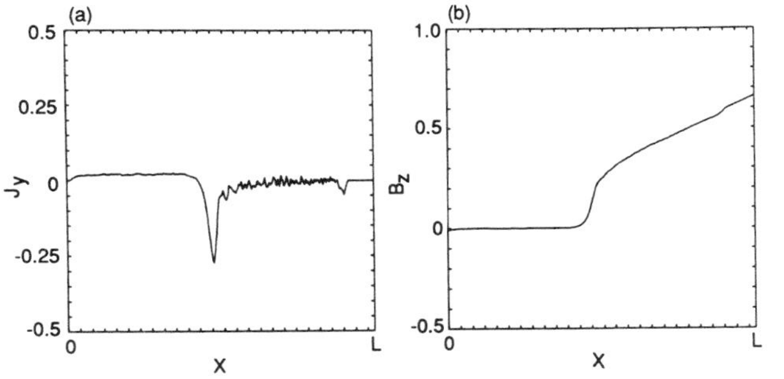


Fig. 10 Results of the one-dimensional simulation in the presence of an ambient plasma in the dipole field. $m_i/m_e$ = 1600 and $\omega_{pe}t$ = 5500. Shown are the (a) current density, Jy, and (b) total magnetic field, $B_z/B_o$.

current layer persists even in the presence of a thermal plasma which can neutralize the charge separation at the current layer. The sharp current layer and the magnetic field ramp are supported by the density discontinuity between the beam and thermal electrons which cannot be easily mixed because they are frozen in their own magnetic fields. The discontinuity is found to be (20 ~ 30) Δ wide in close agreement with the previous results without a thermal plasma and is certainly much narrower than the ion gyroradius.

It is interesting to consider a plasma confined in a magnetic field such as seen in a laboratory plasma to study the

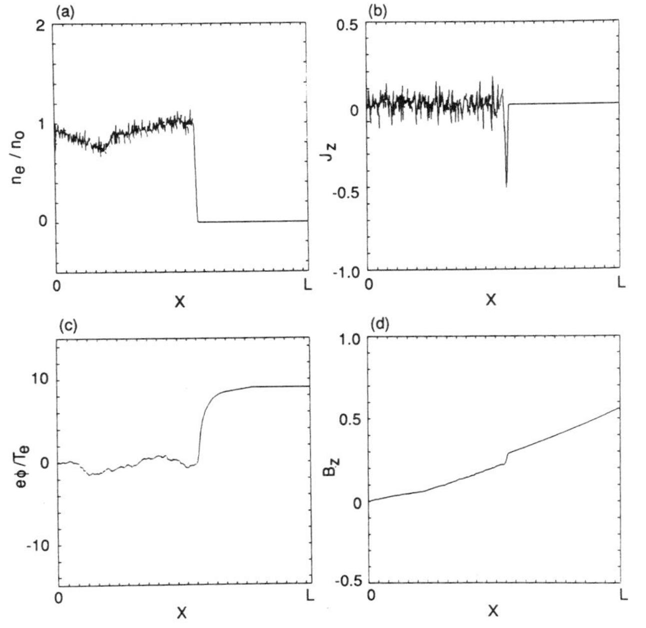


Fig. 11 Results of the one-dimensional simulation for a plasma initially created in a dipole magnetic field at $0 \leq x \leq L/2$. Shown are the (a) electron density, $n_e$, (b) current density, Jy, (c) electrostatic potential, $\phi$, (d) total magnetic field, $B_z/B_o$.

development of the current layer. In order to study this case, a thermal plasma is created at t = 0 in the same dipole magnetic field between x = 0 and x = L/2 with a uniform density. Since the plasma pressure is larger than the magnetic field pressure at t = 0, the plasma thermally expands to the right reaching an equilibrium.

Shown in Fig. 11 are the results from such a simulation using $m_i/m_e$ = 1600 and $T_e/T_i$ = 1 at $\omega_{pe}t$= 4000 where the system is near equilibrium. The electron density, (a), has a sharp discontinuity at the right boundary near x = 280Δ. The ion density follows the electron density very closely. The particles reaching at the left wall, x = 0, are elastically reflected in this model and, therefore, all the particles are confined between the wall and the magnetic field. The current profile, (b), the potential jump, (c), and the corresponding magnetic field (d) indicate the presence of a sharp current layer between a plasma and the magnetic field. The width of the layer is again close to the hybrid gyroradius and is much narrower than the ion gyroradius. Note that in this example the initial dipole field is stronger than the earlier simulations since the plasma density initially created in the magnetic field is larger. It is possible that such a sharp boundary layer can be generated in space when a plasma is created in a magnetic field. One such example may be a comet, both natural and artificial such as the AMPTE mission, where a plasma is generated by the solar radiation.

## Conclusions

We have shown three different examples of plasma simulations in which a boundary layer between a collisionless plasma and a magnetic field is self–consistently studied by means of one–dimensional electromagnetic particle simulations. It is found that in all cases the width of the current layer is much narrower than the ion gyroradius and is close to the hybrid gyroradius even in the presence of a neutralizing thermal plasma. Such a thermal plasma can neutralize the charge separation at the current layer, however, a sharp discontinuity and the current layer persist whose width is much narrower than the ion gyroradius.

While the present results are only in one–dimension, it is found that the plasma can be unstable with respect to ion–ion two–stream instability. The effects of such an instability is thermalization of the beam ions in velocity space and has little effect on the structure of the current layer in real space. The simulations presented here can be used as an equilibrium in two- and three-dimensions in order to study stability of such current layer both analytically and by numerical simulations [Berchem and Okuda, 1990]. In addition to the drift cyclotron or lower hybrid drift waves, $\omega \gtrsim \Omega_i$ in two-dimensions, we expect low frequency drift waves, drift mirror and drift compressional waves, $\omega << \Omega_i$, [Hasegawa 1969; 1971; Kadomtsev, 1965] can cause substantial broadening of the current layer [Cheng and Okuda, 1977].

Acknowledgments. The author acknowledges the collaboration with William Dorland during the course of this work. This work is supported by the NSF Grant ATM–8914792 and DOE Contract DE-AC02-76-CHO3073. Part of the computing was carried out on the Cray Y-MP at the San Diego Supercomputer Center supported by the National Science Foundation.

## References


Berchem, J., and C. T. Russell, The thickness of the magnetopause current layer: ISEE 1 and 2 observations, J. Geophys. Res., 87, 2108, 1982.

Berchem, J., and H. Okuda, A two–dimensional particle simulation of the magnetopause current layer, J. Geophys. Res., 95, 8133, 1990.

Cheng, C. Z., and H. Okuda, Formation of convective cells, anomalous diffusion, and strong plasma turbulence due to drift instabilities, Phys. Rev. Lett., 38, 708, 1977.

Elphic, R. C., The bow shock and the magnetopause, Rev. of Geophys., 25, 3, 510, 1987.

Haerendel, G., and G. Paschmann, Interaction of the solar wind with the dayside magnetosphere, in Magnetospheric Plasma Physics, edited by A. Nishida, p. 49, D. Reidel, Dosdvecht, Holland, 1982.

Hasegawa, A., and H. Okuda, Computer experiment on Ferraro–Rosenbluth problem, J. Phys. Soc. Japan, 23, 906, 1967.

Hasegawa, A., Drift mirror instabilities in the magnetosphere, Phys. Fluids, 12, 2642, 1969.

Hasegawa, A., Drift wave instabilities of a compressional mode in a high β plasma, Phys. Rev. Lett., 27, 11, 1971.

Kadomtsev, B. B., Plasma Turbulence, p. 95, Academic, New York, 1965.

Lin, A. T., J. M. Dawson, and H. Okuda, Application of electromagnetic particle simulation to the generation of electromagnetic radiation, Phys. Fluids 17, 1995, 1975.

Longmire, C. L., Elementary Plasma Physics, p. 90, Interscience Publications, New York, 1963.

Okuda, H., R. Horton, M. Ono, and M. Ashour–Abdalla, Propagation of a nonrelativistic electron beam in a plasma in a magnetic field, Phys. Fluids, 30, 200, 1987.

Paschmann, G., The earth's magnetopause, in Achievements of the International Magnetospheric Study, compiled by B. Battrick and E. Rolfe, p. 53, Rep. ESA SP-217, Noordivijk, The Netherlands, 1984.

Sestero, A., Charge separation effects in the Ferraro-Rosenbluth cold plasma sheath model, Phys. Fluids, 8, 739, 1965.

Stringer, T. E., Electrostatic instabilities in current–carrying and counter–streaming plasmas, J. Nucl. Energy Pt C, 6, 267, 1964.

# ROLE OF SMALL SCALE PROCESSES IN GLOBAL PLASMA MODELLING

G. Ganguli

Space Plasma Physics Branch
and

Y.C. Lee[(a)], P.J. Palmadesso
Special Project for Nonlinear Sciences
Plasma Physics Division, Naval Research Laboratory
Washington D.C. 20375

Abstract. Waves in space plasmas are important not only because they act as a signature of local plasma conditions, and therefore have value as a diagnostic tool, but also because they can profoundly affect the macroscopic evolution of space plasmas by modifying transport properties ("anomalous transport"). Recent results from numerical simulations using generalized fluid codes [S. Ganguli and P.J. Palmadesso, J. Geophys. Res., 92, 8673, 1987], indicate that anomalous transport due to the current driven ion cyclotron instability plays an important role in the Magnetosphere-Ionosphere coupling process. In this regard, a new class of nonlocal instabilities sustained by shears in the ambient flows is of special interest since these instabilities can be excited at lower altitudes (unlike the current driven ion cyclotron instability) and can provide local heat sources. Hence, the velocity shear driven instabilities may be very important, especially to ion upflow processes such as the 'ion cleft fountain'. Shears in the flow velocities, both parallel and perpendicular to the ambient magnetic field, are often encountered in space plasmas. The free energy residing in the velocity shears is capable of driving various instabilities and influencing others and thereby influencing macroscopic plasma behavior such as transport. Recent observations indicate significant correlations of turbulence with shears in the flow velocities. In this paper we develop a general nonlocal kinetic theory of electrostatic waves that can be excited in a magnetized warm plasma with nonuniform plasma flows along and across the ambient magnetic field, a typical scenario often encountered in magnetospheric and ionospheric plasmas. We discuss the various instabilities as limiting cases of the general theory and examine the interaction of shears in the flows parallel and perpendicular to the ambient magnetic field.

(a) Permanent Address: Dept. of Physics, University of Maryland, College Park, MD 20742

Modeling Magnetospheric Plasma Processes
Geophysical Monograph 62


## Introduction

The interaction of plasmas in the earth-sun system involves highly complex phenomenon where various spatial and temporal scale sizes interplay in an intricate fashion. The importance of large scale processes in the solar wind-magnetosphere-ionosphere interaction is obvious and needs no further comment. The study of micro- and mid-scale processes is, however, equally vital to the resolution of major outstanding questions in space physics. It is well known that turbulent small scale electric and magnetic fields, generated via a host of plasma instabilities, are prevalent throughout much of the space medium. Generation, propagation and evolution of small scale waves in space plasmas can be important in two major ways. First, waves act as a signature of local plasma conditions, and therefore have value as a diagnostic tool. Second, and most importantly, plasma instabilities can have a profound effect on the macroscopic evolution of the space plasma by modifying transport properties of the plasma, either in a direct, coherent fashion or through the effects of plasma turbulence ("anomalous transport"). One of the most obvious examples of the significance of anomalous transport is the earth's bow shock, which exists in the absence of significant classical dissipation mechanisms, i.e., Coloumb collisions. However, there are several other regions of the magnetosphere where plasma microprocesses can be important. Plasma instabilities may be responsible for (1) the viscous interaction of the solar wind and magnetospheric plasma at the magnetopause; (2) the enhancement of magnetic merging rates required for fast magnetic field line reconnection in the magnetotail; and (3) the ion heating and acceleration in the topside ionosphere, e.g., the ion cleft fountain region. Thus, by these examples, a clear need is established for the study of microscopic processes in the earth's magnetosphere since these processes feed back to impact the macroscopic plasma behaviour; and hence the knowledge of plasma microprocesses are vital for global plasma modelling.

Space plasmas are often found to sustain equilibrium flows both parallel and transverse to the ambient magnetic field. In general, these flows are nonuniform. We define shear to be the spatial gradient of the flow, $d\mathbf{V}(x)/dx = \mathbf{V}'(x)$ where $\mathbf{V}(x)$ is the flow velocity. Shear in the flow velocity is a source of free energy which can be tapped to support oscillations in a fluid. Also, the interaction of velocity shears with other sources of free energy can affect the wave propagation properties in a medium. For example, it has been well known [Rayleigh, 1896] that shear in the transverse flow velocity can support a long wavelength and low frequency Kelvin-Helmholtz (KH) instability; and more recently we [Ganguli et al., 1985a; 1988a; Nishikawa et al., 1988] have shown that a shorter wavelength and higher frequency inhomogeneous energy density driven instability (IEDDI) can also be sustained by it. While the KH instability is sustained by the gradient of the shear (i.e. $d^2V_E/dx^2$, where $V_E$ is the flow transverse to the ambient magnetic field) the IEDDI is sustained by the inhomogeneity in the energy density introduced by the localized nature of the Doppler shift. It has been shown by D'Angelo [1965] that a shear in the flow along the ambient magnetic field can lead to yet another instability with the real frequency $\omega_r \sim k_z V_d$, where $V_d$ and $k_z$ are the flow velocity and the wave vector along the ambient magnetic field (assumed here to be in the z direction) and the growth rate $\gamma$ is proportional to the shear in the parallel flow, $dV_d/dx$. Recently, the collisional limit of this instability has been investigated by Basu and Coppi [1988; 1989]. Other parallel current driven instabilities of interest to space plasmas were discussed by Kindel and Kennel [1971]. Also, when both ions and electrons are magnetized, shear can couple the flow along the magnetic field with the flow across it. Interesting physics is associated with this coupling, as for example, small shears in the transverse flow can reduce the threshold current necessary for the current driven ion cyclotron instability (CDICI) and change the spectrum and the propagation characteristics of these waves [Ganguli and Palmadesso, 1988; Ganguli et al., 1989a; Earle et al., 1990]. This is a case where the flows along and across the magnetic field interact synergistically to provide a larger growth rate for the instability. Conversely, small shear in the transverse flow can interfere and hamper the growth of the instability driven by the shear in the flow along the magnetic field [Ganguli et al., 1989b].

The scale size of the shear (L), is not always restricted to be larger than the ion gyro radius ($\rho_i$) as has been assumed so far in the examples discussed above. Flows with scale sizes of the order of the electron gyro radius ($\rho_e$) occur both naturally, associated with bow shocks [Scudder et al., 1986] and boundary layers [Romero et al. 1990], and in controlled laboratory experiments [Mostovych et al., 1989]. A high frequency ($\omega_r \sim \omega_{LH}$, the lower hybrid frequency) and short wavelength ($k_y L \sim 1$) instability is found to be associated with such flows. For such short temporal and spatial scale sizes the ion response is essentially unmagnetized while the electron response remains magnetized. Under these conditions, it can theoretically be shown that such a high frequency short wavelength instability is sustained by the shear in the flow of the electron fluid [Ganguli et al., 1988b; 1989c] across the ambient magnetic field.

Whenever $\mathbf{V}'(x)$ (the shear frequency) is much smaller than the frequency of interest it is sufficient to ignore the shear and consider the flow to be uniform. This will not be an interesting limit for this paper. Also when $\mathbf{V}'(x)$ is comparable to the frequency of interest it is sometimes sufficient to employ a local analysis as have been used in some previous investigations [D'Angelo, 1965; Basu and Coppi, 1988; 1989]. Although formally questionable, the local limit is reasonably good for estimating the growth rates of a number of instabilities. However, Mikhailovskii [1974] has discussed a high frequency ($\omega_r \sim \omega_{pe}$, the electron plasma frequency) instability driven by streaming electrons with an inhomogeneous velocity profile. This instability is critically dependent on nonlocal effects and is not accessible via a local treatment. Also, in a number of other applications the local approach can lead to inaccurate and dubious conclusions. For example, the transverse-flow-related instabilities (such as KH and IEDDI) and the interaction of these with parallel-flow-related and other instabilities cannot be investigated if the treatment is local.

Accurate stability analysis is essential for identification, classification and prediction of the onset of various instabilities. For examining the role of inhomogeneous flows a theory containing nonuniform flows both along and across the ambient magnetic field and preserving the coupling between the flows is necessary. Such a theory will provide more reliable conclusions when the resulting disturbance is a consequence of a number of driving sources acting in unison and will still be capable of isolating the constituent instabilities as limiting cases. This in turn leads to a better and more accurate knowledge of the nonlinear stage of the plasma system. The nonlinear consequences such as anomalous resistivity, viscosity, diffusion, heating etc. directly impact the macro properties of the plasma environment such as large scale plasma transport etc. [S. Ganguli and Palmadesso, 1987], and determine the profile of the observed spectrum. Hence, improved and accurate knowledge of the nonlinear stage is crucial for the analysis and interpretation of the observed data. A complicating factor in a general study of this nature is the fact that unlike the instability driven by the shear in the parallel flow [D'Angelo, 1965], the transverse-flow-related instabilities [Rayleigh, 1896, Ganguli et al., 1985a,1988a] have no local limit. These instabilities are inherently nonlocal in nature and hence the dispersion conditions can no longer be simple algebraic relations as typically obtained from a local analysis. Instead, the nonlocal dispersion conditions are eigenvalue conditions which, in general, are integrodifferential equations. Consequently, it becomes even more difficult to study the nonlinear evolution of these instabilities analytically. For accurate investigation into the nonlinear stages of these instabilities we will have to rely on numerical simulations. In the following we will develop a general nonlocal kinetic theory for the electrostatic waves sustained by flows and shears in these flows, since these are often encountered in the natural setting and may play a vital role in the global plasma dynamics.

## Theory

The equilibrium we wish to study is shown in Fig. 1. The ambient magnetic field $B_0$, is uniform and is in the z direction. A nonuniform d.c. electric field $E_0(x)$ is localized in the x direction, and leads to a cross field sheared flow in the y direction. Also, there is a magnetic field aligned flow $V_d(x)$ which is nonuniform. The scale size associated with the transverse flow is

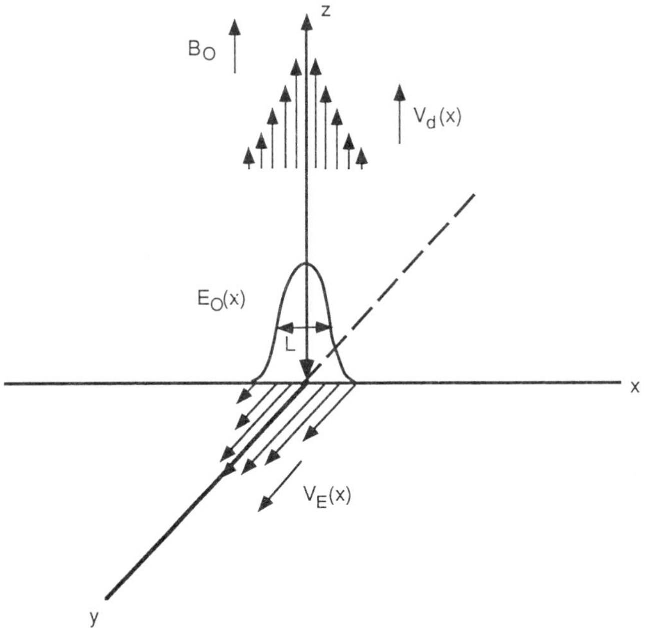


Fig. 1. The general equilibrium configuration, $V_d \neq 0$ and $V_E \neq 0$.

defined to be L while that with the parallel flow is defined to be $L_{||}$. The plasma is warm and collisionless. For obtaining the dispersion condition for this situation we shall generalize our previous theory [Ganguli et al., 1988a] to include a sheared flow along the magnetic field. The constants of the motion are (i) the total energy, $H = (v_x^2+v_y^2+\bar{v}_z^2)/2 + e\Psi(x)/m$, where $\bar{v}_z = v_z - V_d(\xi)$, $E_0(x) = -\partial\Psi(x)/\partial x$ and (ii) the guiding center position, $\xi = x + [v_y - V_E(\xi)]/\Omega$ where $V_E(\xi) = -cE_0(\xi)/B_0$ and $\Omega = eB_0/mc$, is the gyrofrequency. The appropriate initial distribution function constructed from the constants of the motion is

$$f_o(\xi, H) = N \exp\left[-\beta H\right] g(\xi), \qquad (1)$$

where $N = n_0(\beta/2\pi)^{3/2}$, $\beta = 1/v_t^2$, $v_t$ is the thermal velocity, and

$$g(\xi) = \exp\left[\beta\{e\Psi(\xi)/m + V_E^2(\xi)/2\}\right]\eta(\xi)^{-1/2}, \qquad (2)$$

where $\eta(\xi) = 1 + V_E'(\xi)/\Omega$. The quantity $\eta$ parameterizes the magnitude of the velocity shear in the transverse flow. Note that there are two crucial parameters in this problem: (i) $\eta$ and (ii) $\varepsilon = \rho_i/L$ where $\rho_i = v_t/\Omega_i$ is the ion gyroradius. We will allow $\eta$ to be arbitrary but positive while assuming $\varepsilon \ll 1$. Also, in this paper we shall restrict ourselves to the case where $L_{||}$ is either of the order of L or larger and $L > \rho_i$. The equilibrium distribution can be expressed as

$$f_o(\xi,H) = \frac{N\exp(-\frac{\beta}{2}w_\perp^2)\exp(-\frac{\beta}{2}\bar{v}_z^2)}{\sqrt{\eta(\xi)}}, \qquad (3)$$

where we have expanded the x dependence of (3) around $\xi$ and neglected terms of $O(\varepsilon^3)$ and higher. Here $w_\perp^2$ is

$$w_\perp^2 = v_x^2 + \eta(\xi)u_y^2 + \frac{V_E''(\xi)}{\Omega^2}\left(u_y\langle u_y^2\rangle - \frac{u_y^3}{3}\right), \qquad (4)$$

where $u_y = v_y - \langle v_y\rangle$ and "< >" indicates time average. Note that small terms of order $\varepsilon^2$ have to be retained in the formalism in order to recover the low frequency KH instability, as will become clear in the following.

Integrating (3) over all velocities we can show that the equilibrium density distribution, $n = n_0\{1 + o(\varepsilon^2)\}$, is uniform to order $\varepsilon$. It is possible to devise a distribution function with density uniform to any desired higher order in $\varepsilon$, but this is not necessary here. For generality, in the following we shall consider a nonuniform equilibrium density profile, i.e. $n_0 = n_0(\xi)$.

Now using the definitions

$$\phi(r',\ t') = \exp\{-i(\omega t' - k_y y')\}\phi(x'), \quad (5)$$

$$\phi(x') = \int dk_x'\ \exp(ik_x' x')\phi_k(k_x') \quad , \quad (6)$$

where $\phi$ is the electrostatic potential for the perturbed electric field and linearizing the Vlasov equation, we obtain the perturbed distribution function

$$f_1(x,\ \underline{v}) = -\beta\frac{e}{m}f_o\left[\int dk_x'\ \exp\{i(k_x' x)\}\phi_k(k_x') + i\int dk_x'\phi_k(k_x')(\omega - k_y V_g(\xi) - k_z V_d(\xi))\int_{\infty}^{t} dt' A\right], \quad (7)$$

defining $\tau = t' - t$,

$$A(t', x', y', z') = \exp\left\{i(k_x' x' + k_y(y' - y) + k_z(z' - z) - \omega\tau)\right\}, \quad (8)$$

$$V_g(\xi) = \frac{1}{\beta\eta(\xi)f_o\Omega}\frac{\partial f_o}{\partial\xi} = V_E(\xi) - \frac{V_E''(\xi)\rho^2}{2\eta^2} + \frac{\varepsilon_n\rho\Omega}{\eta} + \frac{\bar{v}_z V_d'(\xi)}{\eta\Omega} \quad . \quad (9)$$

Here $\varepsilon_n = \rho/L_n$, and $L_n = ((dn_0(\xi)/d\xi)/n_0(\xi))^{-1}$ is the scale length associated with the equilibrium density profile and $V_d'(\xi) = dV_d(\xi)/d\xi$ is the shear in the flow along the magnetic field. Thus, following the methods described earlier [Ganguli et al., 1988a], we obtain the perturbed distribution function $f_1(x,v)$, by replacing $v_z$ by $\bar{v}_z$, $(\omega - k_y V_g)$ by $(\omega - k_y V_g - k_z V_d)$ in Eq. (8) of Ganguli et al., [1988a] and including the term of the order $V_d'(\xi)$ in the expression for $V_g$ in Eq. (10) of Ganguli et al., [1988a]. In the limit of $V_d \to 0$, the above expressions reduce to the corresponding expressions of Ganguli et al., [1988a]. Also, the orbits remain unaffected except for the inclusion of $V_d$ in the z component. With these changes the derivation proceeds exactly as Ganguli et al., [1988a]. Thus, it is not necessary to repeat the details of the derivation here.

For applications in the auroral region we may assume that the shear in the transverse flow is much smaller than the ion gyro frequency as reported [Earle et al., 1990; Basu et al., 1988], so that $\eta \sim 1$ and $k_x' \sim k_x$ can be assumed. These assumptions will break down for applications to strong shear cases such as transverse shocks [Scudder et al., 1986] and the plasma sheet boundary layer-lobe interface [Romero et al., 1990] where the shear frequency is much larger than the ion gyrofrequency. Hence our theory will have to be adequately modified as in Ganguli et al., [1988b; 1989c]. Thus, for weak shears but arbitrary temperatures we can obtain the expression for the perturbed density by generalizing Eq. 26 of Ganguli et al., [1988a] to include the effects of a nonuniform flow parallel to the ambient magnetic field,

$$n_{1\alpha}(x) = -\frac{e_\alpha\beta_\alpha}{m_\alpha}\int_{-\infty}^{+\infty} dk_x \exp\left(ik_x x\right)\int_{-\infty}^{+\infty} d\xi\int_0^{+\infty} dw_\perp w_\perp\int_{-\infty}^{+\infty} dv_z\int_{-\infty}^{+\infty} dk_x'\ \phi_k(k_x')N_\alpha(\xi)\exp\left(i(k_x'-k_x)\xi\right)\exp\left(-\frac{\beta}{2}(w_\perp^2+\bar{v}_z^2)\right)\left\{1 - \sum_n \frac{(\omega_1 + \omega_{2\alpha} - \omega_\alpha^*)\ J_n^2(\sigma_\alpha)}{\omega_1 - \omega_{2\alpha} - n\Omega_\alpha - k_z\bar{v}_z} + \sum_n \frac{\omega_{3\alpha}\left(\frac{\bar{v}_z}{v_{t\alpha}}\right)J_n^2(\sigma_\alpha)}{\omega_1 - \omega_{2\alpha} - n\Omega_\alpha - k_z\bar{v}_z}\right\}. \quad (10)$$

Here $\alpha$ denotes the species, $\sigma_\alpha = k_\perp w_\perp/\Omega_\alpha$ and $\omega - k_y V_{g\alpha} - k_z V_d = \omega_1 + \omega_{2\alpha} - \omega^*_\alpha - \omega_{3\alpha}(\bar{v}_z/v_t)$ where $\omega_1 = \omega - k_y V_E - k_z V_d$, $\omega_{2\alpha} = k_y V''_E \rho_\alpha^2/2$, $\omega^*_\alpha = k_y \varepsilon_n \Omega_\alpha \rho_\alpha$, $\omega_{3\alpha} = k_y V'_d(\xi)\rho_\alpha$ and $N_\alpha(\xi) = n_{0\alpha}(\xi)(\beta/2\pi)^{3/2}$. As before we have replaced $w_\perp^2$ by it's average value $2v_t^2$, in the $\omega_{2\alpha}$ terms appearing in the denominators. Performing the velocity integrals we get,

$$n_{1\alpha}(x) = -\frac{e_\alpha\beta_\alpha}{2\pi m_\alpha}\int_{-\infty}^{+\infty} dk_x \exp(ik_x x)\int_{-\infty}^{+\infty} d\xi n_0(\xi)\int_{-\infty}^{+\infty} dk_x'\ \phi_k(k_x')\ \exp\left(i(k_x'-k_x)\xi\right)\left\{1 + \sum_n \Gamma_n(b_{1\alpha})\left[\left(\frac{\omega_1+\omega_{2\alpha}-\omega_\alpha}{\sqrt{2}|k_z|v_{t\alpha}}\right) Z\left(\frac{\omega_1-\omega_{2\alpha}-n\Omega_\alpha}{\sqrt{2}|k_z|v_{t\alpha}}\right) + \frac{\omega_{3\alpha}}{2|k_z|v_{t\alpha}} Z'\left(\frac{\omega_1-\omega_{2\alpha}-n\Omega_\alpha}{\sqrt{2}|k_z|v_{t\alpha}}\right)\right]\right\}, \quad (11)$$

where $Z'(\zeta) = dZ/d\zeta = -2[1+\zeta Z(\zeta)]$, $\Gamma_n(b) = I_n(b)\exp(-b)$, and $I_n$ are the modified Bessel functions and $Z(\zeta)$ are the plasma dispersion functions. Here $b_1 = \rho_\alpha^2(k_x^2+k_y^2)$. Expanding $\Gamma_n$ in $k_x^2\rho_\alpha^2$ so that $\Gamma_n(b_1) \sim \Gamma_n(b) + \Gamma_n'(b)k_x^2\rho_\alpha^2 + \ldots.$ , where $\Gamma_n' = d\Gamma_n/db$ and $b = (k_y\rho_\alpha)^2$. Since we have consistently retained terms up to $\varepsilon^2$, we neglect the terms $O(k_x^4\rho_\alpha^4)$ and higher which are of higher order in $\varepsilon$. Using this expansion for $\Gamma_n(b_1)$ in (11) we perform the remaining integrals to obtain

$$n_{1\alpha}(x) = -\frac{1}{4\pi e_\alpha\lambda_\alpha^2}\left[-\frac{\rho_\alpha^2}{n_{0\alpha}(x)}\frac{d^2}{dx^2}\left(n_{0\alpha}(x)\sum_n F_{n\alpha}(x)\Gamma_n'(b_\alpha)\right)\right.$$

$$+ \left( 1 + \sum_n F_{n\alpha}(x)\Gamma_n(b_\alpha)\right)\Bigg] \phi(x), \tag{12}$$

where $F_{n\alpha}(x) = A_{n\alpha}(x) - B_{n\alpha}(x)$ and,

$$A_{n\alpha}(x) = \left( \frac{\omega_1 + \omega_{2\alpha} - \omega^*_\alpha}{\sqrt{2}|k_z|v_{t\alpha}}\right) Z\left(\frac{\omega_1 - \omega_{2\alpha} - n\Omega_\alpha}{\sqrt{2}|k_z|v_{t\alpha}}\right) , \tag{13}$$

$$B_{n\alpha}(x) = - \frac{\omega_{3\alpha}}{2|k_z|v_{t\alpha}} Z'\left(\frac{\omega_1 - \omega_{2\alpha} - n\Omega_\alpha}{\sqrt{2}|k_z|v_{t\alpha}}\right) . \tag{14}$$

The quasineutrality condition i.e., $n_{1i} + n_{1e} = 0$, leads to the dispersion relation. Defining $\psi = n_{0i}\phi\Sigma F_{ni}\Gamma'_{ni}(b)$, $b = (k_y\rho_i)^2$, neglecting terms of the order $(\rho_e/\rho_i)^2$ and since we are now interested in $\omega << \Omega_e$, retaining only the n=0 harmonic for the electrons we get,

$$\left(\rho_i^2\frac{d^2}{dx^2} - \frac{1 + \sum F_{ni}(x)\Gamma_n(b) + \tau\left(1 + F_{0e}(x)\right)}{\sum F_{ni}(x)\Gamma'_n(b)}\right)\psi(x) = 0, \tag{15}$$

where $\tau = T_i/T_e$ and $\Gamma_0(b_e) \sim 1$ has been assumed. Eq. 15 is the general eigenvalue condition. The eigenvalues of (15) will determine the stability of the plasma system which includes inhomogeneous flows both along and across the ambient magnetic field and a gradient in the ambient density. Only in simple cases is it possible to obtain analytic solutions of (15). We will employ numerical methods to search for the eigenvalues of (15).

## Results

Before we embark on numerical solutions of the general eigenvalue condition it will be instructive to study the various limiting cases.

### (a) $V_d(x) = 0$, $V_E(x) \neq 0$

This limit, in which a warm magnetized plasma is subjected to a sheared flow transverse to the ambient magnetic field, was investigated in detail earlier by Ganguli et al., [1988a], and the equilibrium condition is depicted in Fig. 2. A nonuniform d.c. electric field $E_0(x)$ in the x direction initiates a sheared **E X B** flow in the y direction. Recently, there have been numerous observations, widely distributed in the auroral region, of shears in the transverse flows and associated broadband electrostatic turbulence [Mozer et al., 1977; Temerin et al., 1981; Heelis et al., 1984; LaBelle et al., 1986; Basu et al., 1986; Moore et al., 1986a,b; Waite et al., 1986; Basu et al., 1988; Earle et al., 1989]. If we set $V_d$ and $V_d'$ equal to zero then $B_{n\alpha} = 0$ and the Doppler shift due to the $V_d$ vanishes from $\omega_1$. Assuming negligible density gradient (i.e. $\omega^* = 0$) the general eigenvalue condition (15) reduces to,

$$\left(\rho_i^2\frac{d^2}{dx^2} - \frac{1 + \sum A_{ni}(x)\Gamma_n(b) + \tau\left(1 + A_{0e}(x)\right)}{\sum A_{ni}(x)\Gamma'_n(b)}\right)\psi(x) = 0. \tag{16}$$

Neglecting $\omega^*$ and $\omega_{2e}$ compared to $\omega_1 = \omega - k_y V_E$ we get $A_{ni} = \{(\omega_1 + \omega_{2i})/\sqrt{2}|k_z|v_i\} Z\{(\omega_1 - \omega_{2i} - n\Omega_i)/\sqrt{2}|k_z|v_i\}$ and $A_{0e} = \{\omega_1/\sqrt{2}|k_z|v_e\} Z\{\omega_1/\sqrt{2}|k_z|v_e\}$. Now if we take the low frequency (n=0) and long wavelength ($k_z\rho_i \sim 0$, $k_y\rho_i \sim 0$) limit, then $\Gamma_0 \sim 1 - b$, $\Gamma_0' \sim -1$ and the Z functions can be expanded for large arguments. This gives $A_{0e} \simeq -1$ and $A_{0i} \simeq -\{1 + 2\omega_{2i}/(\omega_1 - \omega_{2i})\}$. Substituting these in (16) we see that,

$$\left\{\rho_i^2\frac{d^2}{dx^2} - \left(\frac{1 - (1 + \frac{k_y V_E'' \rho_i^2}{\omega - k_y V_E} - k_y^2\rho_i^2) + \tau(1-1)}{(1)}\right)\right\}\psi(x) = 0. \tag{17}$$

From (17) we see that in this (fluid) limit the order unity terms cancel out and the terms of the order $\varepsilon^2$ become the leading terms and the general eigenvalue condition reduces to the eigenvalue condition for the classical KH instability,

$$\left\{ \frac{d^2}{dx^2} - k_y^2 + \frac{k_y V_E''(x)}{\omega - k_y V_E(x)} \right\} \psi(x) = 0. \tag{18}$$

Now it is clear that although $\varepsilon$ is small we have to keep terms to $O(\varepsilon^2)$ in our formalism in order to recover the KH instability which is a low frequency long wavelength mode. Conversely, for

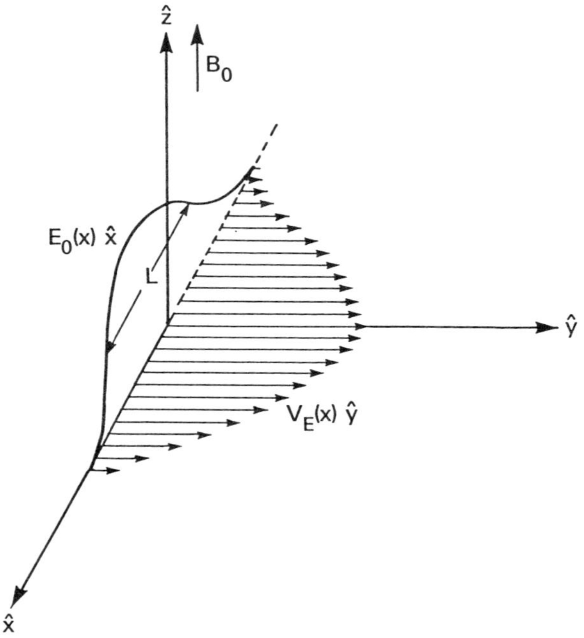


Fig. 2. The equilibrium configuration when $V_d = 0$ but $V_E \neq 0$.

higher frequencies ($n \neq 0$) and shorter wavelengths ($k_y \rho_i \sim 1$) the order unity terms do not cancel out. Now we may neglect $\omega_{2i} \sim O(\varepsilon^2) << \omega_1$ so that $A_{ni} = \{\omega_1/\sqrt{2}|k_z|v_i\}Z\{(\omega_1 - n\Omega_i)/\sqrt{2}|k_z|v_i\}$. With this $A_{ni}$, (16) becomes the eigenvalue condition for the Inhomogeneous Energy Density Driven Instability (IEDDI) [Ganguli et al., 1985a].

The eigenvalue condition (16) was solved numerically [Ganguli et al., 1988a] for the eigenvalues for both the KH and the IEDDI branches. We reproduce the results in Fig. 3. The d.c. electric field profile is $E_0(x) = E_0 \mathrm{sech}^2(x/L)$ with $\varepsilon = 0.43$, $\tau = 3.5$, and a mild density gradient $\varepsilon_n = 0.05$. The ion to electron mass ratio $\mu = 100$ was assumed so that we could compare the results with PIC simulations [Nishikawa et al., 1988]. From Fig. 3a we see that when $u = k_z/k_y$ is very small we can reproduce the classical KH results [Drazin and Howard, 1966] for long wavelengths. However, as u is increased the KH instability is rapidly damped. For $u \sim 0.01$ the KH instability was stabilized. For a fixed u we see that the KH instability peaks for a $b = b_m$ and then damps with increasing values of b due to finite Larmour radius stabilization. The instability peaks for $k_y L \sim 1$ for the electric field profile used here but for a tanh(x/L) profile the peak occurs around $k_y L \sim 0.5$. Thus, we have established that the KH instability is a low frequency long wavelength instability for which transverse propagation ($k_z \sim 0$) is necessary.

It is important to realize that the KH instability is sustained by a gradient in the shear (i.e. by the second derivative of the flow, $d^2V_E/dx^2$). It is in this regard that this branch is different from the IEDDI branch for which an explicit dependence on the gradient of shear is not essential.

In the preceding we found that the KH instability suffers from finite Larmour radius stabilization and hence it becomes less important as the plasma becomes more kinetic. Now the IEDDI branch becomes dominant as can be seen from Fig. 3b. In this plot we have kept all the parameters identical to Fig. 3a except that $u = 0.038$ is held

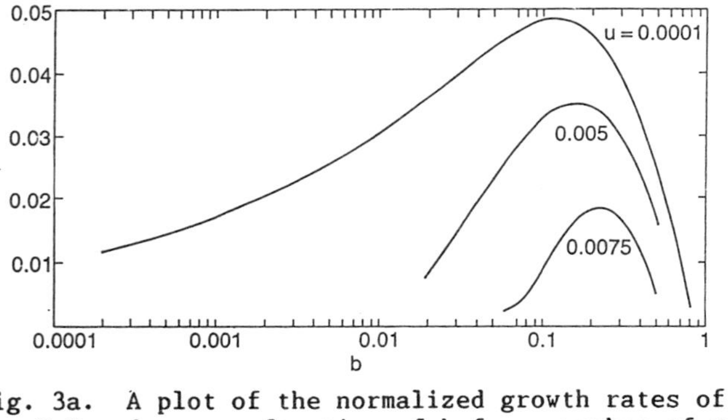


Fig. 3a. A plot of the normalized growth rates of the K-H modes as a function of b for a number of u values. Here $\varepsilon$=0.43, $\mu$=100, and $\tau$=3.5.

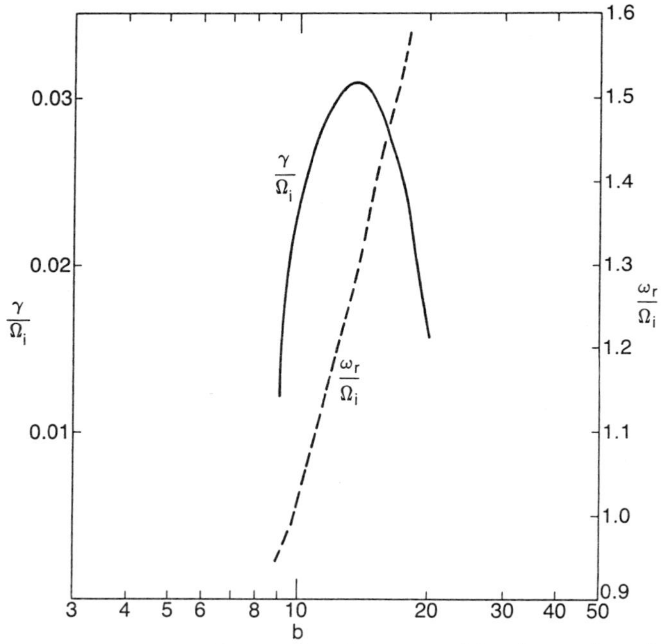


Fig. 3b. A plot of the normalized real and imaginary parts of the eigenfrequency of the IEDDI against b. Here $\varepsilon$=0.43, u=0.038, $\tau$=3.5, $\nabla_E$=0.6, $\mu$=100, $x_{n0}=1.33\rho_i$, and $\varepsilon_n$=-0.05 if $x_{n0} - \rho_i < x < x_{n0} + \rho_i$ and zero otherwise.

constant. We see a high frequency ($\omega_r \sim \Omega_i$) instability peaking around $k_y L \sim 10$. The real frequency of this branch scales as $k_y V_E^0$ and survives for larger angles of propagation. Unlike the KH branch this branch grows faster with increasing density gradient. Nishikawa et al. [1988] have investigated the IEDDI branch through a numerical PIC simulation and have discussed some nonlinear features. However, a number of aspects of the IEDDI, especially it's nonlinear evolution, deserves further investigation.

The IEDDI can be a potential source for ion heating at lower altitudes and may significantly affect plasma transport. In this regard it is of interest to the ion cleft fountain [Lockwood et al., 1985] which appears to be driven by a process which operates at low altitude to heat the ions and initiate a slow upflow. At lower altitudes, bulk parameters such as currents and electron precipitation effects are likely to provide particle heating but it is not entirely clear whether these effects alone are sufficient to initiate the heavy ion upflow. Other processes operating at higher altitude provide additional energization and accelerate the flow. A number of possible mechanisms for high altitude heating exist. Simulations with the generalized fluid model of S. Ganguli and Palmadesso [1987] have shown that at high altitudes ions are heated in the perpendicular direction when the macroscopic effects of plasma microprocesses (e.g., EIC instability and anomalous resistivity in this case) are included in the fluid model. A number of other mechanisms have been suggested for ion heating of ion conics [Crew et al., 1990, and the references therein], which also occur at higher altitude than the ion fountain source region. It is not yet clear, however, what the mechanism is for the low altitude ion heating in the cleft ionosphere. At low altitudes, where cleft ion fountain heating is observed in high density, low velocity plasma, it is difficult to excite current driven instabilities with reasonable levels of field aligned currents. However, our studies of velocity shear driven instabilities suggest that the IEDDI may be able to explain ion heating at low altitudes. Velocity shears, strong enough to excite these modes, have been observed at low altitudes [for example see Basu et al., 1988; Earle et al. 1989]. This is perhaps not so surprising since the velocity shears can be generated in several ways. One such example is the decay of low frequency long wavelength instabilities such as the KH instability etc. [see numerical simulation discussed by Keskinen et al., 1988].

A model for ion heating at lower altitudes by the IEDDI has been discussed by Ganguli et al. [1985b]. The resonance condition for the IEDDI is

$$|\omega_1(x) - k_{||}v_{||} \pm n\Omega_i| \leq \delta\omega,$$

where $\omega_1(x) = \omega - k_yV_E(x)$, $V_E(x)$ is the transverse flow and $\delta\omega$ is the resonance width in the frequency space and is a function of the wave amplitude. For a transverse flow profile that is localized in space we expect that the Doppler shifted frequency $\omega_1(x)$ will vary from a value in excess of $\Omega_i$ to less than $-\Omega_i$. Thus, we expect that there exist one or more points $x_{Rj}$ within the wavepacket at which the condition $|\omega_1(x_{Rj})| \approx \Omega_i$ is satisfied. Since $\delta\omega \sim (ek|E_k|/m_i)^{1/2} \sim kv_i(e\phi_k/T_i)^{1/2}$, where $\phi_k$ is the wave potential [Palmadesso, 1972], and since $e\phi_k/T_i \sim \delta n_i/n_i$ when $T_i \sim T_e$ we can express $\delta\omega \sim kv_i(\delta n_i/n_i)^{1/2}$ where $\delta n_i$ is the amplitude of the wave associated ion density fluctuation. Typical observed values for $\delta n_i/n_i$ can be in the range of 0.1 to 0.5 for fully developed EIC turbulence. Assuming similar values for the IEDDI and since $k_{||} << k_{\perp}$ for the IEDDI we see that $\delta\omega >> k_{||}v_{||}$ for thermal ions ($v_{||} \sim v_i$). Hence $k_{||}v_{||}$ can be neglected. Thus it follows that within a finite sized interval $\Delta x$ around $x_{Rj}$ such that $|\omega_1 - \Omega_1| \leq \delta\omega$, all thermal ions are in cyclotron resonance with the wave. To estimate the size of the resonant regions in x-space, we neglect the small Doppler shift $k_{||}v_{||}$ and expand $V_E$ around $x_{Rj}$.

$$|\omega - k_yV_E(x_{Rj}) - k_y(\partial V_E/\partial x)\Delta x \pm \Omega_i| < \delta\omega.$$

Since $|\omega_1(x_{Rj})| \sim \Omega_i$, we can solve for $\Delta x/L$ as,

$$\Delta x/L \sim (k/k_y)(v_i/V_E)(\delta n_i/n_i)^{1/2},$$

where we have replaced $\partial V_E/\partial x$ by $V_E/L$. Thus in the nonlinear regime we expect to see resonant heating involving all thermal particles within regions of size $\Delta x$ which is a significant fraction of the scale size L, since $k/k_y \sim O(1)$.

It should be noted that the IEDDI branch is not the continuation of the KH branch into the kinetic regime. In fact the KH branch can not evolve to the kinetic regime since it suffers from finite Larmour radius stabilization. The IEDDI should not be mistaken as the kinetic version of the KH instability. In fact the IEDDI has a distinct fluid limit which can be obtained as follows. Since the IEDDI is a higher frequency mode ($\omega_1 \sim \Omega_i$), keep n = 0, ±1 for the ions and $\omega_{2i} << \omega_1$ can be neglected. Consider the case where $k_z \neq 0$ such that the arguments of the ion Z-functions can be expanded for large argument (i.e., $(\omega_1 - n\Omega_i) >> \sqrt{2}|k_z|v_i$) but the argument of the electron Z-function is small (i.e., $\omega_1 << \sqrt{2}|k_z|v_e$). In this limit the general dispersion condition (16) reduces to,

$$\left(\rho_s^2\frac{d^2}{dx^2} - b_s(1 + u^2) + \left(\frac{\omega_1}{\Omega_i}\right)^2 - 1 + \frac{k_z^2\Omega_i^2}{\omega_1^2}\rho_s^2\right)\psi(x) = 0, \tag{19}$$

where $b_s = (k_y\rho_s)^2$ and $\rho_s^2 = 2T_e/m_i\Omega_i^2$. The main difference between the fluid limit of the IEDDI and the KH instability is that while it is essential for the KH instability that $k_z \sim 0$ the opposite is the requirement for the fluid IEDDI. Also note that explicit dependence on the gradient of shear (i.e., $d^2V_E/dx^2$) is not necessary for the IEDDI while it is essential for the KH instability. Detailed analysis of (19) will be provided in a future article.

(b) $V_d \neq 0$, $V_E = 0$.

In this limit a warm magnetized plasma is subjected to a sheared flow $V_d(x)$, along the ambient magnetic field a scenario very common to space plasmas. The case where $V_d$ is uniform was first analyzed by Drummond and Rosenbluth [1962] for the current driven ion cyclotron instability (CDICI) in response to a laboratory experiment by D'Angelo and Motley [1962], and later by Kindel and Kennel [1971] for ionospheric application. Subsequently, there has been an abundance of research in this limit because of it's application potentials to both space and laboratory plasmas. These works were generalized to include a nonuniform flow $V_d(x)$, by Bakshi et al. [1983], but explicit dependence on the velocity shear $V_d'(x)$ was not included. D'Angelo [1965] was first to show that $V_d'$ can sustain a new low frequency long wavelength instability which he termed as the Kelvin-Helmholtz instability since it was related to a nonuniform flow. However, the mechanism leading to this instability is very distinct from the mechanism that excites the KH instability as discussed earlier. Hence, we believe that it is more appropriate to refer to this instability as the D'Angelo instability (DI) since it is very different in character and origin from the KH instability.

The eigenvalue condition for this limit can be obtained from the general eigenvalue condition (15) by setting $V_E = 0$. First let us consider the case where the ions are stationary and the electrons are flowing with respect to the ions (i.e., $V_{di} = 0$ and $V_{di}' = 0$). This makes $B_{ni} = 0$ and $\omega_1 = \omega - k_z V_d$ making $A_{ni} = \{\omega/\sqrt{2}|k_z|v_i\} Z\{(\omega - n\Omega_i)/\sqrt{2}|k_z|v_i\}$, $A_{0e} = \{\omega_1/\sqrt{2}|k_z|v_e\} Z\{\omega_1/\sqrt{2}|k_z|v_e\}$ and $B_{0e} = -\{\omega_3/2|k_z|v_e\} Z'\{\omega_1/\sqrt{2}|k_z|v_e\}$. Thus, $F_{ni} = A_{ni}$ and $F_{0e} = A_{0e}\{1 - V_d'/u\Omega_e\} - V_d'/u\Omega_e$. If we apply the local limit then the eigenvalue condition (15) reduces to an algebraic relation since the operator '$d^2/dx^2$' can now be replaced by $-k_x^2$ and the coordinates can be reoriented so that $k_y = k_\perp$. Consequently, the dispersion relation now is simply,

$$1 + \sum_n \Gamma_n(b)\left(\frac{\omega}{\sqrt{2}|k_z|v_i}\right) Z\left(\frac{\omega - n\Omega}{\sqrt{2}|k_z|v_i}\right) + \tau'\left[1 + \left(\frac{\omega - k_z V_d}{\sqrt{2}|k_z|v_e}\right) Z\left(\frac{\omega - k_z V_d}{\sqrt{2}|k_z|v_e}\right)\right] = 0, \tag{20}$$

where $\tau' = \tau(1 - V_d'/u\Omega_e)$ and $b = (k_\perp \rho_i)^2$. Thus, if $V_d' = 0$ then (18) reduces to the dispersion relation for the CDICI [Drummond and Rosenbluth, 1962], otherwise the $V_d'$ affects the system by effectively reducing the temperature ratio. The effective temperature ratio $\tau'$, can even become negative. This is an interesting domain never explored before. As discussed by Kindel and Kennel [1971], lower $\tau$ reduces the threshold for the CDICI. However, in order for this effect to be important $V_d'/u\Omega_e$ must be comparable to unity. While $V_d'/\Omega_e$ is generally a small number in the auroral region, the propagation angle can also be a small number so that the ratio can become large enough to affect the solutions of (20). For example there is a growing ion cyclotron root ($\omega = (1.03+0.60i)\Omega_i$) for $\tau = 10$, $k_y\rho_i = 44.7$, $u = 0.0015$, $\mu = 1837$, $V_d = 15.0v_i$, if we assume a small $V_d' = 0.0019\Omega_e$. The effective temperature ratio $\tau'$ is negative and is equal to $-2.67$. We defer more detailed analysis of (20) and applications to a future article.

Let us now consider the long wavelength ($k_y\rho_i$, $k_z\rho_i \ll 1$) and low frequency ($\omega \ll \Omega_i$) limit. Since $\omega \ll \Omega_i$ we may keep only $n = 0$ for the ions and consider the case where the ions are flowing with respect to the electrons. We see that $F_{0e} = A_{0e} = (1 - \omega^*_e/\omega)\{\omega/\sqrt{2}|k_z|v_e\} Z\{\omega/\sqrt{2}|k_z|v_e\}$ and $F_{0i} = A_{0i} - B_{0i}$. In this limit the expresssions for $A_{0i} = \{(\omega_1 - \omega^*_i)/\sqrt{2}|k_z|v_i\} Z\{\omega_1/\sqrt{2}|k_z|v_i\}$ and $B_{0i} = -(\omega_{3i}/2|k_z|v_i)Z'\{\omega_1/\sqrt{2}|k_z|v_i\}$. For long wavelengths $\Gamma_0 \sim 1 - b$ and for $\omega_1 \gg |k_z|v_i$ but $\omega \ll |k_z|v_e$ the local dispersion relation reduces to that of D'Angelo [1965],

$$\bar{\omega}_1^2 + (k_y\rho_i)\varepsilon_n\bar{\omega}_1 - (k_z\rho_i)\{(k_z\rho_i) - (k_y\rho_i)\alpha_2\} = 0. \tag{21}$$

Here $\alpha_2 = V_d'/\Omega_i$, $\bar{\omega}_1 = \omega_1/\Omega_i$, and $\tau$ is assumed to be of the order unity. Thus the eigenvalue $\omega_1$ is given by,

$$\bar{\omega}_1 = \frac{1}{2}\left(-(k_y\rho_i)\varepsilon_n \pm ((k_y\rho_i)^2\varepsilon_n^2 + 4(k_z\rho_i)^2(1 - \alpha_2/u))^{1/2}\right). \tag{22}$$

As described by D'Angelo [1965], if $\alpha_2/u$ is sufficiently large then there is instability and the density gradient ($\varepsilon_n$) plays a stabilizing role in this case.

### (c) $V_d \neq 0$, $V_E \neq 0$

Now we study the combined effects of both parallel and transverse flows. For the purpose of this article we will study the kinetic limit only. A fluid formalism for low frequency long wavelength regime for this case has been discussed by Ganguli et al. [1989b]. In this limit we will ignore the low frequency and long wavelength modes such as the KH or D'Angelo instabilities (i.e., terms proportional to $V_E''$ and $V_d'$) and will concentrate on the interaction between the CDICI and the IEDDI. We will apply this to the observations reported by Earle et al., [1990] and Basu et al., [1988]. Since these observations pertain to roughly 500 kms altitude we neglect the ion-neutral collisions but do include the electron-neutral collision. The electron-neutral collision frequency $\nu_{en}$, is introduced by replacing $\omega_1$ in the argument of the electron Z-function by $\omega_1 + i\nu_{en}$ and multiplying the $(1 + \Sigma F_{ni}\Gamma_n)$ in the numerator of (15) by $\{i\nu_{en}/\sqrt{2}|k_z|v_e\} Z\{(\omega_1 + i\nu_{en})/\sqrt{2}|k_z|v_e\}$. Note that here $\omega_1 = \omega - k_y V_E - k_z V_d$. Details of this derivation was explained in Ganguli and Palmadesso [1988] and Ganguli et al., [1989a].

For evaluation of the dispersion relation we will choose typical parameters as encountered by Earle et al., [1990]. We will assume the altitude of observation in the F region to be approximately 500 kms and the plasma is oxygen dominated ($\mu = 29392$). At this altitude $\nu_i/\Omega_i \sim 0.001$ and hence will be ignored while $\nu_e/\Omega_i \sim 2.0$ will be used. Since $k_\perp\rho_i \sim 0.5$ is observed we will seek roots around this value. Typical values of $u = k_\parallel/k_\perp = 0.15$, $\tau = 1.0$, $L = 50\rho_i$ and peak $\bar{V}_d = 20.0$ is chosen for the calculations. These parameters are also very similar to Basu et al. [1988]. In this parameter range the CDICI is stable until $V_d \sim 25v_i$ or larger.

In Figure (4) we first assume idealized profiles for the electric field and the field aligned drift such that $E_\perp(x) = E_0$ and $V_d(x) = V_d$ when $x \leq |L/2|$ and 0 otherwise. We choose $L = 100\rho_i$ (so that $\varepsilon = 0.01$) here and plot the real

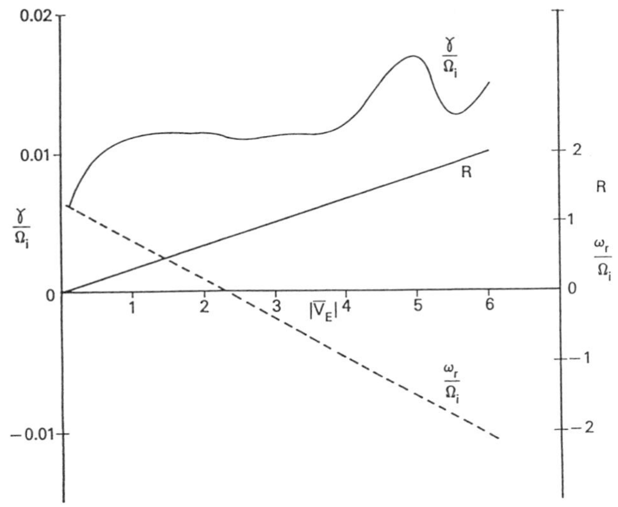


Fig. 4. The normalized real ($\omega_r/\Omega_i$, the dashed curve) and imaginary ($\gamma/\Omega_i$, the top solid curve) parts of the complex eigenvalues are plotted against $|\nabla_E|$. Here b = 0.16, $\tau$ = 1, $\varepsilon$ = 0.01, $\mu$ = 29392, $V_d = 20v_i$, $\nu_i$ = 0, $\nu_e = 2\Omega_i$ and u = 0.15. The values of R as a function of $|\nabla_E|$ is shown in the middle solid curve. Here the velocity profile is piecewise continuous.

frequency ($\omega_r$) and the growth rate ($\gamma$) normalized by the ion gyro frequency as a function of $|\nabla_E|$, the magnitude of the peak value of the $\mathbf{E}_\perp \times \mathbf{B}$ drift velocity normalized by the ion thermal velocity. In the same plot we also show the values of R ($= k_yV_E^0/k_zV_d^0$, where $V_E^0$ and $V_d^0$ are the peak values of the respective flows) for different $|\nabla_E|$. A number of interesting features are demonstrated in this figure. For R << 1 the character of the waves are similar to the CDICI ($\omega_r \geq \Omega_i$) but even a subcritical value of $V_d$ can provide growth. Effectively the threshold value of $V_d$ for the CDICI is lowered. Also, Nishikawa et al., [1990] have demonstrated this feature in a PIC simulation. The real frequency acquires a $k_yV_E$ dependence and decreases to fall under the first cyclotron harmonic as $|\nabla_E|$ is increased. Since the rest of the parameters are held constant, as $V_E$ increases so does R. As $|\nabla_E|$ and hence R increases further the $\omega_r$ changes even more and the character of the waves become increasingly different from the CDICI. For R > 1 we see that $\omega_r$ becomes proportional to $k_yV_E$ which is typical of the IEDDI [Ganguli et al., 1988a]. The transition in the spectral signatures is also to be expected, i.e., from a coherent spectrum of the CDICI to a broadband spiky spectrum of the IEDDI. The coherent spectrum of the CDICI is amply discussed in the literature, and Nishikawa et al. [1990] have discussed the typical spectrum associated with IEDDI and compared it with that of CDICI.

Thus, in the general case where $V_E \neq 0$ and $V_d \neq 0$, we find that the character of the resulting waves is dependent on the value of R, the ratio of the Doppler shifts due to the cross field and the parallel flows. Roughly, it appears that if R << 1 (i.e., $k_yV_E << k_zV_d$) then the waves are more like the CDICI [Drummond and Rosenbluth, 1962], and if R >> 1 (i.e., $k_yV_E >> k_zV_d$) then the wave character resembles the IEDDI [Ganguli et al., 1988a].

So far we have based our discussion on an idealized shear profile. We now relax this condition by choosing a continuous velocity profile,

$$V_E(\xi) = \frac{V_E^0}{C\,\exp\{(\xi^2 - \xi_0^2)/\delta\} + 1 - B}, \qquad (23)$$

where $\xi_0 = L/2\rho_i$, $C = (1 - A)^{-1}$, $A = \exp(-\xi_0^2/\delta)$ and $B = AC$. At $\xi = \xi_0$ we get $V_E(\xi) = V_E^0/2$. Also $V_d(\xi) = V_d^0(V_E(\xi)/V_E^0)$. Here $\delta = (L_1/\rho_i)^2$ is the steepness parameter and L is the extent over which the electric field (and the magnetic field aligned flow) is localized. For $\delta \to 0$ the profile (23) reduces to the idealized square pulse profile used earlier where the peak value goes to zero over just a point. For $\delta = \xi_0^2/\ln(2)$ it becomes a Gaussian while for $\delta \to \infty$ it becomes a Lorentzian. We use $L = 100\rho_i$ and increase $\delta$ to depart from the square pulse profile to a smooth profile typical to the environment of the application. For a typical case of strong shear as encountered by

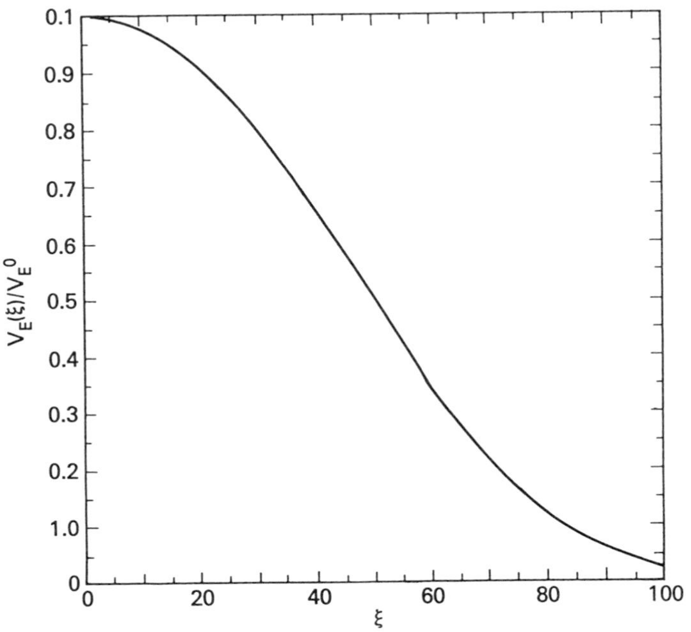


Fig. 5a. A velocity shear profile as given by Eq. (23). Here $\varepsilon$ = 0.01 and $\delta$ = 2500.

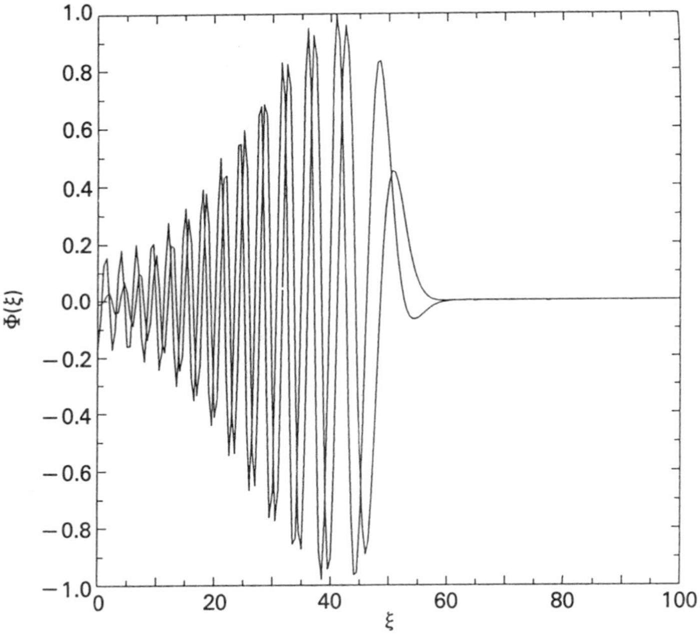


Fig. 5b. Eigenfunction for the shear profile in Fig. (5a). Here $V_E^0 = -5.2v_i$, $u = 0.21$ and rest of the parameters are identical to Fig. (4).

Earle et al., [1990] the peak value goes to zero over $L_1 = 50\rho_i$ making $\delta = 2500$ and the magnitude of the peak value is around $5v_i$. We first choose $V_E^0 = -5.2v_i$, $V_d^0 = 20v_i$ and $\delta = 2500$ in (23) and use it to numerically evaluate the eigenvalue of (15). The rest of the parameters are unchanged. The eigenvalue for $b = 0.171$ is $\omega = (-1.41+0.007i)\Omega_i$. Here we used $u = 0.21$ which makes $R \sim 1.24$ and hence the waves in such a parameter combination are IEDDI-like. The magnitude of the peak shear in the transverse flow in this case is $\alpha_1 = V_E^0/L_1\Omega_i \sim 0.1$. Figure (5a) shows the profile of the shear used while Figure (5b) is the corresponding eigenfunction.

In figure (6) we plot the eigenfunction for the case where we lower the shear and R by using $V_E^0 = -1.5v_i$, $V_d^0 = 30v_i$ and $u = 0.15$. Here $V_E^0/L_1\Omega_i \sim 0.03$ and $R \sim 0.33$. The eigenvalue for $b = 0.17$ is $\omega = (0.697+0.0155i)\Omega_i$. Thus this point is CDICI-like.

## Conclusions

We have developed a general formalism to study electrostatic waves in a magnetized plasma with inhomogeneous flows both along and across the ambient magnetic field. This formalism can explain the small scale processes that are associated with inhomogeneous flows which are widespread in the magnetosphere. S. Ganguli and Palmadesso [1987] have demonstrated that the coupling of the small scale processes associated with the CDICI with large scale plasma modelling can be very important. However, as discussed in this paper, we see that there are a number of other relevant instabilities that are capable of contributing (perhaps substantially) to the overall plasma dynamics and hence affect the magnetosphere-ionosphere coupling. We find that the IEDDI and its combination with the magnetic field aligned flow is a promising candidate for ion heating at lower altitudes and may be able to explain the ion cleft fountain occurring at lower altitudes. This topic is currently under investigation.

In this paper we have limited our analysis to the cases where the scale size of the velocity shear is larger than the ion gyroradius. Interesting results emerge when this scale size becomes smaller than the local ion gyroradius. Such situations have been observed associated with transverse shocks and the plasma sheet boundary layer-lobe interface [Romero et al., 1990]. Strong signatures of broadband electrostatic turbulence are also observed with these structures. Extensions of our theory to the case where the scale size of the shear is smaller than the ion gyroradius, reveal the existence of a number of broadbanded high frequency waves stretching all the way from under the lower hybrid frequency and approaching the electron plasma frequency. One of these branches, around the lower hybrid frequency, has recently been detected

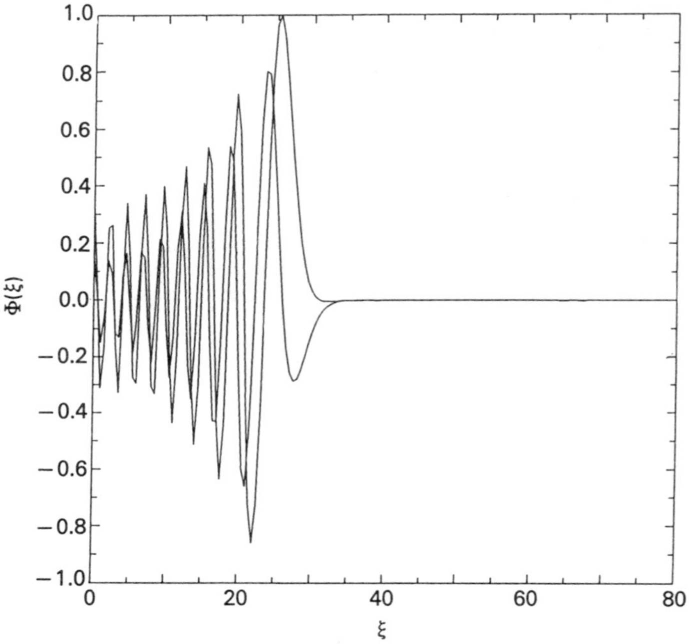


Fig. 6. Eigenfunction similar to Fig. (5b). Here $V_E^0 = -1.5v_i$, $V_d^0 = 30v_i$ and $u = 0.15$ are used while other parameters are identical to Fig. (5).

in a laboratory experiment [Mostovych et al., 1989]. These results will be reported in a forthcoming article.

Acknowledgements

This work is supported by NASA and ONR.

References

Bakshi, P., G. Ganguli and P.J. Palmadesso, "Finite width currents, magnetic shear, and the current driven ion cyclotron instability", Phys. Fluids, 26, 1808, 1983.

Basu, B. and B. Coppi, "Fluctuations associated with sheared velocity regions near auroral arcs", Geophys. Res. Lett., 15, 417, 1988.

Basu, B. and B. Coppi, "Velocity shear and fluctuations in the auroral regions of the ionosphere", J. Geophys. Res., 94, 5316, 1989.

Basu Sunanda, Santimay Basu, E. MacKenzie, P.F. Fougere, W.R. Coley, N.C. Maynard, J.D. Winningham, M. Sugiura, W.B. Hanson, and W.R. Hoegy, "Simultaneous density and electric field fluctuation spectra associated with velocity shears in the auroral oval", J. Geophys. Res., 93, 115, 1988.

Crew, G.B., T. Chang, J.M. Retterer, W.K. Peterson, D.A. Gurnett and R.L. Huff, "Ion cyclotron heated conics: theory and observations", J. Geophys. Res., 95, 3959, 1990.

D'Angelo, N. and R.W. Motley, "Electroststic oscillations near the ion cyclotron frequency", Phys. Fluids, 5, 633, 1962.

D'Angelo, N., "Kelvin-Helmholtz instability in a fully ionized plasma in a magnetic field", Phys. Fluids., 8, 1748, 1965.

Drazin, P.G. and L.N. Howard, "Hydrodynamic stability of parallel flow of inviscid fluid", Advances in Applied Mechanics, V-7, p-1, (Academic Press, New York, 1966).

Drummond, W.E. and M.N. Rosenbluth, "Anomalous diffusion arising from microinstabilities in a plasma", Phys. Fluids, 5, 1507 (1962).

Earle, G.D., G. Ganguli and M.C. Kelly, "Large velocity shears and associated electrostatic waves in the auroral F region", J. Geophys. Res., 94, 15321, 1989.

Ganguli, G., Y.C. Lee and P.J. Palmadesso, "Electrostatic ion cyclotron instability due to a nonuniform electric field perpendicular to the external magnetic field", Phys. Fluids, 28, 761, 1985a.

Ganguli, G., P. Palmadesso and Y.C. Lee, "A new mechanism for excitation of electrostatic ion-cyclotron waves and associated perpendicular ion heating", Geophys. Res. Lett., 12, 643, 1985b.

Ganguli, G., Y.C. Lee and P.J. Palmadesso, "Kinetic theory for electrostatic waves due to transverse velocity shears", Phys. Fluids, 31, 823, 1988a.

Ganguli, G., Y.C. Lee and P.J. Palmadesso, "Electron-Ion hybrid mode due to transverse velocity shear", Phys. Fluids., 31, 2753, 1988b.

Ganguli, G. and P.J. Palmadesso, "Electrostatic ion instabilities in the presence of parallel currents and transverse electric fields", Geophys. Res. Lett., 15, 103, 1988.

Ganguli, G., Y.C. Lee, P.J. Palmadesso and S.L. Ossakow, "Oscillations in a plasma with parallel currents and transverse velocity shears", in Physics of Space Plasmas (1988), SPI Conference Proceedings and Reprint Series, edited by T. Chang, G.B. Crew and J.R. Jasperse, 8, pp. 231, Scientific Publishers, Inc., Cambridge, MA, 1989a.

Ganguli, G., Y.C. Lee, P.J. Palmadesso and S.L. Ossakow, "D.C. electric field stabilization of plasma fluctuations due to a velocity shear in the parallel ion flow", Geophys. Res. Lett., 16, 735, 1989b.

Ganguli, G., P.J. Palmadesso, Y.C. Lee and J.D. Huba, "High Frequency Waves due to Velocity Shear in the Electron Flow", in Proceedings of the 1989 International Conference on Plasma Physics, New Delhi, India, p197, 1989c.

Ganguli, S.B. and P.J. Palmadesso, "Plasma transport in the auroral return current region", J. Geophys. Res., 92, 8673, 1987.

Keskinen, M.J., H.G. Mitchell, J.A. Fedder, P. Satyanarayana, S.T. Zalesak and J.D. Huba, "Nonlinear evolution of the Kelvin-Helmholtz instability in the high-latitude ionosphere", J. Geophys. Res., 93, 137, 1988.

Kindel, J.M. and C.F. Kennel, "Topside current instabilities", J. Geophys. Res., 76, 3055, 1971.

LaBelle, J., P.M. Kintner, A.W. Yau and B.A. Whalen, "Large-amplitude wave packets observed in the ionosphere in association with transverse ion acceleration", J. Geophys. Res., 91, 7113, 1986.

Lockwood, M, M.O. Chandler, J.L. Horwitz, J.R. Waite, Jr., T.E. Moore and C.R. Chappell, "The cleft ion fountain", J. Geophys. Res., 90, 9736, 1985.

Mikhailovoskii, A.B., Theory of Plasma Instabilities, Vol. I, p. 18, (Consultants Bureau, New York, 1974).

Moore, T.E., M. Lockwood, M.O. Chandler, J.H. Waite,Jr., C.R. Chappell, A. Persoon, and M. Sugiura, "Upwelling $O_+$ ion source characteristics", J. Geophys. Res., 91, 7019, 1986a.

Moore, T.E., J.H. Waite,Jr., M. Lockwood, and C.R. Chappell, "Observation of coherent transverse ion acceleration", in Ion Acceleration in the Magnetosphere and Ionosphere, Geophys. Monogr. Ser., 38, edited by T. Chang, p 50, AGU, Washington, D.C., 1986b.

Mozer, F.S., C.W. Carlson, M.K. Hudson, R.B. Torbert, B. Parady, J. Yatteau, and M.C. Kelly, "Observation of paired electrostatic shocks in polar magnetosphere", Phys. Rev. Lett., 38, 292, 1977.

Mostovych, A.N., B.H. Ripin and J.A. Stamper, "Laser-produced plasma jets: collimation and instability in strong transverse magnetic fields", Phys. Rev. Lett., 62, 2837, 1989.

Nishikawa, K.-I., G. Ganguli, Y.C. Lee and P.J. Palmadesso, "Simulation of ion-cyclotron-like modes in a magnetoplasma with transverse inhomogeneous electric field", Phys. Fluids, 31, 1568, 1988.

Nishikawa, K.-I., G. Ganguli, Y.C. Lee and P.J. Palmadesso, "Simulation of electrostatic turbulence due to sheared flows parallel and transverse to the magnetic field", J. Geophys Res., 95, 1029, 1990.

Palmadesso, P.J., "Resonance, particle trapping, and Landau damping in finite amplitude obliquely propagating waves", Phys. Fluids., 15, 2006,1972.

Lord Rayleigh, Theory of Sound, Vol-II, Chapter 21, (MacMillan and Co., London, 1896), Reprinted 1940.

Romero, H, G. Ganguli, P. Palmadesso and P.B. Dusenbery, "Equilibrium Structure of the Plasma Sheet Boundary Layer-Lobe Interface", Geophys. Res. Lett., 17, 2313, 1990.

Scudder J.D., A. Mangeney, C. Lacombe, C.C. Harvey, T.L. Aggson, R.R. Anderson, J.T. Gosling, G. Paschmann and C.T. Russell, "The resolved layer of a collisionless, high $\beta$, supercritical, quasi-perpendicular shock wave 1. Rankine-Hugoniot geometry, currents, and stationarity", J. Geophys. Res., 91, 11019, 1986.

Temerin, M.C., C. Cattell, R.L. Lysak, M.K. Hudson, R.B. Torbert, F.S. Mozer, R.D. Sharp and P.M. Kintner, " The small-scale structure of electrostatic shocks", J. Geophys. Res., 86, 11278, 1981.

Waite, J.H., Jr., T.E. Moore, M.O. Chandler, M. Lockwood, A. Persooon, and M. Sugiura, "Ion energization in upwelling events", in Ion Acceleration in the Magnetosphere and Ionosphere, Geophys. Monogr. Ser., 38, edited by T. Chang, p 61, AGU, Washington, D.C., 1986.

# The Magnetotail

# MODELING OF THE QUASI-STEADY MAGNETOTAIL

Joachim Birn

Los Alamos National Laboratory, Los Alamos, New Mexico 87545

Karl Schindler

Ruhr-Universität, 4630 Bochum 1, Federal Republic of Germany

*Abstract.* Our three-dimensional theory of the quiet magnetotail is reviewed and updated. The most advanced formulation of the theory now allows one to solve the magnetohydrodynamic (MHD) equations including field-aligned flow by reduction to a set of ordinary differential equations and an ordinary integral. These solutions represent lowest order solutions of an asymptotic expansion of the MHD equations for small electric field and weak time dependence. Applications to the magnetotail configuration are presented, which include the following properties in a self-consistent way: Flaring of the tail in $y$ and $z$, associated with the presence of $B_y$ and $B_z$; variation of plasma sheet and current sheet thickness with $x$ and $y$, associated with variations of $B_z$; field-aligned currents at the boundary between plasma sheet and lobes with the signature of "region 1" currents, i.e., toward the Earth on the dawn side and away on the dusk side, associated with the decrease of the tail flaring with distance from the Earth; net cross-tail magnetic field. Particular steady models with field-aligned flow demonstrate the important role of the flow in allowing or producing configurations with a transition from a thick closed plasma sheet to a much thinner distant open current sheet and for the possibility of thick plasmoids propagating in near equilibrium through the thin distant current sheet.

## 1. Introduction

The description of the quasi-equilibrium structure of the magnetophere is the basis for an understanding of its quiet and average properties, including particle behavior and wave propagation. It is also necessary for a study of its stability and as a possible initial state of a dynamic evolution. At present, models that include the near-Earth region and the dayside are either empirical, representing the magnetic field only [e.g., Tsyganenko, 1987, 1989], or highly idealized [e.g., Fuchs and Voigt, 1979; Voigt and Wolf, 1985; see also a recent review by Voigt and Wolf, 1988], or they result from numerical simulations of a temporal evolution with limited spatial resolution and are not necessarily steady states [e.g., Fedder and Lyon, 1987; Walker and Ogino, 1988, 1989]. The elongated structure of the magnetotail, however, has made it possible to apply asymptotic methods to gain self-consistent solutions of the steady state (i.e., $\partial/\partial t = 0$) or static (i.e., $\mathbf{v} = 0$ and $\partial/\partial t = 0$) magnetohydrodynamic equations, appropriate for the description of the quiet tail [e.g., Schindler, 1972; Birn et al., 1975, 1977].

The purpose of this paper is to present our most advanced formulation of the theory and the most general solutions describing the quasi-steady magnetotail structure. It is based largely on a summary and update of recent work of the authors [Schindler and Birn, 1986; Birn, 1987, 1989, 1990a,b] on magnetotail theory and modeling.

A magnetospheric model is usually considered as an exact or approximate solution of the time-independent or weakly time-dependent magnetohydrodynamic (MHD) equations, which can be represented in the following form, using common notation and MKS units:

$$\partial\rho/\partial t = -\nabla\cdot\rho\mathbf{v}, \tag{1}$$

$$\partial\rho\mathbf{v}/\partial t = -\nabla p - \nabla\cdot(\rho\mathbf{v}\mathbf{v}) + \mathbf{j}\times\mathbf{B}, \tag{2}$$

$$\mathbf{E} + \mathbf{v}\times\mathbf{B} = 0, \tag{3}$$

$$\partial(p/\rho^\gamma)/\partial t = -\mathbf{v}\cdot\nabla(p/\rho^\gamma), \tag{4}$$

$$\nabla\cdot\mathbf{B} = 0, \tag{5}$$

$$\nabla\times\mathbf{B} = \mu_o\mathbf{j}, \tag{6}$$

$$\partial\mathbf{B}/\partial t = -\nabla\times\mathbf{E} \tag{7}$$

The most important assumption included in (1)–(7) is that of isotropic particle distribution functions (in the frame of the bulk flow), which implies a scalar pressure as well as an adiabatic convection law (4), following from the absence of heat flux. For three-dimensional isotropic distributions the adiabatic index $\gamma$ should equal 5/3. In addition, collisions are neglected, except for their possible implicit role in isotropizing the pressure.

Usually, for the description of quiet magnetosphere structures, the inertia terms in (2) are neglected, which leads to the quasi-static force balance

$$\nabla p = \mathbf{j}\times\mathbf{B}. \tag{8}$$

This approximation is valid for flow speeds that are small in comparison to typical Alfvén wave or sound speeds. In this case Equations (8), (5), and (6) represent the magnetostatic equations, which remain from (1)–(7) for $\partial/\partial t = 0$, $\mathbf{v} = 0$, and $\mathbf{E} = 0$.

As mentioned above, three-dimensional solutions of (1)–(7), or just of the magnetostatic subset, are not available yet for the entire magnetosphere, except from numerical simulations. Empirical models including the near-Earth magnetosphere inside of $10 - 20R_E$ [e.g.,

Modeling Magnetospheric Plasma Processes
Geophysical Monograph 62

Tsyganenko, 1987, 1989], fulfill (5) and (6) by construction, but satisfy the force balance (2) or (8) only to the extent that the assumed magnetic field expressions and the data set used for determination of free parameters in the field expressions are consistent with it [see, e.g., Spence et al., 1987]. For the magnetotail, however, fairly general solutions of (1)-(7) can be derived [e.g., Birn, 1987, 1990b], due to the fact that the characteristic length scale $L_z$ for variations in the $z$ direction perpendicular to the plasma sheet is much smaller than the length scales in the other space directions, except for regions in the vicinity of the magnetopause. In the following section we will present the most recent advances in solving (1)-(7), which now provide a method to reduce these equations to a set of ordinary differential equations and an ordinary integral in the case of $\partial/\partial t = 0$ and $\mathbf{E} = 0$, when the flow is along the magnetic field in a certain frame of reference [Birn, 1990b] or vanishes completely [Birn, 1987]. Particular solutions, illustrating typical properties of the self-consistent models, will be presented in sections 3 and 4. Solutions for $\mathbf{E} \neq 0$ and $\partial/\partial t \neq 0$, which can be gained for the quasistatic case, will be discussed in section 5.

## 2. General Solution for $\mathbf{E} = 0$ and $\partial/\partial t = 0$

For $\mathbf{E} = 0$ Equation (3) implies that the flow velocity is parallel to the magnetic field, i.e.

$$\mathbf{v} = m\mathbf{B}/\sqrt{\mu_o \rho}, \tag{9}$$

where $m$ denotes the Alfvén Mach number. We will represent the magnetic field $\mathbf{B}$ by Euler potentials $\alpha$ and $\beta$, i.e.,

$$\mathbf{B} = \nabla\alpha \times \nabla\beta, \tag{10}$$

which solves Eq. (5). Equation (1) with $\partial/\partial t = 0$ now yields

$$\mathbf{B} \cdot \nabla(m\sqrt{\rho}) = 0$$

or

$$m\sqrt{\rho} = f(\alpha, \beta), \tag{11}$$

where $f$ is a quantity that is constant on field lines and thus a yet unspecified function of $\alpha$ and $\beta$. In a similar way, we find from Equation (4):

$$p/\rho^\gamma = g(\alpha, \beta). \tag{12}$$

Equations (11) and (12) can be combined to

$$pm^{2\gamma} = F(\alpha, \beta), \tag{13}$$

where

$$F(\alpha, \beta) = f^{2\gamma} g. \tag{14}$$

Using Eqs. (6) and (9) to replace $\mathbf{j}$ and $\mathbf{v}$ in (2), we find

$$\nabla(p + \frac{1}{2\mu_o} B^2) = \frac{1}{\mu_o} \nabla \cdot \left\{(1 - m^2)\mathbf{BB}\right\}. \tag{15}$$

Our basic set of equations has now been reduced to (15) together with (13) and (5) or (10), which are to be solved for $p, m$, and $\mathbf{B}$. The density $\rho$ then follows from (11) or (12), and the velocity $\mathbf{v}$ from (9).

The further solution of Eq. (15) with (13) and (5) is based on the assumption that the configuration varies predominantly in the $z$ direction with weak variations in the other two space directions. Using a smallness parameter $\epsilon$ to describe the weak variation in $x$ and $y$, we can find the scaling

$$\partial/\partial z,\ B_x,\ B_y = O(1), \qquad \partial/\partial x,\ \partial/\partial y,\ B_z = O(\epsilon) \tag{16}$$

with $p$ and $m$ being also of order unity. The scaling of $B_z$ in (16) is found from consistency with Eq. (5). A possible additional constant contribution to $B_z$ of arbitrary magnitude is chosen to vanish, consistent with the magnetotail properties. The asymptotic theory allows $B_y$ to be of order unity, although the average $B_y$ in the magnetotail is about as small as $B_z$.

We introduce further the total pressure $\hat{p}$ by

$$\hat{p} = p + \frac{1}{2\mu_o} B^2. \tag{17}$$

From Eq. (15), we now find that

$$\frac{\partial \hat{p}}{\partial z} = O(\epsilon^2). \tag{18}$$

Neglecting $\epsilon^2$ compared to unity, we obtain the frequently used pressure balance equation

$$p + \frac{1}{2\mu_o} B^2 = \hat{p}(x, y), \tag{19}$$

which has thus not been altered by the presence of the parallel flow. We rewrite the other two components of (15) introducing the new variable $\tau$ by

$$\tau = \int ds/B, \tag{20}$$

where the integral extends along field lines from an arbitrary lower boundary. Using this variable, which represents the volume of a magnetic flux tube of unit flux, we can write

$$\frac{dh}{d\tau} \equiv \mathbf{B} \cdot \nabla h, \tag{21}$$

where $h$ is an arbitrary quantity, and $d/d\tau$ denotes the derivative along a field line, i.e., for constant $\alpha$ and $\beta$

$$\frac{dh}{d\tau} \equiv \frac{\partial h(\alpha, \beta, \tau)}{\partial \tau}. \tag{22}$$

Equation (15) with (5) now yields

$$(1 - m^2)\frac{dB_x}{d\tau} - B_x \frac{dm^2}{d\tau} = \mu_o \frac{\partial \hat{p}}{\partial x}, \tag{23}$$

$$(1 - m^2)\frac{dB_y}{d\tau} - B_y \frac{dm^2}{d\tau} = \mu_o \frac{\partial \hat{p}}{\partial y}. \tag{24}$$

From (14) and (19), we find

$$\frac{dm^2}{d\tau} = -\frac{1}{\gamma} \frac{m^2}{p} \frac{dp}{d\tau}, \tag{25}$$

$$\frac{dp}{d\tau} = B_x \frac{\partial \hat{p}}{\partial x} + B_y \frac{\partial \hat{p}}{\partial y} - \frac{1}{\mu_o} \left(B_x \frac{dB_x}{d\tau} + B_y \frac{dB_y}{d\tau}\right). \tag{26}$$

If (26), (23), and (24) are inserted into (25) and $p$ is eliminated using (13), we obtain

$$\frac{dm^2}{d\tau} = \frac{\mu_o \sigma^2 m^2}{1 - m^2 - \sigma^2(B_x^2 + B_y^2)} \left( B_x \frac{\partial \hat{p}}{\partial x} + B_y \frac{\partial \hat{p}}{\partial y} \right), \tag{27}$$

where

$$\sigma^2 = \frac{1}{\mu_o \gamma} \frac{m^2}{p} = \frac{m^{2+2\gamma}}{\mu_o \gamma F} \tag{28}$$

with

$$\frac{dx}{d\tau} \equiv \mathbf{B} \cdot \nabla x = B_x, \tag{29}$$

$$\frac{dy}{d\tau} \equiv \mathbf{B} \cdot \nabla y = B_y. \tag{30}$$

We now have a complete set of explicit first order differential equations for $x, y, B_x, B_y$, and $m^2$, given by (23), (24), (27), (29), and (30) with an arbitrary choice of $\hat{p}(x,y)$ and $F(\alpha,\beta)$. This set can be integrated by standard numerical procedures. The solution follows in the form $x(\tau;\alpha,\beta)$, $y(\tau;\alpha,\beta)$, which gives $\beta = \beta(x,y,\alpha)$ if $\tau$ is eliminated. We note that special care has to be taken when $m^2 = 1$ or the denominator on the right hand side of (27) vanishes. When $m^2 = 1$ everywhere, only rather restricted solutions with $p + B^2/2\mu_o = \hat{p} = const$ are possible, as implied by (15), with $\rho = \rho(\alpha,\beta)$, $p = P(\alpha,\beta)$, following from (11) and (12), and, consequently, $B = B(\alpha,\beta)$.

For further integration, we use the representation of $\mathbf{B}$ by Euler potentials, Eq. (10), and obtain

$$B_x^2 + B_y^2 = \left| \nabla \beta \right|_\alpha^2 \left( \frac{\partial \alpha}{\partial z} \right)^2 \tag{31}$$

where

$$|\nabla \beta|_\alpha^2 = \left( \frac{\partial \beta(x,y,\alpha)}{\partial x} \right)^2 + \left( \frac{\partial \beta(x,y,\alpha)}{\partial y} \right)^2. \tag{32}$$

We now solve (31) for $\partial\alpha/\partial z$, inserting $B_x^2 + B_y^2$ from the pressure balance (19). If $\partial\alpha/\partial z$ is inverted to gain $\partial z(x,y,\alpha)/\partial\alpha$, we can then find $z$ as a function of $x, y$, and $\alpha$ by integration:

$$z = \pm \int |\nabla \beta|_\alpha \frac{d\alpha}{\sqrt{2\mu_o(\hat{p} - p)}}. \tag{33}$$

In (33) $p$ is a function of $x, y$, and $\alpha$, given by (13), with $m^2$ and $\beta$ expressed as functions of $x, y$, and $\alpha$ through the solutions of our set of ordinary differential equations (23), (24), (27), (29), and (30). A free function of $x$ and $y$ is contained in the yet undetermined lower boundary of the integral (33).

The number of differential equations can be reduced using an integral which can be found from (23), (24), and (27):

$$\frac{\gamma}{\gamma - 1} m^{2(1-\gamma)} + m^{2(2-\gamma)} - m^4 \hat{p}/F = R(\alpha,\beta) \quad \text{for } \gamma \neq 1 \tag{34}$$

and

$$m^2 + \ln m^2 - m^4 \hat{p}/F = R(\alpha,\beta) \quad \text{for } \gamma = 1. \tag{34a}$$

(A further reduction can be obtained, if the variable $\tau$ is replaced, for instance, by $x$ through (29).)

In the static case, $\mathbf{v} = \mathbf{0}$ or $m = 0$, one finds the simplified set of equations:

$$dB_x/d\tau = \mu_o \partial\hat{p}/\partial x \tag{35}$$

$$dB_y/d\tau = \mu_o \partial\hat{p}/\partial y \tag{36}$$

together with (29), (30), and (33), where $p$ is now a function of $\alpha$ and $\beta$ only, to be determined from boundary conditions or other constraints.

## 3. Special Static Solutions

### 3.1 $\hat{p} = \hat{p}(r)$ *where* $r^2 = (x-a)^2 + (y-b)^2$

It is particularly easy to derive static solutions when the total pressure function can be written as a function of $r$ only in a cylindrical coordinate system $r$, $\phi$, $z$ with an arbitrary center at $x = a$, $y = b$ [Birn, 1979, 1987]. In this case one finds $\mathbf{B} \cdot \nabla(rB_\phi) = 0$, i.e., $rB_\phi$ is constant along any field line. For the average magnetotail configuration it is reasonable to assume symmetry around the equatorial plane. This implies

$$B_\phi \equiv 0, \tag{37}$$

that is, field lines must lie in planes $\phi = const$. In this case, the coordinate $\phi$ can be chosen as one of the Euler potentials, say,

$$\beta = \phi. \tag{38}$$

For suitable choices of $p = P(\alpha,\beta)$, the final solution can be given in analytic form. For instance, for

$$p = p_N \exp(-2\lambda(\beta)\alpha) \tag{39}$$

the magnetic field and the pressure are given by [Birn, 1987]

$$B_r = -\hat{B} \tanh\zeta, \tag{40}$$

$$B_z = \frac{z}{r} \frac{d}{dr}(r\hat{B}) \tanh\zeta - L\frac{d\hat{B}}{dr}, \tag{41}$$

$$p = \hat{p}(r)/\cosh^2\zeta, \tag{42}$$

where

$$\hat{B}(r) = \sqrt{2\mu_o \hat{p}(r)}, \tag{43}$$

$$L(r,\phi) = [\lambda(\phi) r \hat{B}(r)]^{-1}, \tag{44}$$

and

$$\zeta = z/L(r,\phi). \tag{45}$$

The solution (40)–(45) still contains two free functions, $\hat{B}(r)$, the lobe magnetic field strength, and $\lambda(\phi)$, a function that determines the variation of the characteristic scale length $L$, equivalent to a current sheet half-thickness, across the tail. Both functions can be reasonably well adjusted to observations. A particular model is shown in Figure 1. It is adjusted to tail observations between 20 and $60R_E$ and extended to $140R_E$. The figure shows magnetic field lines in three different planes $\phi = const$ with the dashed line indicating the scale length $L$ and the

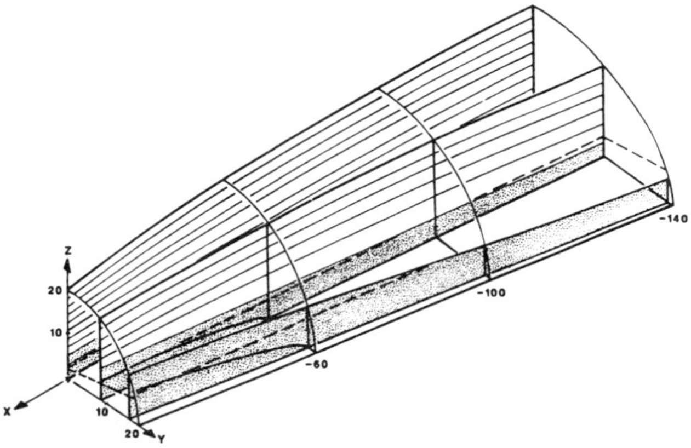


Fig. 1 Static magnetotail solution with $\hat{p} = \hat{p}(r)$, where $r^2 = (x-a)^2 + y^2$, and $B_\phi = 0$ after Birn [1987]. Magnetic field lines are shown in three different planes $\phi = const$. The dashed lines represent the scale length $L(x, y)$ for variations with $z$, equivalent to a current sheet half-thickness, and the shaded region indicates the region of closed field lines, that is, the plasma sheet.

shaded region representing the region of closed (that is, plasma sheet) field lines. This particular example does not include a distant neutral line to terminate the closed field line region. With a different choice for $\hat{B}(r)$, however, a distant neutral line can be modeled also. In the quasi-static theory, without including the dynamics of magnetic reconnection which may occur at a neutral line, the location of the neutral line follows from the form of $\hat{B}(r)$, which is determined by the force balance between the magnetotail and the solar wind plasma. It is associated with a minimum of $\hat{B}(r)$, corresponding to the location where the tail flaring and the influence of the solar wind ram pressure cease [Wiechen and Schindler, 1988; Schindler et al., 1989].

Despite the restricting assumptions used for the derivation of (40)–(45), this solution includes several properties necessary for a quantitatively satisfying magnetotail model: A flaring of the tail with both $y$ and $z$, which can be adjusted; the presence of a normal field component $B_z$ in the neutral sheet with the balance of the corresponding Lorentz forces by pressure gradients; a variation of the current sheet and plasma sheet thickness along and across the tail, associated with a variation of $B_z$, which can also be adjusted. The major restriction is the absence of field-aligned curents, which follows from the assumed form for $\hat{p}$ and the consequence that field lines lie in planes $\phi = const$. This condition will be released in the following subsection.

*3.2 Extended tail solutions with $j_\parallel \neq 0$*

For more general functions $\hat{p}(x, y)$ analytic solutions seem no longer possible, and Equations (35), (36), (29), (30), and (33) have to be integrated numerically. A particular solution, obtained by Birn [1989], is shown in Figure 2. The figure shows, in a perspective view from the tail, magnetic flux surfaces formed by field lines that originate from a circle of $20R_E$ radius at $x = -20R_E$, extending out to $x = -120R_E$. If the magnetopause were a flux surface (that is, a tangential discontinuity) and if the present model would be valid out to the magnetopause and if we further assume that the magnetopause is indeed a circle of $20R_E$ radius at $x = -20R_E$, then this surface would in fact represent the magnetopause shape of the model. The most prominent feature of this particular solution, representative of others, is the bulging of the tail in the low-latitude plasma sheet region. This bulging is due to a difference in the degree of flaring of field lines in their equatorial projections between plasma sheet and lobes, corresponding to a shear of the magnetic field, that is, a rotation of the equatorial projection of **B** with altitude $z$ above the equatorial plane. Such a rotation of the field direction is indeed observed by Candidi et al. [1990]. It is directly related to the presence of field-aligned currents which produce such a shear. In the present model, it is the consequence of just two boundary conditions, namely an adjustment of the tail flaring in the near tail between 20 and $60R_E$ to the observed flaring and the requirement that at large distances field lines should become aligned with the tail axis, which implies a decrease of the flaring with distance, as is also observed [Slavin et al., 1985]. We should note that, in contrast to the deformation of interior flux surfaces, a bulging of the magnetopause is not an inevitable implication of the observed properties and the presence of field-aligned currents. Several of our assumptions break down in the vicinity of the low-latitude magnetopause. The gradient scale length in $y$ becomes small at the magnetopause and fast flow is present in the low-latitude boundary layer. Furthermore, the magnetopause need not be closed.

The distribution of the field-aligned currents at $x = -20R_E$, associated with the magnetic shear, is shown more explicitly in Figure 3. The top part of the figure shows lines of constant current density $j_\parallel$ in a north-dusk sector of the tail, while the bottom part shows the profile of $j_\parallel$ as a function of $z$ for constant $x$ and $y$. The current densities are normalized by $j_N$, the peak cross-tail current density at $x = -20R_E$ and $y = 0$. The contour lines in the top panels represent $j_\parallel$ values at multiples of 2.5% of $j_N$. The signature of these currents is that of region 1 field-aligned currents flowing toward the Earth on the dawnside and away on the duskside [Iijima and Potemra, 1976]. The total current $I_\parallel$ in each direction is about $4 \times 10^5$ A, using a lobe field strength $B_N = 25$ nT at $x = -20R_E$. The magnitude of $I_\parallel$ increases toward the Earth. This indicates that not only contributions from further downtail are conserved, but that new contributions from the near flanks are added closer to the Earth. We do, therefore, expect that these field-aligned currents actually reach the Earth and very likely become even larger due to additional contributions from the morning and evening flanks not included in our model. These values therefore seem consistent with observations of the region 1 currents that persist even during quiet times [Iijima and Potemra, 1976].

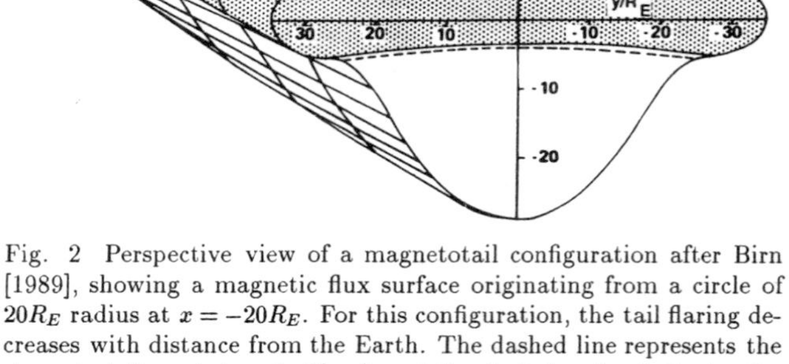

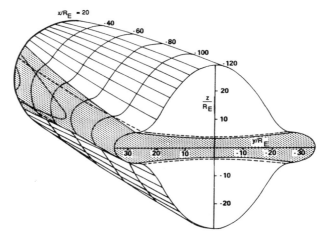

Fig. 2 Perspective view of a magnetotail configuration after Birn [1989], showing a magnetic flux surface originating from a circle of $20R_E$ radius at $x = -20R_E$. For this configuration, the tail flaring decreases with distance from the Earth. The dashed line represents the characteristic scale length $L(x, y)$, and the shaded region indicates the closed plasma sheet region, which extends to infinity in this particular model.

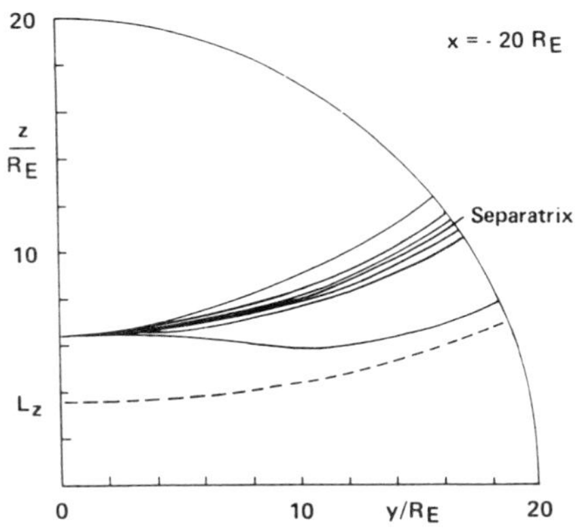


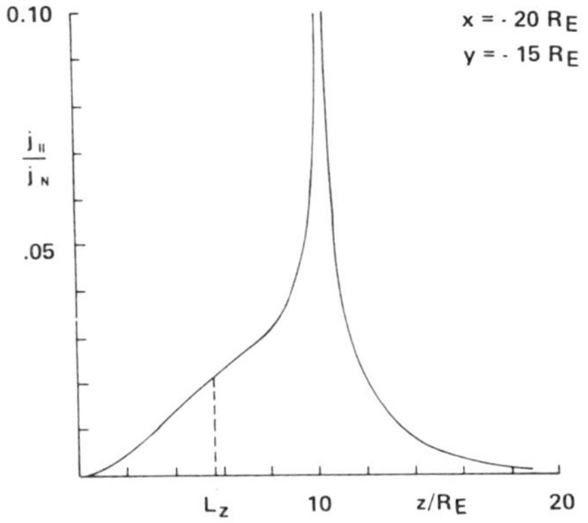


Fig. 3 Distribution of the field-aligned current density at $x = -20R_E$ for the configuration of Figure 2. The top panel shows contours of constant $j_\parallel$ with its magnitude increasing monotonically toward the separatrix from either side. The bottom panel shows a cut through the distribution of $j_\parallel$ at $y = -15R_E$. The values of $j_\parallel$ are normalized by $j_n$, the peak cross-tail current density at $x = -20R_E$. The dashed line indicates again the characteristic scale length $L$.

*3.3 Solutions with net cross-tail magnetic field*

The symmetry of the magnetotail models of sections 3.1 and 3.2 with respect to the midnight meridian and the equatorial planes is a reasonable assumption for the average magnetotail. For a particular instantaneous model, however, the absence of a net cross-tail magnetic field component $B_{yN}$, which is required for such symmetry, is a rather singular case [e.g., Fairfield, 1979; Lui, 1984; Tsurutani et al., 1984]. Two-dimensional ($\partial/\partial y = 0$) models, which imply that $B_y$ is constant on field lines, have been discussed, e.g., by Hilmer and Voigt [1987]. In the three-dimensional case, again, numerical integration of Equations (35), (36), (29), (30), and (33) is required. Such solutions were obtained by Birn [1990a]. Figure 4 shows a representative example, again in the form of a magnetic flux surface originating from $r = 20R_E$ at $x = -20R_E$, drawn in a perspective view from the tail. The magnitude of the net cross-tail field component $B_{yN}$ at $x = -20R_E$ is 2% of the lobe field strength, which is rather typical [Fairfield, 1979]. Note that, in contrast to the two-dimensional ($\partial/\partial y = 0$) models, $B_y$ is no longer constant along field lines in a three-dimensional model.

Figure 4 demonstrates two new characteristic effects. The first one is a separation of the two lobe regions from each other, increasing with

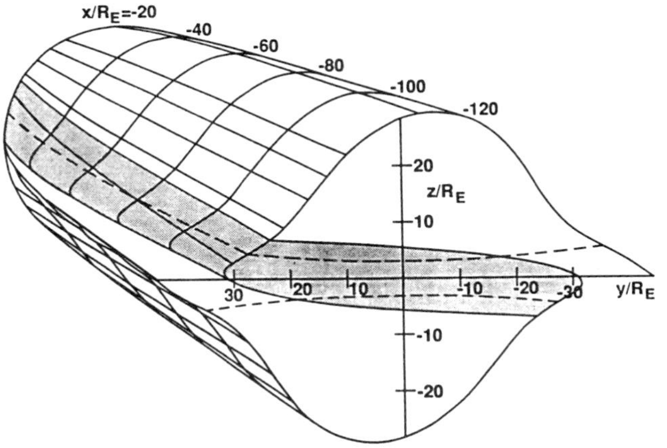


Fig. 4 Perspective view of a magnetotail configuration with a net cross-tail magnetic field component $B_{yN}$ after Birn [1990a]. As in Figure 2, the configuration is represented by a magnetic flux surface originating from a circle of $20R_E$ radius at $x = -20R_E$. The dashed lines indicate again the scale length $L$ and the shaded region the region of closed field lines with both ends inside of the $20R_E$ circle at $x = -20R_E$.

distance down the tail. This can also be found in the simpler two-dimensional models [e.g., Hilmer and Voigt, 1987; Voigt and Hilmer, 1987]. It is a direct consequence of the fact that the net $B_y$ field persists through the lobes. Whether this effect corresponds to a similar behavior of the magnetopause, depends on to what extent the magnetopause is open in these regions. If we postulate a symmetric magnetopause surface, Figure 4 suggests an asymmetric opening of such magnetopause, on the north-dawn and south-dusk sides for our present choice of positive $B_{yN}$. Such an asymmetric opening was inferred already by Cowley [1981] it is supported by observations of $B_y$ related asymmetries in the presence of magnetosheath plasma in the lobes [Hardy et al., 1976; Gosling et al., 1985].

The second effect demonstrated by Figure 4 is a tilt of the closed plasma sheet region, defined by field lines with both ends inside of the circle $r = \sqrt{y^2 + z^2} = 20R_E$ and indicated by shading, against the neutral sheet (defined by $B_x = 0$), which, by assumption, is identical with the equatorial plane $z = 0$. This effect is caused by field lines which have one end at $x = -20R_E$ inside the circle $r = 20R_E$, cross the equatorial plane, but do not return to within the $20R_E$ circle at $x = -20R_E$, extending rather to larger distances in $|y|$. These field lines are likely to be interconnected with the magnetosheath field through the low-latitude boundary region at their other end, so that they become topologically similar to the lobe field lines. A quantitative estimate of this effect is provided by Figure 5, which shows the boundary of the closed field line region in the equatorial plane for various magnitudes of the net $B_{yN}$, indicated by the parameter $b_N$ (representing $B_{yN}$ normalized by a characteristic lobe field strength). The dashed line represents a distant neutral line or separator, which terminates the closed field line region for $b_N = 0$. For increasing $b_N$ the region of closed field lines shrinks, despite the fact that $B_z$ remains positive everywhere earthward of the dashed line. While for $b_N = 0.027$ (corresponding to 2.7% of the typical lobe field), which is typical for the average net $B_y$ field [Fairfield, 1979], the effect is not very strong, it becomes appreciable for $b_N = 0.107$, which is not extremely unusual.

The possible existence of field lines which cross the neutral sheet with positive $B_z$, but are nevertheless open at one end, is of interest for the interpretation of observed properties in the vicinity of

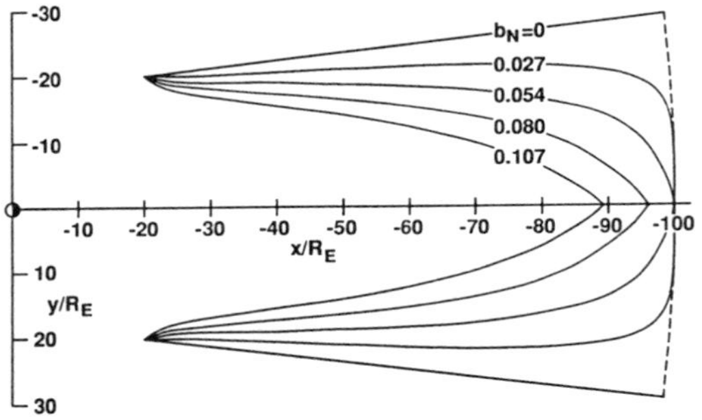


Fig. 5 Boundaries of the closed field line region in the equatorial plane after Birn [1990a] for various net cross-tail fields $b_N$, normalized by the typical lobe field strength. The dashed line represents the distant neutral line or separator, which defines the tailward boundary of the closed field line region for $b_N = 0$. Note that the distribution of $B_z$ in the equatorial plane does not change significantly for the different cross-tail fields, and that $B_z$ is positive earthward of the dashed line in every case.

the low-latitude magnetopause. Depending on the recent history of such field lines and on the location where they are interconnected with the magnetosheath, one expects different plasma properties. If the field lines have become interconnected with the magnetosheath quite recently, one would expect to see a mixture of magnetospheric and magnetosheath plasma, as typically observed in the low-latitude boundary layer [e.g., Eastman and Hones, 1979]. If they have been interconnected for quite a while, one would expect to see only magnetosheath like plasma, as is typical for the plasma mantle [Rosenbauer et al., 1975], or no plasma at all, if the interconnection location is far away. Such density drops inside the low-latitude magnetopause are indeed observed [Gosling et al., 1986]. In any case, the suggested interpretation would require a strong dependance of the occurrence of the observed phenomenon on the magnitude of the net $B_y$ (or the interplanetary $B_y$). This will allow one to find out from statistical data studies whether the presence of such type of open field lines indeed plays a role in influencing the low-latitude boundary properties.

## 4. Steady Flow Solutions

The quasi-static approximation, neglecting the inertia effects in the force balance, is usually a good assumption for the closed plasma sheet region during quiet times. This approximation, however, seems no longer valid in the more distant tail near and beyond the termination of the closed field line region, where fast flows seem to prevail most of the time [Zwickl et al., 1984]. Since these flows are typically strongly field-aligned, it seems justified to apply the theory presented in section 2. One possible application is the description of the termination region itself. Further posssible applications of that theory include the motion of a plasmoid through the distant magnetotail [observed, e.g., by Hones et al., 1984], which in the frame of the moving plasmoid may be considered as the steady flow of a plasma around it, and magnetopause structures, which are often found to have an associated "de Hoffmann-Teller" frame, in which the electric field vanishes [e.g., Papamastorakis et al., 1989; Sonnerup et al., 1990].

A particular example, representing the steady motion of a plasmoid through the distant tail current sheet, was derived by Birn [1990b], using an incompressible two-dimensional model. This example is represented in Figure 6, showing magnetic field lines, which are also the flow lines in the frame moving with the plasmoid. In this frame, the plasma inside the plasmoid is at rest. This is a consequence of the

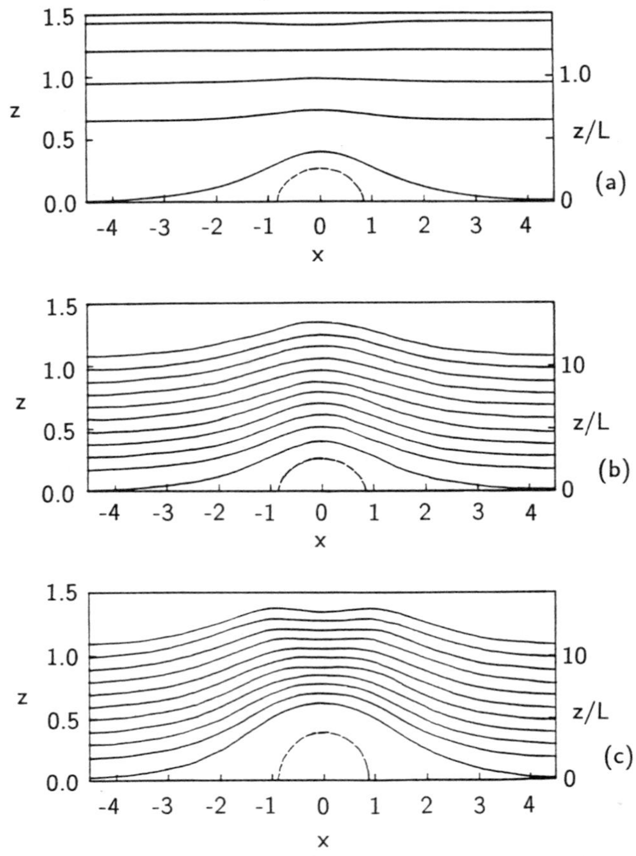


Fig. 6 Magnetic field lines of a two-dimensional magnetic island or plasmoid after Birn [1990b]. In the frame of the moving plasmoid these lines are also flow lines of the steady plasma flow around the plasmoid. Panel (a) represents the static case without flow, panels (b) and (c) cases with nearly Alfvénic flow ($m = 0.995$) around the plasmoid. Panels (a) and (b) correspond to 20% pressure enhancement in the center of the plasmoid as compared to the pressure at $z = 0$ for $|x| \to \infty$, while panel (c) corresponds to 70% pressure enhancement. The horizontal scales are normalized by an arbitrary length $L_x$, the vertical scales on the left hand side by a characteristic length $L_c$, defined by the current sheet half-thickness for $|x| \to \infty$ in the static case, and on the right hand side by the actual current sheet half-thickness $L$ in each case.

symmetry around $z = 0$, which implies that $m = 0$ at $z = 0$ and, via Equation (11), $m = 0$ along each closed plasmoid field line. The three panels of Figure 6 correspond to three different parameter sets with the top panel representing the static case without flow, while the two bottom panels correspond to fast, nearly Alfvénic flow ($m = 0.995$ on open field lines around the plasmoid). The first two panels correspond to 20% enhancement of the total pressure $\hat{p}$ at the center of the plasmoid as compared to $\hat{p}$ at large distances, while the last panel corresponds to 70% pressure enhancement. The $x$ coordinate is normalized by an arbitrary length $L_x$ in each case. The $z$ coordinate on the left hand side of each panel is scaled by a length $L_c$, which represents the current sheet half-width far away from the plasmoid (large $|x|$) in the absence of flow. The vertical scales on the right hand sides are normalized by the actual current sheet half-width $L$ for $|x| \to \infty$

in the presence of the flow. These scales indicate the characteristic difference in the island thickness between the static case (top panel) and the two fast flow cases (two bottom panels). In the static case, the island width is comparable to the current sheet scale $L$, while the islands are much thicker for the two fast flow cases. Observations indicate that the ambient current layer or plasma sheet in the far tail is much thinner than the plasmoid. This follows, for instance, from the fact that the plasmoid is typically entered from the tail lobes [see, e.g., Baker et al., 1987]. In the present model, this requires values of $m$ close to 1 if the pressure enhancement in the center of the plasmoid is only moderate. Observations of magnetic pressure enhancement outside the plasmoid, which should be equivalent to the enhancement of the total pressure, show typical increases of 20-70% [Slavin et al., 1984]. It can easily be shown that much larger pressure enhancements are necessary to produce a thick island in the static case. These results thus demonstrate the necessity of including inertia effects in a realistic description of plasmoids in the distant magnetotail.

Although the model of Figure 6 was constructed to illustrate plasmoid properties, it may also be used to illustrate possible properties of the quasi-steady magnetotail near the termination of the closed field line region. This region is found observationally as the transition region from slow or no flow in the thick closed plasma sheet to fast flow in the typically much thinner more distant current sheet and the adjacent lobes [Zwickl et al., 1984]. If we consider only the right half or less of the lower two panels in Figure 6, these sections may be similar to the plasma sheet termination region. There is no flow in the closed field line region, but fast flow in the regions surrounding it and extending down the tail. Again we can conclude that the transition from a thick closed field line region to a thin current sheet further down the tail should be associated with fast plasma flow.

## 5. Solutions for $\mathbf{E} \neq 0$ and $\partial/\partial t \neq 0$

For the quasi-static case general solutions for $\mathbf{v}$ and $\mathbf{E}$ can be derived for any (weakly) time-dependent sequence of equilibria $\mathbf{B}(\mathbf{r}, t)$ and $p(\mathbf{r}, t)$, if boundary conditions are imposed only at one point on each magnetic field line [Schindler and Birn, 1978]. Using an Euler representation $\alpha(\mathbf{r}, t)$, $\beta(\mathbf{r}, t)$ for the magnetic field, the electric field can be represented by

$$\mathbf{E} = \frac{\partial \beta}{\partial t}\nabla\alpha - \frac{\partial \alpha}{\partial t}\nabla\beta - \nabla\Phi(\alpha, \beta, t) \tag{50}$$

where $\Phi(\alpha, \beta, t)$ is arbitrary and has to be obtained from boundary conditions or other appropriate assumptions. The velocity perpendicular to $\mathbf{B}$ is given by the $\mathbf{E} \times \mathbf{B}$ drift, following from (3),

$$\mathbf{v}_\perp = \mathbf{E} \times \mathbf{B}/B^2. \tag{51}$$

The parallel component of the flow velocity can be determined from the adiabatic law (4) and the continuity equation (1), which can be combined to

$$\mathrm{D}p/\mathrm{D}t \equiv \partial p/\partial t + \mathbf{v} \cdot \nabla p = -\gamma p \nabla \cdot \mathbf{v}. \tag{52}$$

Using our curvilinear coordinate system $\alpha$, $\beta$, and $\tau$, and splitting the velocity into its parallel and perpendicular components, we can express (52) by

$$\frac{d}{d\tau}\left(\frac{v_\parallel}{B}\right) + \nabla \cdot \mathbf{v}_\perp = -\frac{1}{\gamma p}\frac{\mathrm{D}p}{\mathrm{D}t}, \tag{53}$$

where $d/d\tau$ again denotes the derivative along a field line, i.e., for constant $\alpha$ and $\beta$. Equation (53) can be integrated along the field lines:

$$\frac{v_\parallel}{B} = \left.\frac{v_\parallel}{B}\right|_o - \frac{\tau}{\gamma p}\frac{\mathrm{D}p}{\mathrm{D}t} - \int \nabla \cdot \mathbf{v}_\perp d\tau. \tag{54}$$

This equation can also be expressed in the following form [Schindler and Birn, 1978]:

$$\frac{v_\parallel}{B} = \left.\frac{v_\parallel}{B}\right|_o - \frac{\tau}{\gamma p}\frac{\mathrm{D}p}{\mathrm{D}t} - \mathbf{v}_\perp \cdot \nabla\tau + \frac{\partial}{\partial\alpha}\int\frac{\partial\alpha}{\partial t}d\tau + \frac{\partial}{\partial\beta}\int\frac{\partial\beta}{\partial t}d\tau \tag{55}$$

with

$$\frac{\mathrm{D}p}{\mathrm{D}t} = \frac{\partial P(\alpha, \beta, t)}{\partial t} + \frac{\partial\Phi}{\partial\alpha}\frac{\partial P}{\partial\beta} - \frac{\partial\Phi}{\partial\beta}\frac{\partial P}{\partial\alpha}. \tag{56}$$

The term $v_\parallel/B|_o$ in (54) and (55) represents a function of $\alpha$, $\beta$, and $t$ to be determined by boundary conditions.

The general solution for $\mathbf{E}$ and $\mathbf{v}$ for the given fields $\mathbf{B}(\mathbf{r}, t)$ and $p(\mathbf{r}, t)$ thus follows from (50), (51), and (54) with two free functions of $\alpha$, $\beta$, and $t$, following from prescribing the potential $\Phi$ and the parallel flow at one point on each field line. Such a boundary condition causes no obvious problem for lobe field lines that extend to large distances. For closed field lines, however, the ionosphere generally provides boundary conditions at both ends, which are not necessarily consistent with each other. Furthermore, boundary conditions for $\Phi$ are generally coupled to the magnetospheric field through the closure of magnetospheric field-aligned currents in the ionosphere. The inclusion of ionospheric boundary conditions thus restricts the possible field solutions in the magnetosphere. A basic approach to such a self-consistent closed chain of equations has been formulated by Vasyliunas [1970] (see also Schindler and Birn [1978]). A major piece of this chain, but without the self-consistent magnetospheric magnetic field calculation, is the basis of the "Rice Convection Model" [Wolf, 1970, 1974; Harel and Wolf, 1976]. For the magnetotail part, only particle and entropy conservation have been included through a boundary condition for $v_\parallel$ [Schindler and Birn, 1982; Birn and Schindler, 1983].

If north-south symmetry is imposed, then $v_\parallel = 0$ at the equatorial plane, and (55) can be used directly to calculate $v_\parallel$ at the ionosphere if the integrals are extended from the equatorial plane to the ionosphere. Using the average magnetospheric configuration, $\partial/\partial t = 0$, and typical electric fields, Equation (55) implies an outflow into the ionosphere for earthward convection which by far exceeds the average particle precipitation. This is another way of formulating the "pressure inconsistency" [e.g., Erickson and Wolf, 1980; Schindler and Birn, 1986], which essentially states that the average observed magnetic field and pressure in the magnetosphere are inconsistent with a steady convection, as associated with typically observed ionospheric electric potentials, and the observed losses of particles from the closed field line region into the ionosphere, if adiabatic convection with $\gamma = 5/3$ is assumed. The more common way of formulating this inconsistency is based on a different representation of (52), using mass conservation (1),

$$\frac{\mathrm{D}}{\mathrm{D}t}(pV^\gamma) = 0, \tag{57}$$

where $V$ represents the total volume of a magnetic flux tube of unit flux extending to the ionosphere,

$$V = \int d\tau = \int ds/B, \tag{58}$$

and flux tubes are assumed to be closed at the ionosphere, i.e., outflow and inflow are neglected. In a steady state Equation (57) implies that $pV^\gamma = const$ in each plane of convecting flux tubes. Using the average magnetic field, Equation (57) predicts a much larger pressure increase toward the Earth than is observed [Erickson and Wolf, 1980]. Relations between $p$ and $V$ (or, rather, the density $\rho$) that are derived from observations have yielded values of $\gamma$ less than 5/3, even as low as 2/3 [Schindler and Birn, 1982; Huang et al., 1989], while Baumjohann and Paschmann [1989] found $\gamma = 1.4$, still consistent with $\gamma = 5/3$ within the experimental uncertainties. We should add, however, that the statistically derived indices $\gamma$ may not be very meaningful for properties of convection, if the scatter in the data, reflecting variations in the properties of plasma entering the plasma sheet, is larger than the typical variation of $p$ and $\rho$ due to the adiabatic or non-adiabatic convection across the region sampled.

Several possibilities exist to solve the pressure inconsistency: (1) While particle losses into the ionosphere do not seem to be sufficient to account for the discrepency for typical convection electric fields of 0.1–1mV/m, they may be sufficient to allow for steady convection when the electric field is smaller. (2) Kivelson and Spence [1988] (see also Spence and Kivelson [1990]) suggested that preferred losses of more energetic particles through the flanks of the magnetotail during earthward convection could account for the observed pressure deficiency; it is, however, not proved yet that this effect is quantitatively sufficient and that the observed particle distributions indeed show the reduction in high energy particles closer to the Earth, which corresponds to this mechanism. (3) If steady loss mechanisms are not sufficient, the conclusion must be that the average magnetosphere configuration is indeed not the result of a steady convection, but rather must include non-adiabatic dynamic processes, which lead to a loss of particles and/or energy [Erickson and Wolf, 1980; Schindler and Birn, 1982, 1986]. This conclusion leads to a plausible scenario for the development of magnetospheric substorms: Adiabatic convection, driven by the solar wind electric field which couples to the magnetosphere for southward IMF via frontside magnetic reconnection, deforms the magnetotail in a time-dependent way, governed by Equation (57), until a stability threshold is exceeded. The resulting instability of tearing type leads to the severance and tailward ejection of plasma and energy from the plasma sheet in the form of a plasmoid, as suggested, e.g., by Hones [1977] and simulated by Birn and Hones [1981]. Schindler and Birn [1982] and Birn and Schindler [1983] showed for explicit sequences of equilibria, representing the possible result of such forced convection, that they evolved toward instability, using criteria for current driven microinstabilities as well as for a collisionless ion tearing mode. A slightly different approach by Erickson [1984] and Hau et al. [1989], based on the attempt to find steady state solutions consistent with $pV^{5/3} = const$ at the expense of giving up consistency with the average magnetic field, indicates that, even if steady states are approached during the quasi-static evolution, they also would tend to lead to instability, because they develop a deep minimum of the equatorial $B_z$ near the inner edge of the plasma sheet, which would favor the onset of a tearing instability.

## 6. SUMMARY AND DISCUSSIONS

We have presented the most updated version of our general quasi-steady magnetotail theory. This version allows one to calculate general magnetotail configurations by solving ordinary differential equations and ordinary integrals with the free choice of boundary functions as in the general boundary value problem. The main assumptions are (1) isotropic pressure, (2) ideal MHD, (3) weak electric fields and slow temporal variations, measured by some small parameter $\delta$, and (4) the so-called "tail approximation," using the characteristic magnetotail property that spatial variations in $x$ and $y$ are smaller, by some factor $\epsilon$, than variations in $z$. The solutions that can be obtained by this theory are of order zero in the parameter $\delta$ but include the first order in $\epsilon$. For quasi-static cases, when the inertia terms are negligible, the first order terms in $\delta$ can also be found under certain conditions, as discussed in section 5.

Particular solutions of the quasi-static equations were presented that include the following three-dimensional effects present in the actual magnetotail: Flaring of the tail in both $y$ and $z$, associated with the presence of $B_z$ and $B_y$; variation of the plasma sheet or current sheet thickness with $x$ and $y$, associated with a variation of $B_z$ with $x$ and $y$; field-aligned currents of region 1 type, which are found to be associated with the decrease of the tail flaring for larger distances; a net cross-tail magnetic field component $B_{yN}$, which may be chosen variable in $z$ at a given location. In addition, a particular steady model with field-aligned flow was presented, which indicates the important role of fast plasma flow for the transition of a thick closed plasma sheet to the more distant, much thinner current sheet and for the possibility of thick plasmoids moving in near equilibrium through the thin distant current sheet.

Refinements of the quasi-static theory which allow for pressure anisotropy are presently restricted to two-dimensional models [Bird and Beard, 1972; Cowley, 1978; Nötzel et al., 1985]. Such anisotropies are probably not important for the average large scale quiet magnetotail structure [Stiles et al., 1978; Nötzel et al., 1985], but they may play a role in the near-Earth region around $x \approx -10R_E$, where thin current sheets may form in the substorm growth phase which seem to be associated with pressure anisotropy [Mitchell et al., 1990].

Another refinement of the present theory concerns the assumption of ideal MHD. This assumption is typically a good one for large scale magnetospheric structures as considered here [see, e.g., Vasyliunas, 1976]. It may break down, however, within localized regions, particularly at the reconnection sites in the near and far tail. While near-Earth reconnection is presumably part of a dynamic process, which is beyond our present approach, the distant reconnection process may occur in more steady fashion, which, in priciple, should be included in a quasi-steady theory. Our present approach does not incorporate non-ideal effects. It constitutes, however, the basis for an exterior solution to the reconnection problem, which has to be matched to an interior solution for the non-ideal "diffusion region" in a typical boundary layer approach to the reconnection problem [e.g., Petschek, 1964; Vasyliunas, 1975; Priest and Forbes, 1987].

*Acknowledgments.* This work was supported by the U.S. Department of Energy through the Office of Basic Energy Sciences, by NASA, and by the Deutsche Forschungsgemeinschaft through its Sonderforschungsbereich program.

## References

Baker, D. N., R. C. Anderson, R. D. Zwickl, and J. A. Slavin, Average plasma and magnetic field variations in the distant magnetotail associated with near-Earth substorm effects, *J. Geophys. Res.*, *92*, 71, 1987.

Baumjohann, W., and G. Paschmann, Determination of the polytropic index in the plasma sheet, *Geophys. Res. Lett.*, *16*, 295, 1989.

Birn, J., Self-consistent magnetotail theory: General solution for the quiet tail with vanishing field-aligned currents, *J. Geophys. Res.*, *84*, 5143, 1979.

Birn, J., Magnetotail equilibrium theory: The general three-dimensional solution, *J. Geophys. Res.*, *92*, 11,101, 1987.

Birn, J., Three-dimensional equilibria for the extended magnetotail and the generation of field-aligned current sheets, *J. Geophys. Res.*, *94*, 252, 1989.

Birn, J., The distortion of the magnetotail equilibrium structure by a net cross-tail magnetic field, *J. Geophys. Res.*, *95*, 8019, 1990a.

Birn, J., Stretched three-dimensional plasma equilibria with field-aligned flow, *Phys. Fluids B*, in press, 1990b.

Birn, J., and E. W. Hones, Jr., Three-dimensional computer modeling of dynamic reconnection in the geomagnetic tail, *J. Geophys. Res.*, *86*, 6802, 1981.

Birn, J., and K. Schindler, Self-consistent theory of three-dimensional convection in the geomagnetic tail, *J. Geophys. Res.*, *88*, 6969, 1983.

Birn, J., R. Sommer, and K. Schindler, Open and closed magnetospheric tail configurations and their stability, *Astrophys. Space Sci.*, *35*, 389, 1975.

Birn, J., R. R. Sommer, and K. Schindler, Self-consistent theory of the quiet magnetotail in three dimensions, *J. Geophys. Res.*, *82*, 147, 1977.

Bird, M. K., and D. B. Beard, The selfconsistent geomagnetic tail under static conditions, *J. Geophys. Res.*, *73*, 907, 1968.

Candidi, M., S. Orsini, M. Stokholm, and R. C. Elphic, On the structure of the tail magnetic field, *J. Geophys. Res.*, *95*, 7929, 1990.

Cowley, S. W. H., The effect of pressure anisotropy on the equilibrium structure of magnetic current sheets, *Planet. Space Sci.*, *26*, 1037, 1978.

Cowley, S. W. H., Magnetospheric asymmetries associated with the y-component of the IMF, *Planet. Space Sci.*, *29*, 79, 1981.

Eastman, T. E., and E. W. Hones, Jr., Characteristics of the magnetospheric boundary layer and magnetopause layer as observed by Imp 6, *J. Geophys. Res.*, *84*, 2019, 1979.

Erickson, G. M., On the cause of X-line formation in the near-Earth plasma sheet: Results of adiabatic convection of plasma sheet plasma, in *Magnetic Reconnection in Space and Laboratory Plasmas, Geophys. Monogr. Ser.*, vol. 30, edited by E. W. Hones, Jr., p. 296, Am. Geophys. Union, Washington, D.C., 1984.

Erickson, G. M., and R. A. Wolf, Is steady convection possible in the Earth's magnetotail?, *Geophys. Res. Lett.* *7*, 897, 1980.

Fairfield, D. H., On the average configuration of the geomagnetic tail, *J. Geophys. Res.*, *84*, 1950, 1979.

Fedder, J. A., and J. G. Lyon, The solar wind–magnetosphere–ionosphere current–voltage relationship, *Geophys. Res. Lett.*, *14*, 880, 1987.

Fuchs, K., and G.-H. Voigt, Self-consistent theory of a magnetospheric **B**-field model, in *Quantitative Modeling of Magnetospheric Processes, Geophys. Monogr. Ser.*, vol. 21, edited by W. P. Olson, p. 86, AGU, Washington, D. C., 1979.

Gosling, J. T., D. N. Baker, S. J. Bame, W. C. Feldman, and R. D. Zwickl, North-south and dawn-dusk plasma asymmetries in the distant tail lobes: ISEE 3, *J. Geophys. Res.*, *90*, 6354, 1985.

Gosling, J. T., M. F. Thomsen, S. J. Bame, and C. T. Russell, Accelerated plasma flows at the near-tail magnetopause, *J. Geophys. Res.*, *91*, 3029, 1986.

Hardy, D. A., J. W. Freeman, and H. K. Hills, Plasma observations in the magnetotail, in *Magnetospheric Particles and Fields*, edited by B. M. McCormac, p. 89, D. Reidel, Hingham, Mass., 1976.

Harel, M., and R. A. Wolf, Convection, in *Physics of Solar Planetary Environments*, vol. 2, edited by D. J. Williams, p. 617, Am. Geophys. Union, Washington, D.C., 1976.

Hau, L.-N., R. A. Wolf, G.-H. Voigt, and C. C. Wu, Steady state magnetic field configurations for the Earth's magnetotail, *J. Geophys. Res.*, *94*, 1303, 1989.

Hilmer, R. V., and G.-H. Voigt, The effects of a magnetic $B_y$ component on geomagnetic equilibria, *J. Geophys. Res.*, *92*, 8660, 1987.

Hones, E. W., Jr., Substorm processes in the magnetotail: Comments on "On hot tenuous plasmas, fireballs, and boundary layers in the Earth's magnetotail," by L. A. Frank, K. L. Ackerson, and R. P. Lepping, *J. Geophys. Res.*, *82*, 5633, 1977.

Hones, E. W., Jr., D. N. Baker, S. J. Bame, W. C. Feldman, J. T. Gosling, R. D. Zwickl, J. A. Slavin, E. J. Smith, and B. T. Tsurutani, Structure of the magnetotail at $220 R_E$ and its response to geomagnetic activity, *Geophys. Res. Lett.*, *11*, 5, 1984.

Huang, C. Y., C. K. Goertz, L. A. Frank, and G. Rostoker, Observational determination of the adiabatic index in the quiet time plasma sheet, *Geophys. Res. Lett.*, *16*, 563, 1989.

Iijima, T., and T. A. Potemra, The amplitude distribution of field-aligned currents at northern high latitudes observed by Triad, *J. Geophys. Res.*, *81*, 2165, 1976.

Kivelson, M. G., and H. E. Spence, On the possibility of quasi-static convection in the quiet magnetotail, *Geophys. Res. Lett.*, *15*, 1541, 1988.

Lui, A. T. Y., Characteristics of the cross-tail current in the Earth's magnetotail, in *Magnetospheric Currents, Geophys. Monogr. Ser.*, vol. 28, edited by T. A. Potemra, p. 158, AGU, Washington, D.C., 1984.

Mitchell, D. G., D. J. Williams, C. Y. Huang, L. A. Frank, and C. T. Russell, Current carriers in the near-Earth cross-tail current sheet during substorm growth phase, *Geophys. Res. Lett.*, *17*, 583, 1990.

Nötzel, A., K. Schindler, and J. Birn, On the cause of approximate pressure isotropy in the quiet near-Earth plasma sheet, *J. Geophys. Res.*, *90*, 8293, 1985.

Papamastorakis, I., G. Paschmann, W. Baumjohann, B. U. Ö. Sonnerup, and H. Lühr, Orientation, motion, and other properties of flux transfer event structures on September 4, 1984, *J. Geophys. Res.*, *94*, 8852, 1989.

Petschek, H. E., Magnetic field annihilation, *NASA Spec. Publ., SP-50*, 425, 1964.

Priest, E. R., and T. G. Forbes, New models for fast steady state magnetic reconnection, *J. Geophys. Res.*, *91*, 5579, 1986.

Rosenbauer, H., H. Grunwaldt, M. D. Montgomery, G. Paschmann, and N. Sckopke, Heos 2 plasma observations in the distant polar magnetosphere: The plasma mantle, *J. Geophys. Res.*, *80*, 2723, 1975.

Schindler, K., A self-consistent theory of the tail of the magnetosphere, in *Earth's Magnetospheric Processes*, edited by B. M. McCormac, p. 200, D. Reidel, Dordrecht-Holland, 1972.

Schindler, K., and J. Birn, Magnetospheric physics, *Physics Reports*, *47*, 109, 1978.

Schindler, K., and J. Birn, Self-consistent theory of time-dependent convection in the Earth's magnetotail, *J. Geophys. Res.*, *87*, 2263, 1982.

Schindler, K., and J. Birn, Magnetotail theory, *Space Sci. Rev.*, *44*, 307, 1986.

Schindler, K., D. N. Baker, J. Birn, E. W. Hones, Jr., J. A. Slavin, and A. B. Galvin, Analysis of an extended period of Earthward plasma sheet flow at $\sim 220\ R_E$: CDAW 8, *J. Geophys. Res.*, *94*, 15,177, 1989.

Slavin, J. A., E. J. Smith, B. T. Tsurutani, D. G. Sibeck, H. G. Singer, D. N. Baker, J. T. Gosling, E. W. Hones, Jr., and F. L. Scarf, Substorm associated traveling compression regions in the distant tail: ISEE-3 geotail observations, *Geophys. Res. Lett.*, *11*, 657, 1984.

Slavin, J. A., E. J. Smith, D. G. Sibeck, D. N. Baker, R. D. Zwickl, and S.-I. Akasofu, An ISEE 3 study of average and substorm conditions in the distant magnetotail, *J. Geophys. Res.*, *90*, 10875, 1985.

Sonnerup, B. U. Ö., I. Papamastorakis, G. Paschmann, and H. Lühr, The magnetopause for large magnetic shear: Analysis of convection electric fields from AMPTE/IRM, *J. Geophys. Res.*, *95*, 10,541, 1990.

Spence, H. E., and M. G. Kivelson, The variation of the plasma sheet polytropic index along the midnight meridian in a finite width magnetotail, *Geophys. Res. Lett.*, *17*, 591, 1990.

Spence, H. E., M. G. Kivelson, and R. J. Walker, Static magnetic field models consistent with nearly isotropic plasma pressure, *Geophys. Res. Lett.*, *14*, 872, 1987.

Stiles, G. S., E. W. Hones, Jr., S. J. Bame, and J. R. Asbridge, Plasma sheet pressure anisotropies, *J. Geophys. Res.*, *83*, 3166, 1978.

Tsurutani, B. T., D. E. Jones, R. P. Lepping, E. J. Smith, and D. G. Sibeck, The relationship between the IMF $B_y$ and the distant tail

($150 - 238 R_E$) lobe and plasmasheet $B_y$ fields, *Geophys. Res. Lett.*, *11*, 1082, 1984.

Tsyganenko, N. A., Global quantitative models of the geomagnetic field in the cislunar magnetosphere for different disturbance levels, *Planet. Space Sci.*, *35*, 1347, 1987.

Tsyganenko, N. A., A magnetospheric field model with a warped tail current sheet, *Planet. Space Sci.*, *37*, 5, 1989.

Vasyliunas, V. M., in *Particles and Fields in the Magnetosphere*, edited by B. M. McCormac, p. 60, D. Reidel, Dordrecht, Holland, 1970.

Vasyliunas, V. M., Theoretical models of magnetic field line merging, 1, *Rev. Geophys.*, *13*, 303, 1975.

Vasyliunas, V. M., An overview of magnetospheric dynamics, in *Magnetospheric Particles and Fields*, edited by B. M. McCormac, p. 99, D. Reidel, Dordrecht, Holland, 1976.

Voigt, G.-H., and R. A. Wolf, On the configuration of the polar cusps in Earth's magnetosphere, *J. Geophys. Res.*, *90*, 4046, 1985.

Voigt, G.-H., and R. A. Wolf, Quasi-static magnetospheric MHD processes and the "ground state" of the magnetosphere, *Rev. Geophys.*, *26*, 823, 1988.

Voigt, G.-H., and R. V. Hilmer, The influence of the IMF $B_y$ component on the Earth's magneto-hydrostatic magnetotail, in *Magnetotail Physics*, edited by A. T. Y. Lui, p. 91, Johns Hopkins University Press, Baltimore, MD, 1987.

Walker, R. J., and T. Ogino, Field-aligned currents and magnetospheric convection—a comparison between MHD simulations and observations, in *Modeling Magnetospheric Plasma*, *Geophys. Monogr. Ser.*, vol. 44, edited by T. E. Moore and J. H. Waite, p. 33, AGU, Washington, D.C., 1988.

Walker, R. J., and T. Ogino, Global magnetohydrodynamic simulations of the magnetosphere, *IEEE Trans. Plas. Sci.*, *17*, 135, 1989.

Wiechen, H., and K. Schindler, Quasi-static theory of the Earth's magnetotail, including the far tail, *J. Geophys. Res.*, *93*, 5579, 1988.

Wolf, R. A., Effects of ionospheric conductivity on convective flow of plasma in the magnetosphere, *J. Geophys. Res.*, *75*, 4677, 1970.

Wolf, R. A., Calculations of magnetospheric electric fields, in *Magnetospheric Physics*, edited by B. M. McCormac, p. 167, D. Reidel, Hingham, Mass., 1974.

Zwickl, R. D., D. N. Baker, S. J. Bame, W. C. Feldman, J. T. Gosling, E. W. Hones, Jr., D. J. McComas, B. T. Tsurutani, and J. A. Slavin, Evolution of the Earth's distant magnetotail: ISEE-3 electron plasma results, *J. Geophys. Res.*, *89*, 11007, 1984.

# PLASMA TRANSPORT IN THE EARTH'S MAGNETOTAIL

A. T. Y. Lui

Applied Physics Laboratory Johns Hopkins University
Laurel, Maryland 20723-6099

*Abstract.* An accurate assessment of plasma transport within and across the various plasma domains of the magnetotail is essential in the development of a global magnetospheric model. We summarize here the observed transport in the magnetotail during geomagnetic quiet and substorm periods. We note that the transport characteristics of magnetotail domains, determined to a great extent by the entry of solar wind and ionospheric plasmas to the magnetotail, are modulated by the solar and geomagnetic activities. During quiet conditions, plasma transport in the distant tail (>100 $R_E$ downstream) is dominated by tailward flowing plasma originated from the solar wind. However, the traditional idea of the tail region closer in being populated mostly by plasma of solar wind origin (transported earthward from the distant tail region) is now in question since it grossly underestimates the strengths of the ionospheric sources. The neglect of upwelling ionospheric plasmas in previous magnetotail studies has also contributed to much confusion in the determination of plasma transport associated with plasma sheet thinning during the substorm expansion phase. Comprehensive studies have now revealed that plasma transport in the central plasma sheet within ~ 20 $R_E$ downstream is almost exclusively sunward and that fast plasma transport (>400 km/s in speed) within the plasma sheet is basically infrequent and transient (mostly lasting for <1 min), even during highly disturbed geomagnetic conditions. This characteristic may be intrinsic to the true nature of the substorm process; the substorm expansion phase may well be a conglomeration of transient substorm intensifications, much like lightning in a thunderstorm. Finally, we emphasize that a complete picture of plasma transport within the magnetotail relies to some extent on future ability to provide accurate measurements of the very low-energy (<50 eV) part of the plasma population which is directly traceable to ionospheric sources.

## Introduction

The maturity of a research discipline is often measured by its success in providing accurate predictions. Naturally, one goal in magnetospheric research is to forecast the magnetospheric environment with reasonable precision. The accomplishment of this endeavor relies considerably on the development of a suitable global magnetospheric plasma model. An essential step in achieving such a model is a thorough assessment of plasma transport within and across the various magnetospheric plasma domains. This topic is intimately related to two other fundamental issues. The question of transport is closely linked with the determination of origins and sinks of the magnetotail particle population. In addition, plasma transport is commonly associated with dynamic processes in which particles are energized. Therefore, a clear picture of transport may not only assess the relative importance of ionospheric and solar wind sources but also pinpoint where energization and loss processes are located within the system. Further details of plasma transport such as the shape of the velocity distribution function may also offer clues to these processes.

Although the significance of plasma transport has long been recognized and substantial efforts have been made on this subject, a consensus description of plasma transport in the magnetotail has not emerged. There are a number of factors attributing to this poor understanding. Early investigations of plasma transport dealt mainly with magnetospheric convection which addresses only motions perpendicular to the magnetic field. We now recognize that many significant plasma transport phenomena manifest via motions along the magnetic field. Another factor is the lack of suitable instruments to monitor the very low-energy (< 50 eV) component of the plasma due to spacecraft charging. The complexity of dynamic processes in this region, coupled with the variability on the strengths of the two competing plasma sources (the solar wind and the ionosphere) for magnetospheric particles, is yet another factor. Nevertheless, there is progress made towards the resolution of some key issues on the global plasma transport in the magnetotail during both quiet and substorm times. In this paper, we shall review some generally accepted results, some previous controversial matters, some recent advances, as well as some unresolved issues on this subject. The main emphasis here will be on observational results rather than theory/modelling work since the latter often trails behind the former.

The magnetotail plasma is often treated as a single fluid. The first velocity moment of the distribution function then gives the bulk flow, which is an important parameter in the consideration of the overall momentum of the population. Due to the large mass difference between electrons and ions (by at least three orders of magnitude), the bulk flow is almost always dominated by ions. Thus, ion transport will be emphasized, with only a brief mention of electron transport in exceptional situations such as the entry of solar wind electrons into the tail lobe.

## Plasma Sources and Sinks

An overview on the plasma sources and sinks in the magnetotail system is instructive before considering the question of transport. Figure 1 is a block diagram illustrating the dominant sources and sinks for the magnetospheric plasma. It is well known that there are two main natural plasma sources.

The magnetopause is an imperfect barrier to the oncoming solar wind, thus permitting the solar wind to become a plasma source for the magnetosphere. The exact entry process (or processes) is still under intense investigation. Among the proposed mechanisms are magnetic reconnection (Dungey, 1961), plasma diffusion (Eviatar and Wolf, 1968) or impulsive penetration (Lemaire, 1977; Heikkila, 1982; Lundin, 1984)

Modeling Magnetospheric Plasma Processes
Geophysical Monograph 62

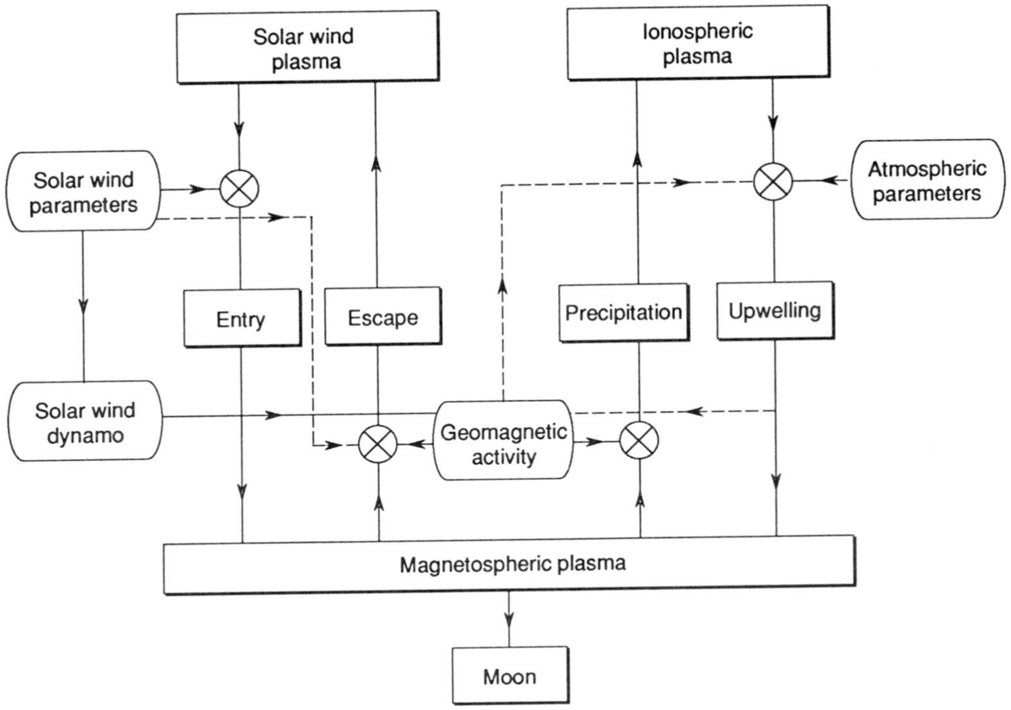


Fig. 1. A block diagram to illustrate the major sources and sinks of magnetospheric plasmas. The solar wind parameters regulate the entry of solar wind particles as well as govern the efficiency of the solar wind magnetosphere dynamo which powers geomagnetic activities. The upwelling of ionospheric plasma is governed by geomagnetic activities and atmospheric parameters determined also by solar activity. Geomagnetic activities also control the rate of particle loss due to precipitation mechanisms operative at that time.

across the magnetopause, gradient drift across the flanks of the magnetosphere (Stevenson and Comstock, 1968; Pfitzer and Olson, 1973), and entry via gyroviscosity effects (Stasiewicz, 1990). Observations have clearly indicated that the entry efficiency of the solar wind plasma is regulated by the intrinsic parameters of the solar wind itself. For instance, the plasma mantle, a domain governed by plasma of solar wind origin, is found to be thicker during southward interplanetary magnetic field (IMF) while the low latitude boundary layer, a domain related to solar wind plasma entry near the equatorial region, is found to be thinner at this IMF orientation (Sckopke et al., 1976; Mitchell et al., 1987).

The other source, also highly variable, is the ionosphere. Although particles from the ionosphere are of low energy and therefore contribute very little to the total pressure or force balance, they contribute significantly to the number density and thus the determination of the characteristic speeds (e.g., the Alfven and the fast mode speeds) in the medium (Moore et al., 1987; Chappell et al., 1987). The source strength depends on atmospheric parameters such as the scale heights and the abundance of various ionospheric ions (affected by the solar EUV radiation (White, 1977)). There is strong evidence that the composition of the near-earth magnetosphere is dependent on the solar cycle, with the richer ionospheric ion content during high solar activity (Lennartsson, 1987). It is further regulated by the level of geomagnetic activity which, in turn, is modulated by the external solar wind condition. Interestingly enough, the geomagnetic activity may also depend on the amount of upwelling of ionospheric ions based on the idea that ionospheric oxygen ions may instigate the onset of a plasma instability (e.g., tearing mode) responsible for the initiation of magnetospheric substorms (Baker et al., 1982). This may provide a positive feedback loop by which ionospheric plasma can be very effectively pumped into the magnetosphere.

There are two major and one minor sinks of magnetospheric plasma. Magnetospheric particles can exit the system through either the frontside magnetopause or downstream in the tail direction. Particle trajectory calculations have shown that some of the magnetospheric particles exit the magnetosphere on the dayside. The amount of particle escape on the frontside magnetopause depends on the magnetopause location, the magnetic field configuration, and the global convection electric field. Early modelling on the shape of the magnetosphere based on hydrodynamic consideration has predicted reasonably well the dayside magnetopause location and its dependence on the solar wind dynamic pressure.

The major loss to the ionosphere is through precipitation of particles in the polar region, mostly in the auroral oval. Particles residing in the loss cone will impact the neutral atmosphere at the ionospheric level and are lost through collisions with the neutrals. This particle drainage can be sustained by the refilling of the loss cone through pitch angle diffusion induced in wave-particle interaction or can be enhanced by parallel electric fields established by a double layer or by suitable wave modes. A related sink, which arises from high altitude extension of the neutral atmosphere, is the loss due to the charge exchange process between the energetic magnetospheric particles and the geocorona. This relatively minor loss is now explored as a means to probe the dynamic evolution of energetic particle population in the magnetosphere through imaging of the resultant energetic neutral particles (e.g., Roelof, 1987). Another relatively minor loss occurs when the moon traverses across the magnetotail (at a downstream distance of ~60 $R_E$) and absorbs particles along its path. This effect is more pronounced for moons in other planetary magnetospheres (see, e.g., Van Allen, 1984).

## Quiet Time Transport

The early investigation of quiet time transport were focused on the steady state convection in association with a dawn-to-dusk electric field. This global electric field is presumably generated by the interconnection of magnetic fields between the solar wind and the magnetosphere. However, more quantitative studies have cast doubts in the notion of a steady state convection (Erickson and Wolf, 1980; Schindler and Birn, 1982; Lui and Hasegawa, 1986) and the simplicity of a dawn-to-dusk electric field pattern in the entire magnetotail (Rostoker and Bostrom, 1976; Lui and Akasofu, 1980; Candidi and Orsini, 1981). In this section, we shall describe transports which occur irrespective of substorm activity. The emphasis will be on features which are not well represented by the traditional description of convection in each of the magnetotail domains. The main plasma domains in the magnetotail within the magnetopause are the plasma mantle (PM), the low latitude boundary layer (LLBL), the tail lobe, the plasma sheet boundary layer, and the central plasma sheet. For the boundary layers adjacent to the tail magnetopause (i.e., the PM and the LLBL), the transport characteristic is one of their identification signatures and, therefore, is well defined. For other regions, the transport characteristic is less apparent.

### *Magnetopause Boundary Layers*

The PM is a broad layer of magnetosheath-like plasma extending from the polar cusps into the polar regions of the tail lobe (Paschmann et al., 1974; Rosenbauer et al., 1975). The plasma has bulk tailward flow usually between 100–200 km/s and its speed is positively correlated with the solar wind speed. In the near-Earth region surveyed by Heos-2, it has a thickness varying from ~0.5 $R_E$ to >4 $R_E$; the thicker layer occurs at times of a larger southward component of the IMF. The PM is frequently separated from the magnetopause by a narrow gap of 0.1 to 0.2 $R_E$. Its proton density, temperature, and bulk flow all tend to decrease with increasing distance from the magnetopause. The PM is suggested to form by solar wind plasma entering the magnetosphere through the polar cusps. These particles are mirrored at low altitudes in the cusp field lines and travel back to the tail lobe region next to the magnetopause. The thicker PM at southward IMF is consistent with the effect from a stronger convection electric field which brings the PM particles deeper into the tail lobe. Although upwelling of ionospheric ions exists also in the cusp region, their speeds along the field lines are low compared with their convection speeds; hence they are carried towards the equatorial region and away from the PM (Cladis, 1986).

The transport in the LLBL is also dominated by the solar wind source (Hones et al., 1972; Akasofu et al., 1973). Similar to the PM, the LLBL has bulk tailward flow between 100–200 km/s. With increasing depth inside the magnetopause, its number density and bulk speed tend to decrease while its temperature increases (approaching that of the plasma sheet). Unlike the PM, ionospheric ions are frequently detected in the LLBL. These ionospheric ions are believed to be convected from the plasmasphere in the form of "blobs" toward the dayside LLBL (Freeman et al., 1977; Horwitz et al., 1981). Once they enter the LLBL, they are gradually assimilated into the tailward boundary flow (Lundin et al., 1987; Fuselier et al., 1989).

### *Tail Lobe*

The tail lobe is the lowest density region of the tail (typically <0.01 $cm^{-3}$) but it is rich in transport phenomena. The tail lobe is often filled with hot electrons (typically a few hundred eV) originating from the solar wind. These hot electrons were first detected at low altitudes (Winningham and Heikkila, 1974) and named as the polar rain. It is also observed at high altitudes (Yeager and Frank, 1976). The hemisphere in which polar rain appears is controlled by the IMF $B_x$, consistent with the notion that the field lines in the tail lobe interconnect with those in the solar wind, i.e., these electrons fill the northern hemisphere when the IMF $B_x$ is negative (away sector). There is no indication of any accompanying solar wind ions.

There are two main spatial locations in the ionosphere, namely, the cusp/cleft region and the polar cap, which can contribute low-energy ions (<100 eV) to the tail lobe (Shelley et al., 1982; Waite et al., 1985; Yau et al., 1985). The cusp/cleft region is found to be a strong source of various ionospheric ions ($H^+$, $O^+$, $N^+$, $O_2{}^+$, $N_2{}^+$, $NO^+$) and has been referred to as the cleft ion fountain (Lockwood et al., 1985). After being energized at low altitudes, these ionospheric ions travel up to the tail lobe and are dispersed by the convection electric field according to their flow speeds along the magnetic field line, much like a mass spectrometer (Moore et al., 1985).

The other important ionospheric outflow for the tail lobe is the polar wind consisting of ions with initial energies below a few eV. This population eludes detection by most instruments due to spacecraft charging. Only very few measurements of this source are available at high altitudes but they already suggest that this may be a very important source for the magnetospheric plasma (e.g., Nagai et al., 1984; Chappell et al., 1987). Supersonic outflow of polar wind occurs even in closed field lines such as in the trough and plasmasphere regions (Chandler and Chappell, 1986). The variability of its strength with solar and geomagnetic activities is not ascertained, which leads Chappell et al. (1987) to suggest that more statistical measurements from spacecraft with potential control are needed. Chappell et al. also emphasized that the low energy ions from both the cleft ion fountain and the polar wind are not detectable initially by the conventional plasma detectors flown in the magnetotail region. Only after they are energized in the neutral sheet region do they become recognized as part of the plasma population. Figure 2 from Chappell et al. illustrates this important point.

During strongly northward IMF, an unusual convection pattern may occur in the tail lobe. There are observations suggesting that particles in the plasma sheet are convected up into the tail magnetopause through the middle of the tail lobe, thus bifurcating the tail lobe (e.g., Potemra et al., 1984; Huang et al., 1987). This unusual configuration and transport of the magnetotail deserve further investigation.

### *Plasma Sheet*

The term plasma sheet used here encompasses the plasma sheet boundary layer (PSBL), the central plasma sheet (CPS), and the neutral sheet region. In the near-earth tail ($|X_{SM}|$ <20 $R_E$), the plasma flow is typically slow. Extensive statistical analyses of plasma flows from the IRM plasma measurements (Baumjohann et al., 1988, 1989, 1990) show that in both the CPS and the PSBL, the flow speeds are typically below 100 km/s, as indicated in Figure 3 by the statistics performed on ~10 s resolution samples from 4 months of IRM data. There is only ~5% chance of detecting flows in excess of 300 km/s for quiet conditions (AE <200 nT) at the PSBL and ~3% chance at the CPS adjacent to the boundary under the same condition. The dominance of slow flow is consistent with the earlier result from ISEE-1 reported by Huang and Frank (1986). The flow direction is almost exclusively earthward at high speeds (>300 km/s) but this feature gradually disappears at low speeds (<100 km/s) in the PSBL, as shown in Figure 4. In the CPS, there remains a higher probability of earthward flow than other directions even at low speeds. Baumjohann et al. found that there is a tendency for larger flows to align closer to the magnetic field direction in the PSBL. This trend, however, does not appear to hold for the CPS. They also noted that the faster the flow, the

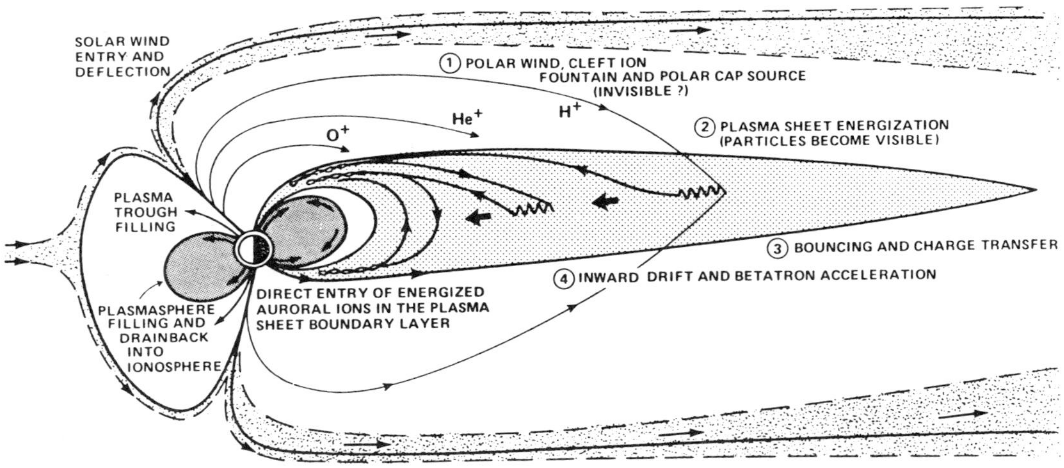


Fig. 2. A schematic diagram to illustrate the transport in the plasma mantle governed by the solar wind entry in the polar cusp/cleft region, and the transport from the ionospheric sources at the polar cusp/cleft (the cleft ion fountain), at the polar cap (the polar wind), at the auroral oval, and at subauroral latitudes of the plasmasphere (after Chappell et al., 1987).

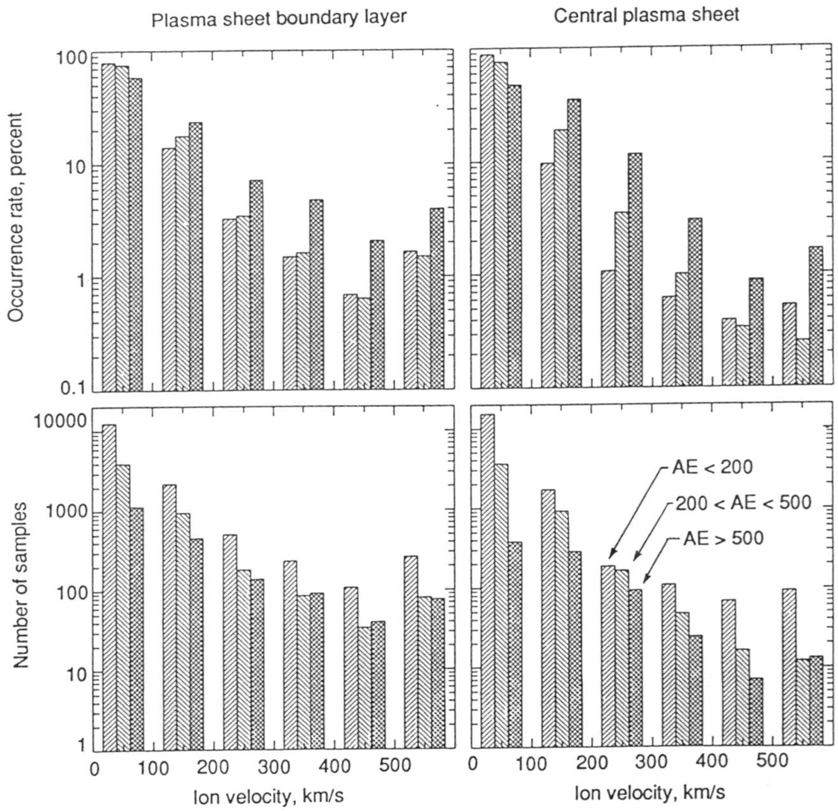


Fig. 3. A statistical survey of plasma flow measurements at ~10 s temporal resolution from the plasma instrument on IRM. Results pertaining to the PSBL are shown in the panels at the left and those to the CPS are shown at the right. Three AE conditions are distinguished. The histograms indicate that ion velocities are typically low for both regions and at any geomagnetic condition. The ion speeds are obtained from survey within the tail region of $X_{SM} \sim -10$ to $-19\ R_E$, and $|Y_{SM}| < 10\ R_E$ (after Baumjohann et al., 1988).

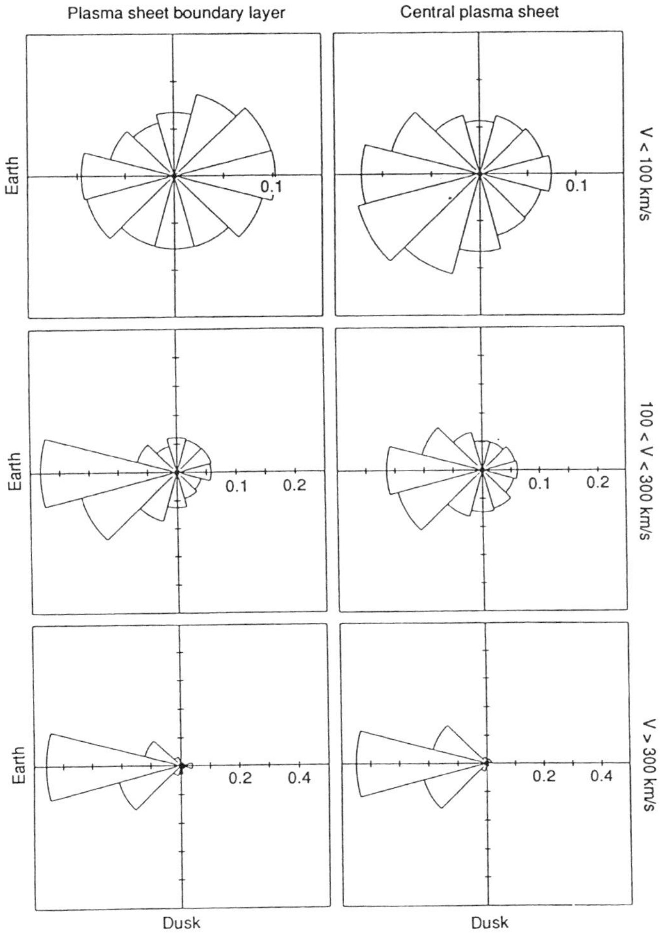


Fig. 4. The statistical distribution of flow direction in the equatorial projection for PSBL (left panels) and the CPS (right panels). Earthward flows are in sectors pointing towards left and duskward flows are in sectors pointing downwards. The flow directions are separated into three ranges of flow speeds (after Baumjohann et al., 1988).

shorter is its duration, with no difference between the PSBL and the CPS. This is confirmed later with 8 months of data at 4.5 s temporal resolution (Baumjohann et al., 1990). Whether earthward convection is steady-state and adiabatic is still actively investigated (Baumjohann and Paschmann, 1989; Huang et al., 1989; Kivelson and Spence, 1988).

Auroral arcs are present even without any substorm activity. In association with the formation of auroral arcs, ionospheric ions are accelerated both parallel and perpendicular to the magnetic field line (Shelley et al., 1972; Whalen et al., 1978) and are detected in the PSBL (Candidi et al., 1982; Orsini et al., 1982). Lennartsson (1987) has shown the general trends that ionospheric ions have their peak intensities in the midnight region of the plasma sheet while the solar wind ions are predominantly near the flanks. Number density of oxygen ions increases with the general magnetic activity. A solar cycle dependence is also found in which plasma sheet oxygen ions tend to be more energetic and are seen more often when the magnetic activity is high. In addition, Candidi and Orsini (1981) were able to use the obliqueness of the ionospheric $O^+$ ion beams with respect to the magnetic field direction to obtain estimates of the electric field in the PSBL. They deduced that the north-south component of the electric field reverses in sign at the interface between the tail lobe and the boundary layer so that the dawn-dusk component of the flow is towards the flanks within the PSBL but towards the midnight meridian in the tail lobe (Orsini et al., 1990).

In the distant tail ($|X_{SM}| > \sim 120\ R_E$), there is no clear indication that the plasma sheet can be separated into two distinct regions of CPS and PSBL, presumably due to the CPS region not extending this far downstream. Due to the partial failure of the plasma instrument on the ISEE-3 prior to its tail mission, plasma flows in the distant tail were inferred from plasma electron measurements. Tailward flow is almost always observed in the distant plasma sheet although the magnetic field normal to the neutral sheet is usually weakly northward, suggesting a closed field topology (Zwickl et al., 1984; Slavin et al., 1985; Heikkila, 1987). This leads to two different interpretations. This situation may arise if the distant neutral line is curved, being closer to the Earth near the midnight sector (Slavin et al., 1985), or if both dawn and dusk LLBL are broadened to occupy the entire tail width (Heikkila, 1987, 1988). In connection with the latter interpretation, we note that the convection electric field would have brought the PM particles down to the plasma sheet at this tail distance and thus they can contribute to the tailward flow. Ionospheric ions, due to their low speeds along the magnetic field line, would have convected down into the neutral sheet and then moved earthward within this downstream distance. Therefore, the tailward flowing plasma at the distant tail is mainly of solar wind origin regardless of which of the two interpretations is adopted. In addition, this is an important loss of tail plasma even without substorm activity.

## Substorm Time Transport

A magnetospheric substorm is the most energetic and complex transient phenomenon arising from the interaction of the magnetized solar wind with the geomagnetic field. Not surprisingly, the associated plasma transport can be extremely complex and, consequently, the subject can be quite controversial in spite of the fact that the largest transport speeds are detected during these episodes of activity (Lui, 1979; Huang, 1987; Baumjohann, 1986, 1988).

It is well documented that there are very significant dimensional changes of the plasma sheet during substorms. In the mid-tail region ($X_{SM} \sim -15$ to $-40\ R_E$), the plasma sheet is observed to start thinning at the onset of substorm expansion phase (Hones et al., 1971; Lui et al., 1975; Dandouras et al., 1986). This thinning phenomenon occurs over a wide local time sector ($|Y_{SM}| > 10\ R_E$) but is more severe in the midnight sector. Thinning is not just a compressional effect since the total pressure is not altered adequately to account for the thickness change (Hones et al., 1971). This leads to an important question relating to the plasma transport at this substorm phase. Where does the magnetotail plasma in the mid-tail region go during the plasma sheet thinning stage? Similarly, the plasma sheet is seen to thicken again during the later stage of a substorm (Akasofu et al., 1971). Where does the magnetotail plasma come from during this recovery of the plasma sheet thickness?

### *Plasma Sheet Thinning Phase*

Detailed data analyses were conducted on the issue of the direction of plasma drainage during plasma sheet thinning in the late 1970's. At that time, two competing models existed, as depicted in Figure 5. The near-earth neutral line model predicts that the plasma is ejected tailward after the formation of a near-earth neutral line at substorm onset (Hones, 1979). The other substorm model, referred to as the current diversion model, predicts earthward plasma flow during plasma sheet thinning. The latter model suggests that a disruption and diversion of the cross-tail current occurs in the midnight sector at the substorm expansion onset. This leads

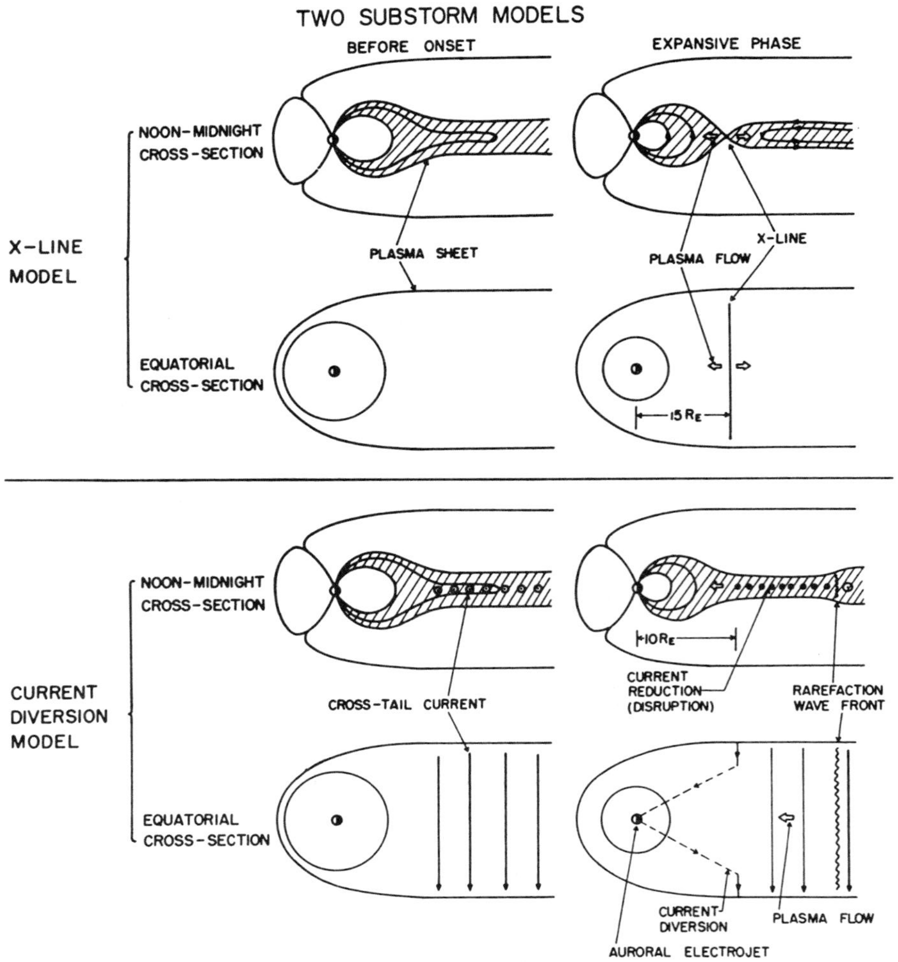


Fig. 5. A schematic diagram illustrating the two substorm models which predict two opposite directions of plasma drainage in the near-earth plasma sheet during the thinning stage.

to a collapse of stretched magnetic field lines (developed during the substorm growth phase) and an earthward injection of plasma in the near-earth tail. This disturbance is communicated tailward via the propagation of a rarefaction wave front which leads to earthward displacement of the magnetotail plasma (Chao et al., 1977; Lui, 1979).

Observational evidence for the current diversion substorm model is found in the studies of IMP-6 data. Figure 6 shows plasma flows observed during plasma sheet thinning based on a study of 74 individual thinning cases. It is found that plasma sheet thinning is associated with earthward flow in ~40 % of the cases, with tailward flow in ~20 %, and with dawnward, duskward or no systematic flow direction in the rest of the cases.

The occurrence of tailward flow during plasma sheet thinning (even when earthward flow occurs twice as often) has been taken as a contradiction to the current disruption model. However, a careful consideration of the situation suggests that tailward flow can arise from a number of reasons that may not indicate the direction of the magnetotail plasma loss during the thinning stage. For instance, it is discussed in a previous section that ions are extracted out from the ionosphere into the magnetosphere in the auroral region, especially over the active auroras seen at substorm intervals (Shelley et al., 1972; Whalen et al., 1978). These ionospheric ions (which include protons and oxygen ions) constitute tailward streaming ion beams in the magnetotail (Lui et al., 1983; Lennartsson and Sharp, 1985; Eastman et al., 1985). The bulk flow is a first moment of the velocity distribution function. Therefore, if the flow contribution from the ionospheric components dominates over that from the magnetotail plasma, then a tailward bulk flow will be obtained regardless of the flow direction of the magnetotail plasma. In other words, this bulk flow direction may not be related at all to the question of where the magnetotail plasma goes during the plasma sheet thinning stage.

Figure 7 shows an example of the multi-component plasma at the PSBL during plasma sheet thinning. Each panel is an energy-angle spectrogram obtained over a 128 s interval by the ISEE-1 plasma instrument. Time progresses from left to right for each row. High intensities (represented by the dark shading) near the middle of the spectrogram indicate a tailward streaming population. A sunward flowing population will show up with high intensities concentrated at the left and right edges of the spectrogram.

If we examine the first row of spectrograms, the first three panels indicate that at the PSBL, there was a low-energy (~50 eV to 5 keV),

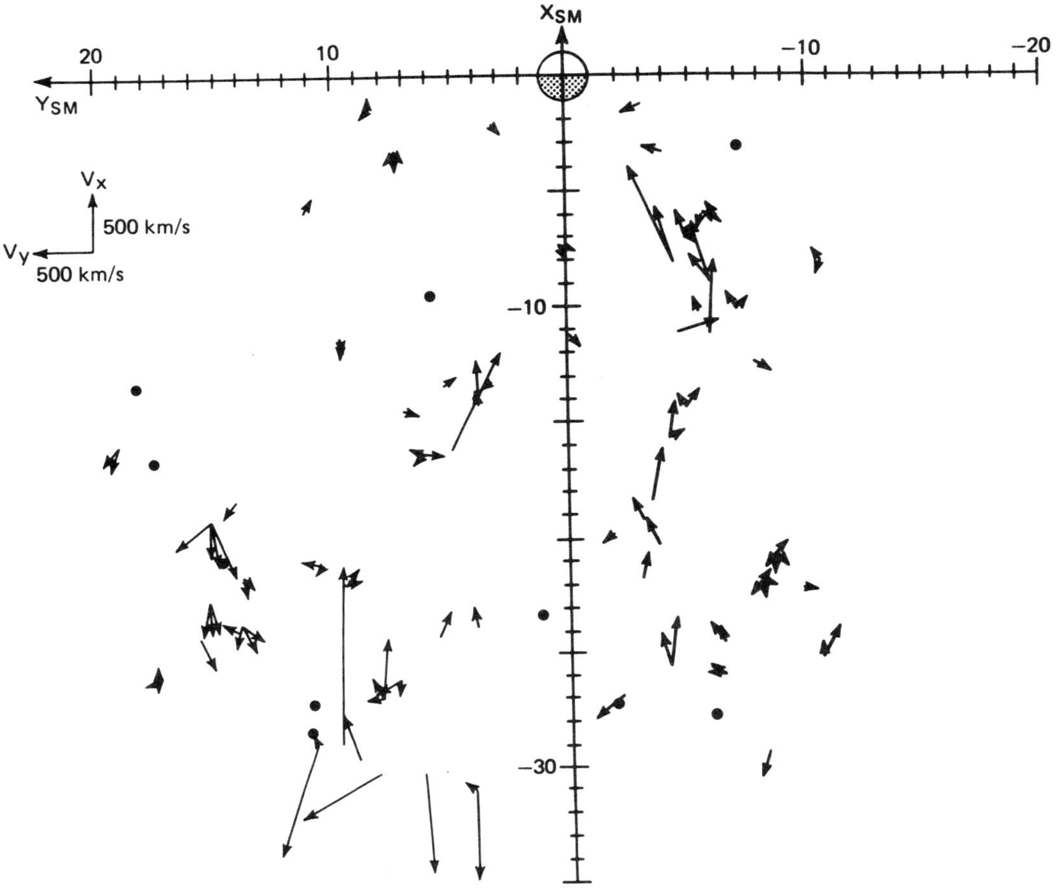


Fig. 6. A pattern of plasma flows constructed from plasma sheet thinning events observed by IMP-6. The predominance of sunward flow is seen for the majority of the thinning events.

tailward flowing component co-existing with a higher energy (~1 to 5 keV), earthward flowing component. The low-energy component is of ionospheric origin, as verified from the ion composition data. Just before the total plasma dropout at the satellite, only the low-energy component was seen. These features are well exhibited by the three cases of plasma sheet thinning shown and exemplifies that the hot magnetotail transport is earthward in these cases in spite of the bulk flow being tailward.

Another source for tailward flow is demonstrated well by Williams (1981). By studying the arrival times of ions at different energies and pitch angles, he found that tailward flowing particles can be due to earthward flowing particles (which were originated at ~80 to 100 $R_E$ downstream) mirroring at a location closer to the Earth as they approach the stronger magnetic field earthward.

The dominance of earthward flow during plasma sheet thinning has been challenged by Hones and Schindler (1979) and Hayakawa et al. (1982). Both studies indicate tailward flow is more pronounced but they did not attempt to separate the ionospheric contribution to the flow computation nor consider the possibility of mirroring of earthward flowing plasma.

*Flow Characteristic of the CPS and Its Implications*

Another means of differentiating the two substorm models is by examining cases in which measurements within the thinned plasma sheet can be made during the entire or a good portion of the substorm interval. This is feasible even though plasma sheet thinning reduces the chance for such occurrence since substorms occur frequently, and there are many satellites flown to provide extensive coverage of the mid-tail region.

Several such cases have been reported (Lui, 1979; Lui et al., 1981) and Figure 8a,b illustrate one such case. The IMP-6 satellite was at a relatively constant downstream distance ($X_{SM} \sim -31\ R_E$) in the midnight sector ($Y_{SM} \sim -6\ R_E$). That the satellite remained in the CPS during most of this interval is indicated by the high number density (> 0.1 $cm^{-3}$), the low magnetic field magnitude (typically below 10 nT), and the frequent change in the magnetic field azimuth angle from near sunward to tailward (indicating crossing of the neutral sheet). There were five substorms during this time span and the spacecraft was within the substorm disturbance region as verified by the local time extent of magnetic disturbance on the ground. During the expansion phase of the first substorm (1530–1610 UT), the bulk flow of plasma in the CPS was sunward, with speeds varying between 40 and 530 km/s. Similarly, plasma flow with sunward component of 100 to 370 km/s was detected in the expansion phase of the second substorm between 1640 and 1750 UT. In fact, the plasma flow during most of this interval had a significant sunward component, indicating sunward flow prevailing in the CPS of the mid-tail region during substorm expansion when the mid-tail plasma sheet thinning occurs.

The result that sunward flow prevails is now confirmed by the compre-

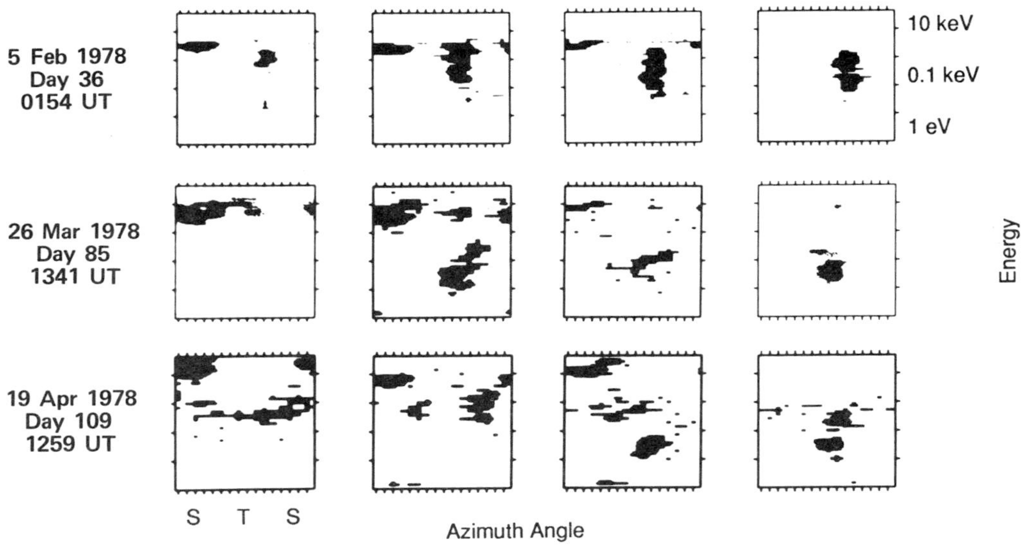


Fig. 7. Plasma velocity distribution represented by the energy-azimuth angle spectrogram observed by ISEE-1 during intervals just prior to plasma sheet thinning at the satellite location. The occurrence of at least two populations is often indicated. The hot component is typically flowing sunward while the cold component (from the ionosphere as indicated by the simultaneous ion composition measurement) is flowing tailward.

hensive survey of IRM plasma measurements within X ~ $-19$ $R_E$. Using 4 months of tail crossing data, Baumjohann et al. (1989) performed a statistical survey on the behavior of ion and electron moments in the CPS. They used the absence of photoelectrons as an identification signature for the CPS region (Baumjohann et al., 1988). With this criterion, they were able to verify that the PSBL is indeed a distinct region of the magnetotail and is not just a gradual transition from the plasma sheet to the tail lobe. Their result also indicates that high speed flows do occur in the CPS but typically last for short durations (often less than 1 min). The plasma analyzer on IRM is able to sample the velocity distribution in only ~4.5 s, faster than any other plasma detector flown to survey the magnetotail so far. Baumjohann et al. concluded from their study that the distribution of plasma flow directions strongly favors sunward flow for velocities above 300 km/s regardless of the value of the AE index (i.e. at all times), indicating, in their own words, "that a near-earth neutral line is rarely, if ever, located inside of X ~ $-19$ $R_E$." They further found that they could separate the inner portion of the CPS from its outer portion by imposing a criterion on the magnetic field strength (<15nT if the z-component is excluded or the z-component is greater than half of the root-mean-square of the other two components).

Another very important and revealing finding from the analyses of Baumjohann et al. is that most of the flow speeds are low (<300 km/s). This result is consistent with the IMP-6 study on flows during plasma sheet thinning as shown in Fig. 6 and the earlier ISEE-1 result (Huang and Frank, 1986). Baumjohann et al. found that the occurrence rate of high speed flows varies from ~0.5 % in the near-earth magnetotail ($-14$ $R_E < X_{SM} < -9$ $R_E$) to ~7% further out at ($-19$ $R_E < X_{SM} < -14$ $R_E$). They have interpreted this result as a possible indication of a high spatial variability of the ion flow, i.e. containment of the fast flows to thin layers close to the neutral sheet and the bursty nature of the flow as due to a high temporal variability of the reconnection rate at a neutral line beyond $X_{SM}$ ~ $-20$ $R_E$.

Another interpretation to these observed features which they have not considered is that the fast but transient earthward flows in the CPS may be due to tailward propagating rarefaction waves generated by current disruption in the near-earth region. Current disruption observed by the CCE spacecraft is a highly transient phenomenon (Lui et al., 1988; Lopez et al., 1989). Each current disruption is associated with earthward injection of particles (transient fast earthward flow) and collapse of stretched magnetic field lines in the near-earth tail. The earthward drainage of plasma initiates a rarefaction wave which restores the local plasma loss by earthward displacement of the plasma from downstream (transient earthward flow at lower speeds). This causes a chain reaction of plasma downstream as the rarefaction wave propagates tailward. This scenario is consistent with the observed tailward propagation of plasma sheet thinning, substorm injection of energetic particles, and dipolarization of the magnetic field configuration (Lui, 1979; Lopez and Lui, 1990).

The substorm expansion phase may consist of a series of these current disruptions in localized regions. This idea is consistent with observations that a substorm expansion phase usually embodies several substorm intensifications (Pytte et al., 1976; Rostoker et al., 1980; Sergeev and Yahnin, 1979; Lui, 1988). Thus, substorm expansion is not due to a process operating continuously. Lui (1988) suggested that substorm expansion is analogous to a thunderstorm, as illustrated in Figure 9. Both are natural phenomena involving electrical discharge. Substorm activity can occur over several hours of local time, but each substorm intensification is very transient and localized. Similarly, a thunderstorm can cover a large region but each lightning stroke is highly transient and localized.

*Plasma Sheet Thickening Phase*

There is a general consensus on the plasma flow pattern during the thickening stage of the plasma sheet near the peak of substorms. Figure 10 from Lui et al. (1977) is a flow pattern derived from IMP-6 study of plasma sheet recovery during substorms. As is rather evident from this figure, sunward plasma flow is predominantly seen, suggesting that the refilling of the thinned plasma sheet was achieved by plasma coming from further down the tail as the spacecraft became engulfed by the expanding

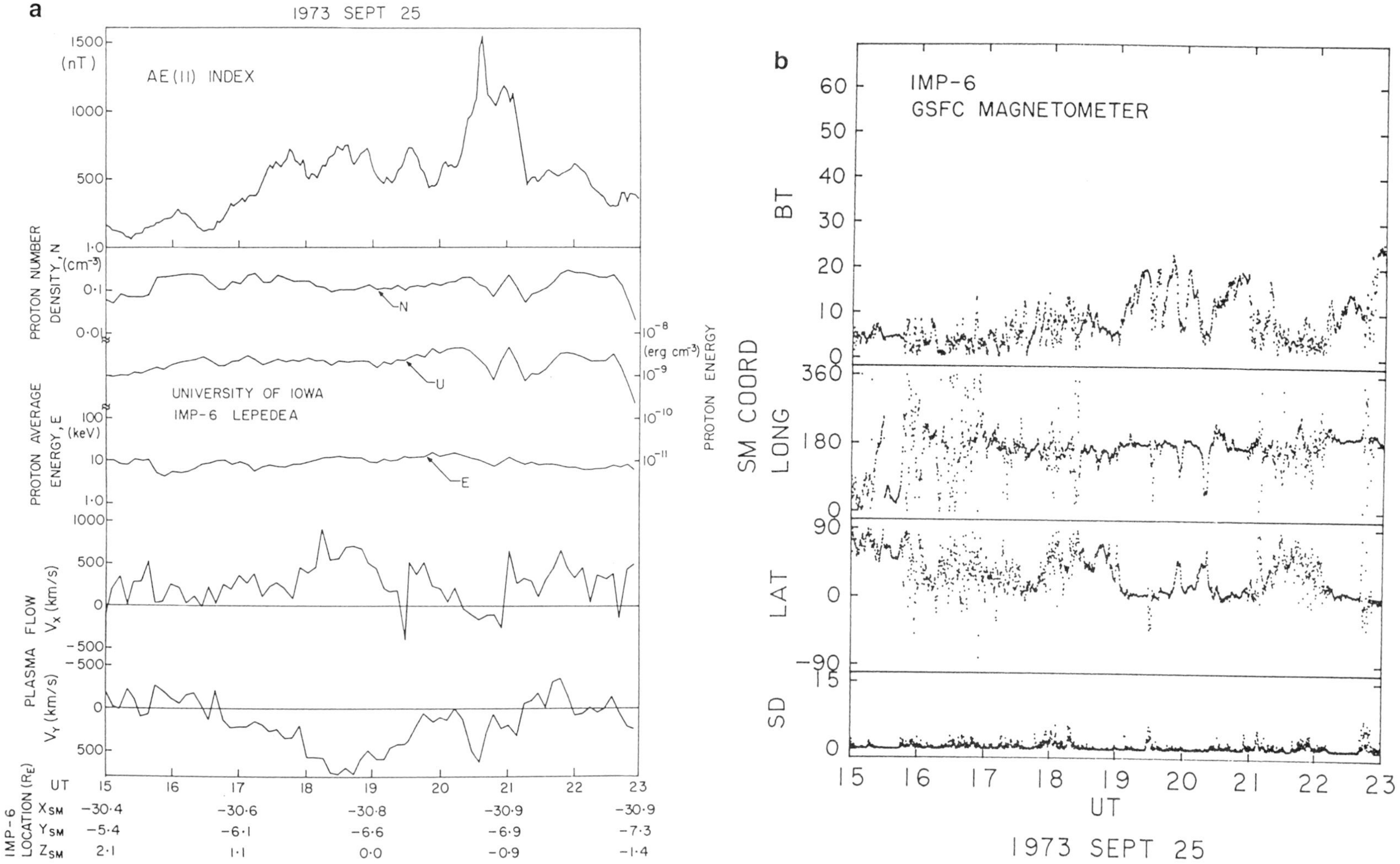


Fig. 8(a) Plasma flow observations from IMP-6 (~6 min resolution measurements) during substorm intervals when the spacecraft remained within the plasma sheet (and sometimes crossed the neutral sheet). The prevalence of sunward flow in the CPS during these substorms are seen. (b) Magnetic field measurements illustrating the proximity of the spacecraft to the neutral sheet during these substorm intervals.

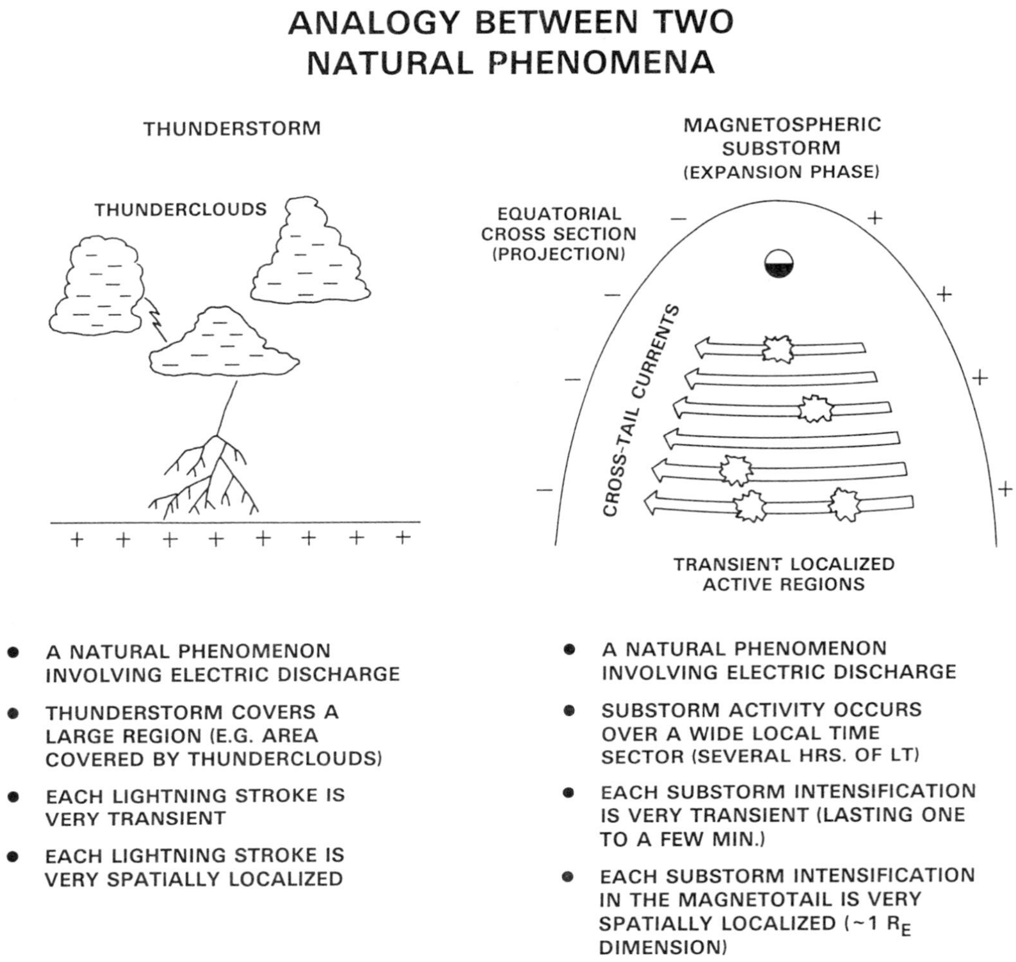


Fig. 9 A schematic diagram to illustrate the analogy between two natural phenomena, namely, the thunderstorm and the magnetospheric substorm. Both are natural phenomena involving electric discharges which are transient and localized in spite of the large coverage in area of such activity.

plasma sheet. The flow magnitude in general is also comparatively higher than that typically found during thinning.

Spatial structures are generally found at the PSBL during plasma sheet thickening but not during thinning. The observed structures are very similar to the geomagnetic mass spectrometer effect found in the tail lobe for the upwelling ionospheric ions. Particles with the fastest speeds along field lines are found at the outermost layer of the PSBL (e. g., Andrews et al., 1981; Takahashi and Hones, 1988).

Close to the time of plasma sheet thickening in mid-tail is a sequence of phenomena in the distant tail observed by the ISEE-3. A sudden northward swing of the magnetic field is often observed in association with spacecraft entry into the plasma sheet (Baker et al., 1984). This is followed by southward swing of the magnetic field accompanied by faster tailward flowing plasma electrons (recall that no measurements of plasma ions were obtained in the distant tail from ISEE-3 due to the partial failure of the plasma instrument on board). It has been inferred from this sequence of phenomena that a plasmoid passes by. Many such events have been found in the ISEE-3 data and corresponding substorm activity in the near-earth region or on the ground were sought. The onsets of these ISEE-3 events were often delayed with respect to substorm onset times. The delay is interpreted as due to the travel time for the near-earth plasmoid to reach the ISEE-3 location.

Unfortunately, it is not entirely clear whether or not the formation of a plasmoid is a necessary consequence of a substorm. There has been no systematic study in this substorm-plasmoid correlation by first selecting substorm onset times and then examining ISEE-3 data for plasmoid signatures when the spacecraft was appropriately situated in the distant tail. In view of the fact that energetic phenomena with southward field at the neutral sheet can occur in the tail outside of substorm intervals (e.g., Lui and Meng, 1979; Sergeev et al., 1986), the link between plasmoid appearance and substorm activity needs further investigation with inclusion of simultaneous measurements at the intermediate distances of 30 to 60 $R_E$ downstream.

## Summary

A brief review is given on the plasma transport during quiet times and substorm periods in the magnetotail. Only a few issues are covered in

FIGURE 10

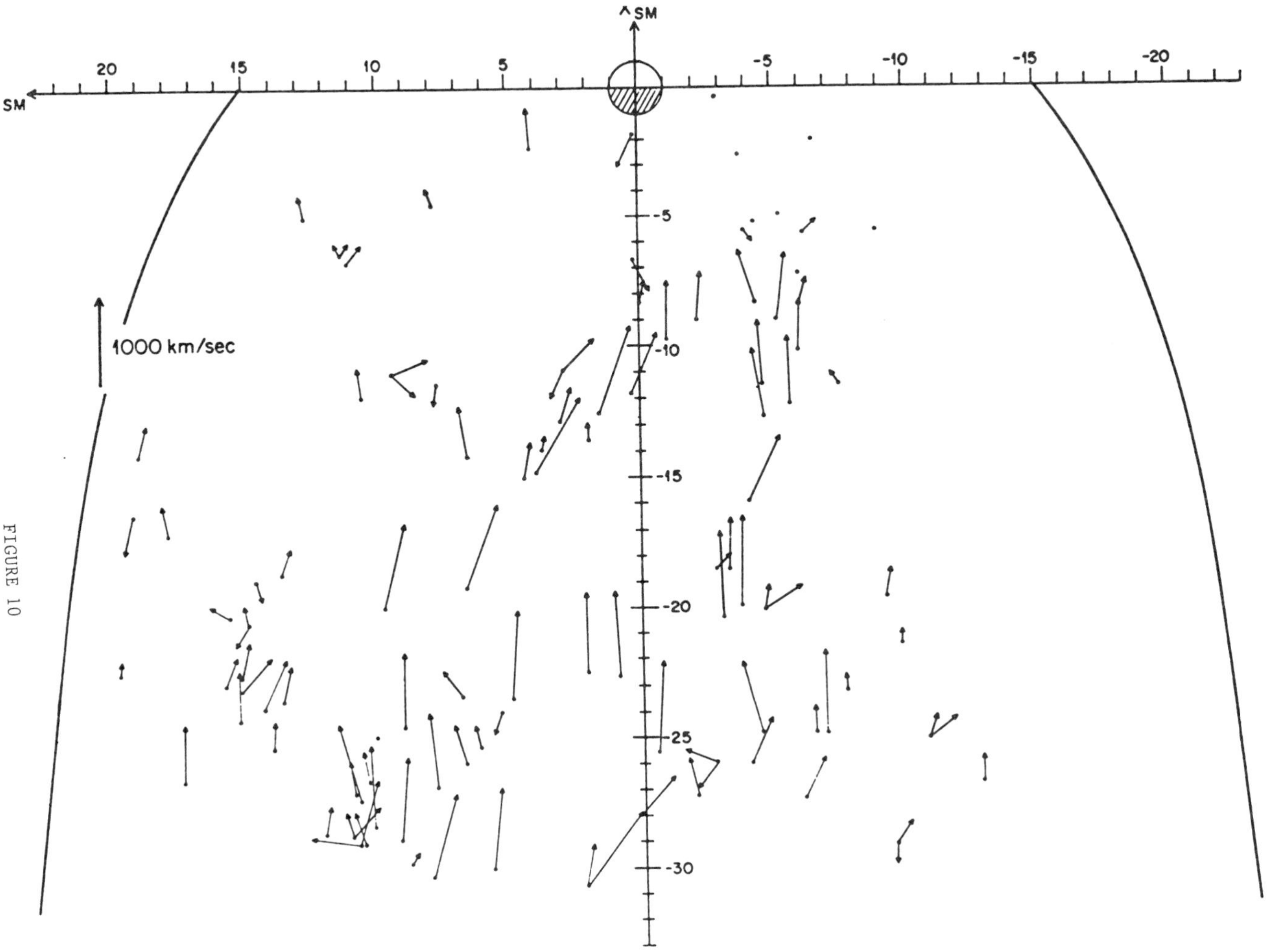


Fig. 10. The plasma flow pattern constructed from plasma flows observed during the thickening of the plasma sheet. The prevalence of sunward flow is very evident.

some detail. We point out that ionospheric sources have contributed significantly to the plasma population in the magnetotail and they have also increased the complexity on the problem of determining the plasma transport occurring during the substorm-associated thinning phase of the mid-tail plasma sheet.

Flows of hot plasma in the central plasma sheet of the mid-tail region is determined to be predominantly earthward. Flows are typically slow and fast flows are very transient (with a duration often less than 1min). Thus, the intrinsic mode of plasma transport in the tail is bursty in nature even during large intensity substorms. This lends support to the suggestion that substorm expansion may consist of a multitude of transient activations in the tail. Each of these activations may bear the same basic physical process (or processes) responsible for the substorm development. Their occurrence frequency may govern the substorm intensity. There are similarities between these activations and lightning in thunderstorms.

A major unknown in our observation of plasma in the tail is the low-energy (< 50 eV) component of the plasma population which originates from the ionospheric sources. Future investigations with instruments capable of monitoring this part of the population, perhaps by controlling the spacecraft potential, are needed to construct a more comprehensive picture on the transport of the entire plasma population in the magnetotail.

*Acknowledgments.* This work is supported by NASA under U. S. Navy contract N00039-87-C-5301 and by the Atmospheric Sciences Section of the National Science Foundation, Grant ATM-9000052 to the Johns Hopkins University.

References

Akasofu, S.-I., E. W. Hones, Jr., M. D. Montgomery, S. J. Bame, and S. Singer, Association of magnetotail phenomena with visible auroral features, *J. Geophys. Res.*, *76*, 5985, 1971.

Akasofu, S.-I., E. W. Hones, Jr., S. J. Bame, J. R. Asbridge, and A. T. Y. Lui, Magnetotail and boundary layer plasmas at a geocentric distance of ~18 $R_E$: Vela 5 and 6 observations, *J. Geophys. Res.*, *78*, 7257, 1973.

Andrews, M. K., P. W. Daly, and E. Keppler, Ion jetting at the plasma sheet boundary: simultaneous observations of incident and reflected particles, *Geophys. Res. Lett.*, *8*, 987, 1981.

Baker, D. N., E. W. Hones, Jr., D. T. Young, and J. Birn, The possible role of ionospheric oxygen in the initiation and development of plasma sheet instabilities, *Geophys. Res. Lett., 11*, 1337, 1982.

Baker, D. N., S. J. Bame, R. D. Belian, W. C. Feldman, J. T. Gosling, P. R. Higbie, E. W. Hones, Jr., D. J. McComas, and R. D. Zwickl, Correlated dynamical changes in the near-earth and distant magnetotail regions: ISEE 3, *J. Geophys. Res., 89*, 3855, 1984.

Baumjohann, W., Some recent progress in substorm studies, *J. Geomag. Geoelectr., 38*, 633, 1986.

Baumjohann, W., The plasma sheet boundary layer and magnetospheric substorms, *J. Geomag. Geoelectr., 40*, 157, 1988.

Baumjohann, W., and G. Paschmann, Determination of the polytropic index in the plasma sheet, *Geophys. Res. Lett., 16*, 295, 1989.

Baumjohann, W., G. Paschmann, N. Sckopke, C. A. Cattell, and C. W. Carlson, Average ion moments in the plasma sheet boundary layer, *J. Geophys. Res., 93*, 11507, 1988.

Baumjohann, W., and G. Paschmann, Determination of the polytropic index in the plasma sheet, *Geophys. Res. Lett., 16*, 295, 1989.

Baumjohann, W., G. Paschmann, and C. A. Cattell, Average plasma properties in the central plasma sheet, *J. Geophys. Res., 94*, 6597, 1989.

Baumjohann, W., G. Paschmann, H. Luhr, Characteristics of high-speed ion flows in the plasma sheet, *J. Geophys. Res., 95*, 3801, 1990.

Candidi, M., S. Orsini, Estimates of the north-south electric field component in the low latitude boundary layer, *Geophys. Res. Lett., 8*, 637, 1981.

Candidi, M., S. Orsini, and V. Formisano, The properties of ionospheric $O^+$ ions as observed in the magnetotail boundary layer and northern plasma lobe, *J. Geophys. Res., 87*, 9097, 1982.

Chandler, M. O., and C. R. Chappell, Observations of the flow of $H^+$ and $He^+$ along magnetic field lines in the plasmasphere, *J. Geophys. Res., 91*, 8847, 1986.

Chao, J. R., J. R. Kan, A. T. Y. Lui, S.-I. Akasofu, A model for thinning of the plasma sheet, *Planet. Space Sci., 25*, 703, 1977.

Chappell, C. R., T. E. Moore, J. H. Waite, Jr., The ionosphere as a fully adequate source of plasma for the Earth's magnetosphere, *J. Geophys. Res., 92*, 5896, 1987.

Cladis, J. B., Parallel acceleration and transport of ions from polar ionosphere to plasma sheet, *Geophys. Res. Lett., 13*, 893, 1986.

Dandouras, J., H. Reme, A. Saint-Marc, J. A. Sauvaud, G. K. Parks, K. A. Anderson, and R. P. Lin, A statistical study of plasma sheet dynamics using ISEE 1 and 2 energetic particle flux data, *J. Geophys. Res., 91*, 6861, 1986.

Dungey, J. W., Interplanetary magnetic field and the auroral zones, *Phys. Res. Lett., 6*, 47, 1961.

Eastman, T. E., L. A. Frank, and C. Y. Huang, The boundary layers as the primary transport regions of the Earth's magnetotail, *J. Geophys. Res., 90*, 9541, 1985.

Erickson, G. M., and R. A. Wolf, Is steady convection possible in the Earth's magnetotail?, *Geophys. Res. Lett., 7*, 897, 1980.

Eviatar, A. and R. A. Wolf, Transfer processes at the magnetopause, *J. Geophys. Res., 73*, 5561, 1968.

Freeman, J. W., Jr., H. K. Hills, T. W. Hill, P. H. Reiff, and D. A. Hardy, Heavy ion circulation in the Earth's magnetosphere, *Geophys. Res. Lett., 4*, 195, 1977.

Fuselier, S. A., W. K. Peterson, D. M. Klumpar, and E. G. Shelley, Entry and acceleration of $He^+$ in the low latitude boundary layer, *Geophys. Res. Lett., 16*, 751, 1989.

Hayakawa, H. A. Nishida, E. W. Hones, Jr., S. J. Bame, Statistical characteristics of plasma flow in the magnetotail, *J. Geophys. Res., 87*, 277, 1982

Heikkila, W. J., Impulsive plasma transport through the magnetopause, *Geophys. Res. Lett., 9*, 159, 1982.

Heikkila, W. J., Neutral-sheet crossings in the distant magnetotail, in Magnetotail Physics, ed. by A. T. Y. Lui, Johns Hopkins University Press, Baltimore, p. 65, 1987.

Heikkila, W. J., Current sheet crossings in the distant magnetotail, *Geophys. Res. Lett., 15*, 299, 1988.

Hones, E. W., Jr., Transient phenomena in the magnetotail and their relation to substorms, *Space Sci. Rev., 23*, 393, 1979.

Hones, E. W., Jr., J. R. Asbridge, and S. J. Bame, Time variations of the magnetotail plasma sheet at 18 $R_E$ determined from concurrent observations by a pair of Vela satellites, *J. Geophys. Res., 76*, 4402, 1971.

Hones, E. W., Jr., J. R. Asbridge, S. J. Bame, M. D. Montgomery, S. Singer, and S.-I. Akasofu, Measurements of magnetotail plasma flow made with Vela 4B, *J. Geophys. Res., 77*, 5503, 1972.

Hones, E. W., Jr., and K. Schindler, Magnetotail plasma flow during substorms: a survey with IMP 6 and IMP 8, *J. Geophys. Res., 84*, 7155, 1979.

Horwitz, J. L., C. R. Baugher, C. R. Chappell, E. G. Shelley, and D. T. Young, Pancake pitch angle distributions in warm ions observed with ISEE-1, *J. Geophys. Res., 86*, 3311, 1981.

Huang, C. Y., and L. A. Frank, A statistical study of the central plasma sheet: implications for substorm models, *Geophys. Res. Lett., 13*, 652, 1986.

Huang, C. Y., L. A. Frank, W. K. Peterson, D. J. Williams, W. Lennartsson, D. G. Mitchell, R. C. Elphic, and C. T. Russell, Filamentary structures in the magnetotail lobes, *J. Geophys. Res., 92*, 2349, 1987.

Huang, C. Y., Quadrennial review of the magnetotail, *Rev. Geophys., 25*, 529, 1987.

Huang, C. Y., C. K. Goertz, L. A. Frank, and G. Rostoker, Observational determination of the adiabatic index in the quiet time plasma sheet, *Geophys. Res. Lett., 16*, 563, 1989.

Kivelson, M. G., and H. E. Spence, On the possibility of quasi-static convection in the quiet magnetotail, *Geophys. Res. Lett., 15*, 1541, 1988.

Lemaire, J., Impulsive penetration of filamentary plasma elements into the magnetospheres of the Earth and Jupiter, *Planet. Space Sci., 25*, 887, 1977.

Lennartsson, W., Dynamical features of the plasma sheet in composition, density and energy, in *Magnetotail Physics*, ed. by A. T. Y. Lui, Johns Hopkins Univ. Press, p. 35, 1987.

Lennartsson, W., and R. D. Sharp, Relative contributions of Terrestrial and solar wind ions in the plasma sheet, *Adv. Space Res., 5*, 411, 1985.

Lockwood, M., M. O. Chandler, J. L. Horwitz, J. H. Waite, Jr., T. E. Moore, and C. R. Chappell, The cleft ion fountain, *J. Geophys. Res., 90*, 9736, 1985.

Lopez, R. E., A. T. Y. Lui, D. G. Sibeck, K. Takahashi, R. W. McEntire, L. J. Zanetti, and S. M. Krimigis, On the relationship between the energetic particle flux morphology and the change in the magnetic field magnitude during substorms, *J. Geophys. Res., 94*, 17105, 1989.

Lopez, R. E., and A. T. Y. Lui, A multisatellite case study of the expansion of a substorm current wedge in the near-earth magnetotail, *J. Geophys. Res., 95*, 8009, 1990.

Lui, A. T. Y., Observations of plasma sheet dynamics during magnetospheric substorms, in *Dynamics of the Magnetosphere*, ed. by S.-I. Akasofu, D. Reidel, p. 563, 1979.

Lui, A. T. Y., What is a magnetospheric substorm expansion made of?, *Trans. Amer. Geophys. Union, 69*, 435, 1988.

Lui, A. T. Y., and S.-I. Akasofu, Estimated north-south component of the electric field in the magnetotail plasma sheet, *Geophys. Res. Lett., 7*, 877, 1980.

Lui, A. T. Y., and A. Hasegawa, Implications of a steady-state magnetospheric convection, *Planet. Space Sci., 34*, 315, 1986.

Lui, A. T. Y., E. W. Hones, Jr., D. Venkatesan, S.-I. Akasofu, and S. J. Bame, Complete dropouts at Vela satellites during thinning of the plasma sheet, *J. Geophys. Res.*, *80*, 4649, 1975.

Lui, A. T. Y., E. W. Hones, Jr., F. Yasuhara, S.-I. Akasofu, and S. J. Bame, Magnetotail plasma flow during plasma sheet expansions: Vela 5 and 6 and IMP 6 observations, *J. Geophys. Res.*, *82*, 1235, 1977.

Lui, A. T. Y., and C.-I. Meng, Relevance of southward magnetic fields in the neutral sheet to anisotropic distribution of energetic electrons and substorm activity, *J. Geophys. Res.*, *84*, 5817, 1979.

Lui, A. T. Y., C.-I. Meng, L. A. Frank, K. L. Ackerson, and S.-I. Akasofu, Temperature variation of the plasma sheet during substorms, *Planet. Space Sci.*, *29*, 837, 1981.

Lui, A. T. Y., T. E. Eastman, D. J. Williams, and L. A. Frank, Observations of ion streaming during substorms, *J. Geophys. Res.*, *88*, 7753, 1983.

Lui, A. T. Y., R. E. Lopez, S. M. Krimigis, R. W. McEntire, L. J. Zanetti, and T. A. Potemra, A case study of magnetotail current disruption and diversion, *Geophys. Res. Lett.*, *15*, 721, 1988.

Lundin, R., Solar wind energy transfer regions inside the dayside magnetopause, II. evidence for an MHD generator process, *Planet. Space Sci.*, *32*, 757, 1984.

Lundin, R., K. Stasiewicz, and B. Hultqvist, On the interpretation of different flow vectors of different ion species in the magnetospheric boundary layer, *J. Geophys. Res.*, *92*, 3214, 1987.

Mitchell, D. G., F. Kutchko, D. J. Williams, T. E. Eastman, L. A. Frank, and C. T. Russell, An extended study of the low-latitude boundary layer on the dawn and dusk flanks of the magnetosphere, *J. Geophys. Res.*, 92, 7394, 1987.

Moore, T. E., C. R. Chappell, M. Lockwood, and J. H. Waite, Jr., Superthermal ion signatures of auroral acceleration processes, *J. Geophys. Res.*, *90*, 1611, 1985.

Moore, T. E., D. L. Gallagher, J. L. Horwitz, and R. H. Comfort, MHD wave breaking in the outer plasmasphere, *Geophys. Res. Lett.*, *14*, 1007, 1987.

Nagai, T., J. H. Waite, Jr., J. L. Green, C. R. Chappell, R. C. Olsen, and R. H. Comfort, First measurements of supersonic polar wind in the polar magnetosphere, *Geophys. Res. Lett.*, *11*, 669, 1984.

Orsini, S., M. Candidi, H. Balsiger, and A. G. Ghielmetti, Ionospheric ions in the near-earth geomagnetic tail plasma lobes, *Geophys. Res. Lett.*, *9*, 163, 1982.

Orsini, S., M. Candidi, M. Stokholm, H. Balsiger, Injection of ionospheric ions into the plasma sheet, *J. Geophys. Res.*, *95*, 7915, 1990.

Paschmann, G., H. Grunwaldt, M. D. Montgomery, H. Rosenbauer, and N. Sckopke, Plasma observations in the high-latitude magnetosphere, in *Correlated Interplanetary and Magnetospheric Observations*, ed. by D. E. Page, p. 249, D. Reidel, Dordrecht, Netherlands, 1974.

Pfitzer, K. A., and W. P. Olson, The entry of low energy charged particles into the magnetosphere, *Trans. Amer. Geophys. Union*, *54*, 432, 1973.

Potemra, T. A., L. J. Zanetti, P. F. Bythrow, A. T. Y. Lui, and T. Iijima, $B_y$ dependent convection patterns during northward interplanetary magnetic field, *J. Geophys. Res.*, *89*, 9753, 1984.

Pytte, T., R. L. McPherron, S. Kokubun, The ground signatures of the expansion phase during multiple onset substorms, *Planet. Space Sci.*, *24*, 1115, 1976.

Roelof, E. C., Energetic neutral atom image of a storm-time ring current, *Geophys. Res. Lett.*, *14*, 652, 1987.

Rosenbauer, H., H. Grunwaldt, M. D. Montgomery, G. Paschmann, and N. Sckopke, Heos 2 plasma observations in the distant polar magnetosphere: the plasma mantle, *J. Geophys. Res.*, *80*, 2723, 1975.

Rostoker, G., and R. A. Bostrom, A mechanism for driving the gross Birkeland current configuration in the auroral oval, *J. Geophys. Res.*, *81*, 235, 1976.

Rostoker, G., S.-I. Akasofu, J. Foster, R. A. Greenwald, Y. Kamide, K. Kawasaki, A. T. Y. Lui, R. L. McPherron, and C. T. Russell, Magnetospheric substorms—definition and signatures, *J. Geophys. Res.*, *85*, 1663, 1980.

Schindler, K. and J. Birn, Self-consistent theory of time-dependent convection in the Earth's magnetotail, *J. Geophys. Res.*, *87*, 2263, 1982.

Sckopke, N., G. Paschmann, H. Rosenbauer, and D. H. Fairfield, Influence of the interplanetary magnetic field on the occurrence and thickness of the plasma mantle, *J. Geophys. Res.*, *81*, 2687, 1976.

Shelley, E. G., R. G. Johnson, and R. D. Sharp, Satellite observations of energetic heavy ions during a geomagnetic substorm, *J. Geophys. Res.*, *77*, 6104, 1972.

Shelley, E. G., W. K. Peterson, A. G. Ghielmetti, and J. Geiss, The polar ionosphere as a source of energetic magnetospheric plasma, *Geophys. Res. Lett.*, *9*, 941, 1982.

Sergeev, V. A., and A. G. Yahnin, The features of auroral bulge expansion, *Planet. Space Sci.*, *27*, 1429, 1979.

Sergeev, V. A., A. G. Yahnin, R. A. Rakhmatulin, S. I. Solovjev, F. S. Mozer, D. J. Williams, and C. T. Russell, Permanent flare activity in the magnetosphere during periods of low magnetic activity in the auroral zone, *Planet. Space Sci.*, *34*, 1169, 1986.

Slavin, J. A., E. J. Smith, and B. T. Tsurutani, An ISEE 3 study of average and substorm conditions in the distant magnetotail, *J. Geophys. Res.*, *90*, 10875, 1985.

Stasiewicz, K., A global model of gyroviscous field line merging at the magnetopause, preprint, 1990.

Stevenson, T. E., and C. Comstock, Particles incident on magnetic field gradients, *J. Geophys. Res.*, *73*, 175, 1968.

Takahashi, K., and E. W. Hones, Jr., ISEE 1 and 2 observations of ion distributions at the plasma sheet-tail lobe boundary, *J. Geophys. Res.*, *93*, 8558, 1988.

VanAllen, J. A., Energetic particles in the inner magnetosphere of Saturn, in *Saturn*, ed. by T. Gehrels, and M. S. Matthews, Univ. of Arizona Press, Tucson, Arizona, 1984.

Waite, J. H., Jr., T. Nagai, J. F. E. Johnson, C. R. Chappell, J. L. Burch, T. L. Killeen, P. B. Hays, G. R. Carignan, W. K. Peterson, and E. G. Shelley, Escape of suprathermal O+ ions in the polar cap, *J. Geophys. Res.*, *90*, 1619, 1985.

Whalen, B. A., W. Bernstein, and P. W. Daly, Low altitude acceleration of ionospheric ions, *Geophys. Res. Lett.*, *5*, 55, 1978.

White, O. R., The Solar Output and Its Variation, Colorado Associated University Press, Boulder, CO, 1977.

Williams, D. J., Energetic ion beams at the edge of the plasma sheet: ISEE 1 observations plus a simple explanatory model, *J. Geophys. Res.*, *86*, 5507, 1981.

Winningham, J. D., and W. J. Heikkila, Polar cap auroral electron fluxes observed with ISIS 1, *J. Geophys. Res.*, *79*, 949, 1974.

Yau, A. W., E. G. Shelley, W. K. Peterson, and L. Lenchyshyn, Energetic auroral and polar ion outflow at DE 1 altitudes: magnitude, composition, magnetic activity dependence, and long-term variations, *J. Geophys. Res.*, *90*, 8417, 1985.

Yeager, D. M., and L. A. Frank, Low energy electron intensities at large distances over the Earth's polar cap, *J. Geophys. Res.*, *81*, 3966, 1976.

Zwickl, R. D., D. N. Baker, S. J. Bame, W. C. Feldman, J. T. Gosling, E. W. Hones, Jr., D. J. McComas, and E. J. Smith, Evolution of the Earth's distant magnetotail: ISEE-3 electron plasma results, *J. Geophys. Res.*, *89*, 11007, 1984.

# PROGRESS IN THE STUDY OF THREE-DIMENSIONAL PLASMOIDS

Michael Hesse and Joachim Birn

Los Alamos National Laboratory, Los Alamos, New Mexico

Results of MHD simulations of three-dimensional plasmoid evolution in the Earth's magnetotail are reviewed. Particular emphasis is on the features that can only be studied in a three-dimensional approach. We find that the magnetic topology of plasmoids in a more realistic three-dimensional system differs considerably from comparatively simple two-dimensional models. The inclusion of cross-tail variations in reconnection associated flow speeds also leads to new conclusions about plasmoid velocities. Magnetic reconnection in three-dimensional systems requires a new approach that has been theoretically investigated previously. Our results are found consistent with conclusions from that approach. They also predict specific features of electron acceleration due to the reconnection process. An important substorm related effect is the diversion of part of the cross-tail current and its diversion into the ionosphere via the substorm current wedge. Our simulation results suggest that the generation of the substorm current wedge is associated with the cross-tail variation of the earthward flows induced by the three-dimensional tearing mode, which also leads to plasmoid formation and ejection. At last, we briefly discuss the impact of an ionospheric coupling model on the plasmoid evolution in our model. We find that the resistive closure of the current wedge currents modifies the evolution slightly, but not enough to be solely responsible for the dynamic evolution of the tail.

## 1. Introduction

Substorm associated plasmoid evolution has been mainly investigated in two-dimensional models in the past [e.g., Birn, 1980; Otto et al., 1990]; an exception is the numerical study by Birn and Hones [1981]. However, theoretical considerations [e.g., Schindler et al., 1988] as well as observations indicating, e.g., strong flows in the cross-tail direction during substorms [e.g., Hones et al., 1976; Hones and Schindler, 1979; Hones et al., 1986] show that the substorm development in the Earth's magnetotail involves strong three-dimensional effects.

Starting from simple considerations such as the cross-tail extent of the region of reconnected field lines that constitute the plasmoid, there are a variety of new problems to be dealt with in a more realistic three-dimensional magnetotail. One of these is the question of the evolution of the magnetic topology. In a two-dimensional model, the severance of the plasmoid from the Earth occurs at one instant of time; see, e.g., the substorm scenario of Hones [1979], and the two-dimensional computer simulations of Otto et al. [1990]. In a three-dimensional system, however, this may be quite different. Birn et al. [1989] found in a relatively simple kinematical model that a plasmoid in a three-dimensional system is not necessarily a closed structure, but rather may be magnetically open with loop-like magnetic field lines. Further, the authors saw strong evidence that the severance of the plasmoid from the closed field line region should involve a gradual change of topology from closed to open, involving a finite amount of time. Hence the question remains whether a self-consistent model produces similar changes during the topological evolution of the plasmoid magnetic field.

Another open question is the mechanism that accounts for the observed plasmoid velocities of several hundreds km/s. This question has been addressed by Slavin et al. [1989] through the use of magnetotail observations. Slavin et al. came to the conclusion that the most prominent contribution to the acceleration is the magnetic slingshot effect, involving the tension forces of interplanetary field lines that wrap around the partially or fully severed plasmoid volume. The same problem was addressed by Otto et al. [1990], who used a two-dimensional MHD model to describe plasmoid evolution. They found that the change in plasmoid velocity was mainly brought about by the pressure gradient prevalent in the quiet magnetotail. In fact, Otto et al. typically saw the plasmoid velocity increase over several hundred percent during the plasmoid evolution. Although at first, these investigations seem sufficient to deal with the acceleration problem it is by no means clear that a more realistic three-dimensional system exhibits the same features. One of the effects that may influence the results is the expected variations of flow velocities in the cross-tail direction, which is taken to be invariant in the two-dimensional models.

Plasmoid evolution has been studied as a prime example of magnetic reconnection [e.g., Schindler, 1974; Hones, 1979; Birn, 1980; Hautz and Scholer, 1987]. This fact should not be changed by a three-dimensional investigation. Indeed Schindler et al. [1988] and Hesse and Schindler [1988] predicted on a theoretical basis that reconnection in systems without zeroes of the magnetic field is not only possible but strongly depends on properties of the electric field component parallel to the magnetic field. In the situation relevant for plasmoid evolution, the parallel electric field should yield a nonnegative value when integrated along field lines passing through the reconnection region. Although the theoretical predictions are quite adequate, it is nevertheless of interest to study the evolution of these quantities in a self-consistent model.

Another intrinsically three-dimensional effect is the deviation of part of the cross-tail current to the ionosphere by means of the so-called substorm current wedge [ McPherron et al., 1973]. Since this effect clearly is incompatible with any kind of translational invariance, a fully three-dimensional model is needed to add to its understanding. The major point of discussion is here, whether the substorm current wedge and the dipolarization of the magnetic field in the central tail region, allegedly associated with the current deviation, can be accounted for in the plasmoid substorm picture.

Further, it is of interest to investigate the changes brought about by the ionospheric closure of the wedge currents. A resistive ionosphere would imply that the current closure generates electric fields which should force convection in the magnetotail, which in turn could modify the tail evolution, up to driving or significantly modifying substorm evolution. For the same reasons as the substorm current wedge, a study of magnetosphere-ionosphere coupling requires an intrinsically three-dimensional model.

Modeling Magnetospheric Plasma Processes
Geophysical Monograph 62

In the following sections, we will review our answers to these questions based on three-dimensional resistive MHD simulations. Due to the complexity of the problem, the study of plasmoid evolution in a three-dimensional resistive MHD model requires numerical integration of the associated differential equations. We integrate the MHD equations in the following form

$$\frac{\partial \rho}{\partial t} + \nabla \cdot (\rho \mathbf{v}) = 0 \tag{1}$$

$$\rho \left( \frac{\partial \mathbf{v}}{\partial t} + \mathbf{v} \cdot \nabla \mathbf{v} \right) = -\nabla p + \mathbf{j} \times \mathbf{B} \tag{2}$$

$$\frac{\partial \mathbf{B}}{\partial t} = \nabla \times (\mathbf{v} \times \mathbf{B}) + \eta \nabla^2 \mathbf{B} \tag{3}$$

$$p = nkT, \quad T = \mathrm{const} \tag{4}$$

$$S = \mu_0 \frac{L_c v_c}{\eta} = 200 \tag{5}$$

All quantities were normalized using the scale length for variations with $z$, $L_z = 2$ R$_E$, the lobe field strength, and the density difference between the center of the plasma sheet and the lobe, all chosen at $y = 0$ and the near-Earth boundary $x = -20$ R$_E$, and suitable combinations of these constants. The computations were done inside a box $0 \geq x \geq x_{min}$, $0 \leq y \leq 10$, $0 \leq z \leq 10$, where $.x_{min} = -60$ or $x_{min} = -120$ depending on the application.

Equation (4) implies the choice of an isothermal pressure law. Independent investigations (not discussed here) including adiabatic effects and Ohmic heating showed that for the specific properties of tail dynamics during the substorm expansion phase discussed below the actual form of (4) is not important. The resistivity was chosen constant in all cases, consistent with Eq. (5).

A fully consistent problem requires the specification of boundary conditions. Symmetry boundary conditions were used at $y = 0$ and $z = 0$. At all other boundaries, except at the distant one $x = -60$, we assumed $\mathbf{v} = 0$, representing solid reflecting walls with an effective field line tying, and held the normal magnetic field component fixed, while vanishing normal derivatives were imposed on the tangential magnetic field components, density and pressure. The distant boundary $x = -60$ was assumed to be open with vanishing normal derivatives of density, velocity, and pressure, and a convective condition $d/dt = 0$ for the tangential magnetic field components in the case of outward flow, while the normal magnetic field component and the tangential magnetic field components for inward flow were held fixed.

Our near-Earth boundary condition is basically equivalent to an ideally conducting ionosphere (if the weak diffusion due to the presence of uniform resistivity is neglected). A more realistic boundary condition takes the convection along the boundary into account, which is generated by electric fields resulting from the closure of field-aligned currents through the resistive ionosphere and mapping back into the tail. This problem will be treated in section 6.

## 2. The Topology of Plasmoids in Three Dimensions

Based on the first two-dimensional models, a plasmoid has been defined to be a volume of closed field lines [e.g., Hones, 1979], that is, loop-like field lines closing on themselves. The border of this volume is formed by a magnetic surface, called a separatrix, made up of field lines that connect to a separator line or X-line. The situation becomes more complicated if the symmetry of the system is relaxed by including a net cross-tail magnetic field component and a dependence on the cross-tail coordinate in the initial state of the simulation. It has been shown in an analytic kinematical model [ Birn et al., 1989] that in such a general system the topology of the plasmoid changes from closed to open, i.e., to a structure that consists of helical field lines rather than closed loops. The first investigation reviewed here is concerned with the self-consistent topological evolution of the magnetic field associated with the plasmoid. In order to study the topology in a self-consistent model, we ran an MHD simulation starting from the equilibrium shown in Figure 1 [ Hesse and Birn, 1990c]. The equilibrium shown is a self-consistent solution of the magnetohydrostatic force balance by a method given by Birn [1987]. The initial state includes a net cross-tail magnetic field component of about 3% of the lobe field strength, consistent with typical observations [ Fairfield, 1979]. Furthermore, the tail model includes a far separator line located at $x = -100$R$_E$. In the initial configuration this line is given by the $B_z = 0$ line in the equatorial plane of the magnetotail model. We integrated the MHD equations (1)–(5) in a tail region extending from $x = -20$ $R_E$to $x = -260$R$_E$.

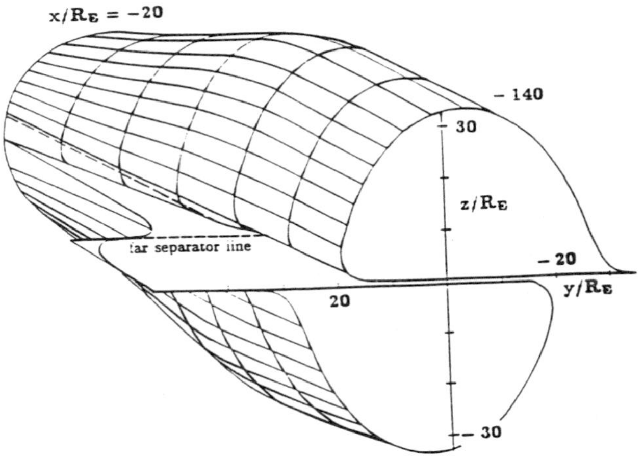


Fig. 1 The MHD equilibrium used as initial state in this investigation. The model shown here exhibits an average cross-tail magnetic field of about 3% of the typical lobe field strength. The equilibrium is shown here for $x \geq$ -120 $R_E$.

In this simulation, we found that the plasmoid formed at $t \approx 95$ (times normalized to a characteristic Alfvén transit time over the distance of the current sheet half-width). Initially, shortly after formation time, identified by the occurrence of negative $B_z$, the plasmoid consists solely of closed field lines that connect to the Earth with either "end." Figure 2 shows a typical plasmoid field line at $t = 130$. In this situation, no field line is topologically preferred, in contrast to the symmetric case. Thus there is no unique topological concept to distinguish plasmoid field lines from the ambient closed field lines in order to define a flux surface including the entire plasmoid.

To investigate the actual evolution of the magnetic field topology we follow Birn and Hesse [1990a,b], using the fact that the plasmoid forms mainly in the region around the equatorial plane of our tail model. This enables us to identify the type of field lines by counting the number of intersections with this plane; if this number is greater than one, the field line is termed a plasmoid field line. Although this identification is not invariant (a tilt of the equatorial plane would change the type of some field lines), it has proved to be useful to study the complexity of the topological substructure of the plasmoid.

Figure 3 displays the actual three-dimensional magnetic field topology in our present case. Intersections of field lines with the equatorial plane of the simulation box are assigned a symbol depending on the topological type of the field line that intersects at this very

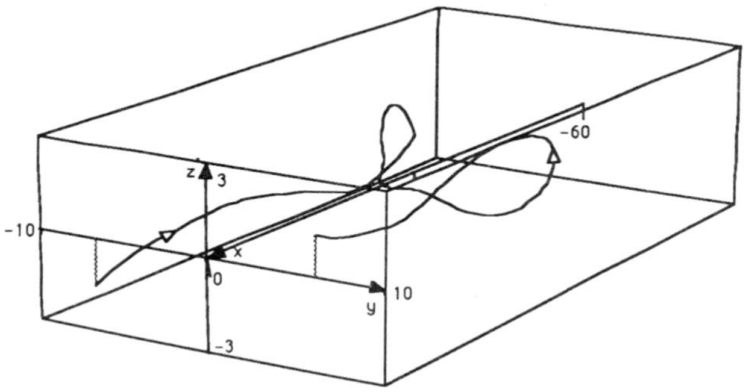


Fig. 2 Example of a typical closed plasmoid field line. This plot illustrates that the plasmoid can still be magnetically connected to the Earth at the earlier times of its evolution.

point. The symbols denote:

- "|": closed field lines, connected with the Earth (more accurately, with the near-Earth boundary) at both ends and intersecting the equatorial plane once
- "#": closed field lines that intersect the equatorial plane more than once ("closed plasmoid field lines")
- "*": half open field lines that intersect the equatorial plane once, connecting to the Earth and the IMF through the sides of the simulation box
- "x": half open field lines that intersect the equatorial plane more than once, ("lobe-like plasmoid field lines") connecting to the Earth on one side and to the IMF on the other through either the tailward end plane or the sides of the simulation box
- "o": open field lines (IMF type) that intersect the equatorial plane once
- "+": "open plasmoid field lines", intersecting the equatorial plane more than once, without connections to the Earth

The plasmoid region, defined by the region of helical field lines, is characterized by four possible symbols: #, *, x, and +. The panels show that the plasmoid initially consists entirely of closed field lines (#), i.e., field lines that connect to the Earth with both ends. The process of severance is found to be similar to that presented by Birn and Hesse [1990a,b], and very much different from two-dimensional models. Instead of an instantaneous severance of the entire plasmoid, the three dimensional model displays a complicated change of topology of the plasmoid field lines, taking a finite amount of time. During intermediate stages of the evolution, the plasmoid exhibits a tangled structure of flux tubes of different magnetic topologies, qualitatively as predicted by the kinematical model of Birn et al. [1989]. The first type of flux that appears to be intermingled with the closed flux tubes is open lobe-like flux (x), resulting from the progression of magnetic reconnection to lobe field lines. At times starting at about $t = 170$, interplanetary type flux ropes (+) appear inside the plasmoid flux rope. Further, another simpler interplanetary flux rope (o) forms and wraps around the plasmoid flux rope, intersecting the equatorial plane earthward from the it. At times later than $t = 200$, the last closed flux tubes vanish from the plasmoid, whereas half-open plasmoid flux tubes that still connect to the Earth at one end have not completely vanished yet.

The impact of an open topology on the particle distributions has been discussed by Birn et al. [1989]. The latter authors concluded that the different magnetic topology should only have an impact on a minor part of the particle distribution. Because of the openness of the plasmoid field lines, the most important effect could be an electron heat flux just outside the plasmoid volume, such as has been observed by Hones et al. [1984].

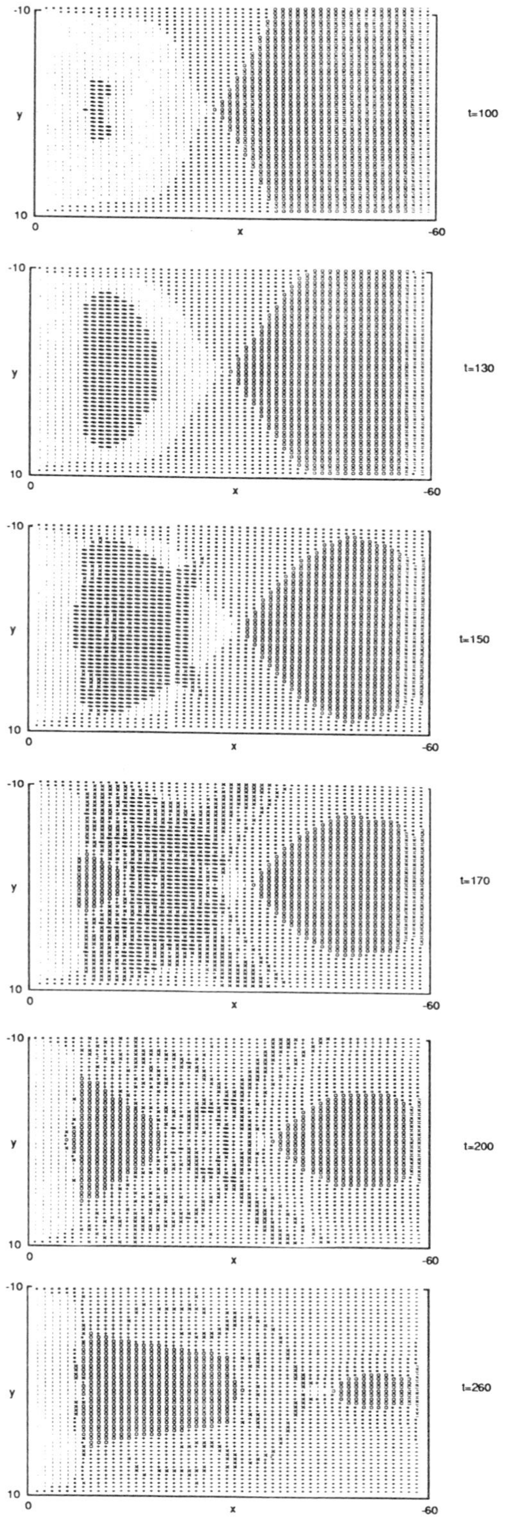


Fig. 3 Topological evolution of the plasmoid. The meaning of the symbols is explained in the text.

## 3. Acceleration Mechanism for Plasmoids

The color representation of Figure 3 is based on the integration of field lines including a count of the number of intersections with the equatorial plane. Although this representation gives a good overview of the topological evolution of the magnetic field, it is less well suited for the identification of the plasmoid volume and the study of plasmoid properties. The investigation of plasmoid properties, such as total forces and total momentum, requires integration of the respective spatial densities. For this purpose, one must determine a well defined volume, which can be identified with the plasmoid or at least a typical part thereof.

For our analysis, we will use the following approach. We consider the part of each helical field line that lies between successive intersections with the equatorial plane. The volume resulting from the set of these sections defines the "body" of the plasmoid. Note that the symmetry of our system with respect to the x-line permits us to restrict the investigation to the northern half of the body, since the southern half is line symmetric to it. Further, it is easy to see that the so-defined body of the plasmoid becomes the entire plasmoid in the symmetric case, which involves only field lines closing on themselves. Therefore, this choice of the body of the plasmoid seems to be the most natural for our purposes. The panels of Figure 4 show the northern surface of the plasmoid body at different times. The southern half of the entire plasmoid body is derived from the northern by mirroring the northern part across the $x$ axis. Note that this implies that the intersection of the body with the equatorial plane is symmetric with respect to the $x$ axis, as indicated in Figure 3. After the formation, which we identify with the appearance of a region of $B_z$ reversal in the equatorial plane, the most prominent features of plasmoid evolution are a growth in volume and a tailward motion. The latter seems to be caused by the action of either the pressure gradient

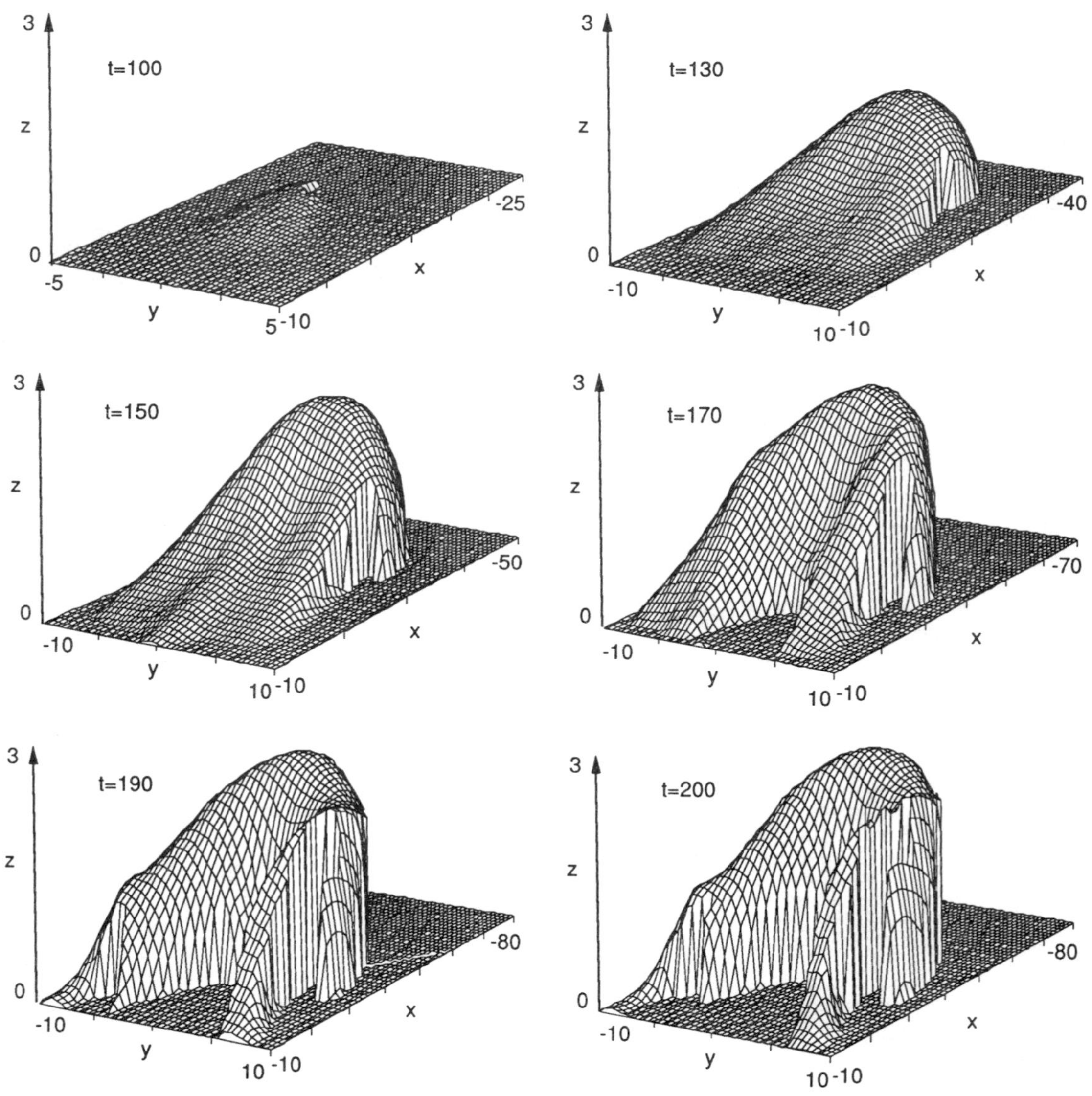


Fig. 4 The northern surface of the plasmoid body during the later evolution.

along the tail axis, or the Lorentz force due to the cross tail current density and the magnetic $B_z$ component. At later times, when the plasmoid becomes detached from the region of closed field lines, it has been suggested that the Lorentz force plays the dominant role (the so-called slingshot mechanism) via the combined effect of the cross-tail current and the open magnetic flux that wraps around the plasmoid [ Slavin et al., 1989]. This argument implies, however, that this effect cannot play the major role initially (there is no open flux wrapping around the plasmoid); instead it seems natural to assume that the pressure gradient dominates here, due to the reduction in the Lorentz force by the reduction of $B_z$ before onset.

To analyze this question and to investigate the temporal evolution of the plasmoid we calculate the volume, the mass, and the $x$ components of the momentum, pressure gradient and the Lorentz force by integration over the plasmoid body. The volume $V$ over which the integrations are performed is taken to be the northern half of the plasmoid body as defined above.

For the analysis of the early time period, we have used the simulation results at every 0.5 Alfvén times, starting from $t = 95$. At each of these times we have determined the position of the northern surface of the plasmoid body, which was then used as the boundary of our integration volume.

The panels of Figure 5 show the temporal evolution of the quantities (6)–(10) in the time range considered. The graphs of the different quantities display a weak increase at formation time (around $t = 95$), which becomes increasingly steep later. An exception of this rule is the integrated Lorentz force, which is initially negative but becomes subsequently positive. This is a consequence of the initial equilibrium, where the (positive) pressure force along the tail axis is balanced by the Lorentz force. At onset, the Lorentz force gets

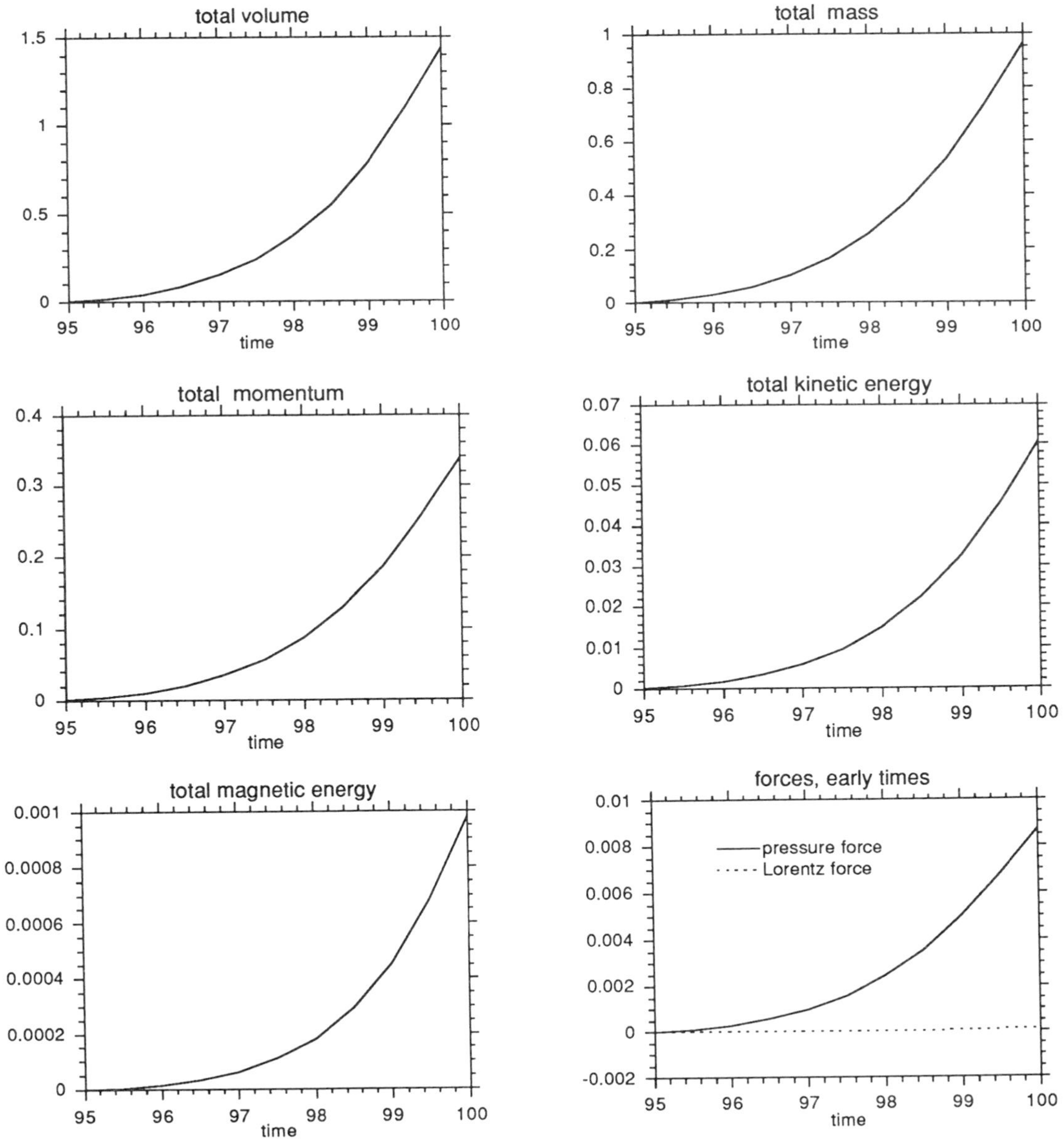


Fig. 5 Typical global plasmoid parameters for the early time range of the evolution.

reduced due to the reduction of $B_z$, which leads to the formation of the near-Earth $B_z = 0$ lines. The total force, however, stays positive at all the times considered. It is obvious that the contribution of the pressure gradient dominates that of the Lorentz force, even after the latter becomes positive and contributes to the tailward acceleration. To find out whether the uncertainty in defining the plasmoid has a significant effect, we did a comparative run without the cross-tail field, for which the plasmoid could be identified in the conventional (but singular) way through closed loops of field lines. The results were quite similar to those discussed here and are therefore not presented in detail.

Figure 6 shows the evolution of the forces acting on the plasmoid body during the later evolution. For the entire evolution period the forces are found positive, that is, tailward. Initially, the pressure force dominates, which seems consistent with results of Otto et al. [1990]. At later times, $t \gtrsim 150$, roughly coinciding with the period of severance of the plasmoid, the Lorentz force becomes dominant and the pressure force drops. Finally, when the plasmoid is nearly fully separated, the Lorentz force drops again to almost the level of the pressure force.

A complete analysis of the plasmoid acceleration process, however, requires a comparison of the total force to the time derivative of the plasmoid momentum. The two panels of Figure 7 show the total force and the time derivative of the momentum in the two evolution regimes. Except in the very last stage the momentum gain $dp/dt$ by far exceeds the total force acting on the plasmoid body. This means that the integrated force cannot account for the change of momentum, particularly during the early evolution, when $dp/dt$ exceeds the force by almost an order of magnitude.

This further implies that some effect other than the action of the

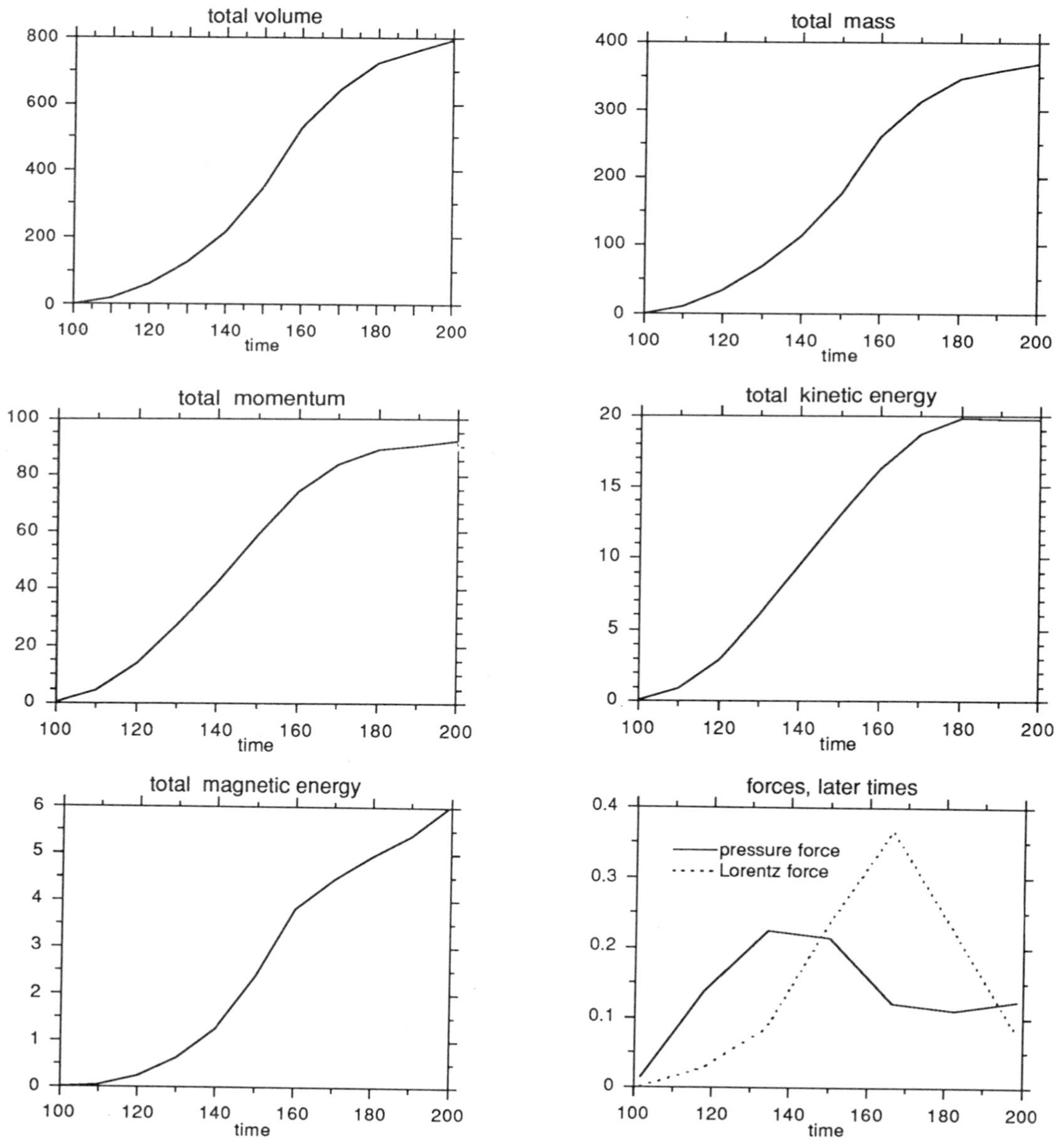


Fig. 6 Typical global plasmoid parameters for the later times of the evolution.

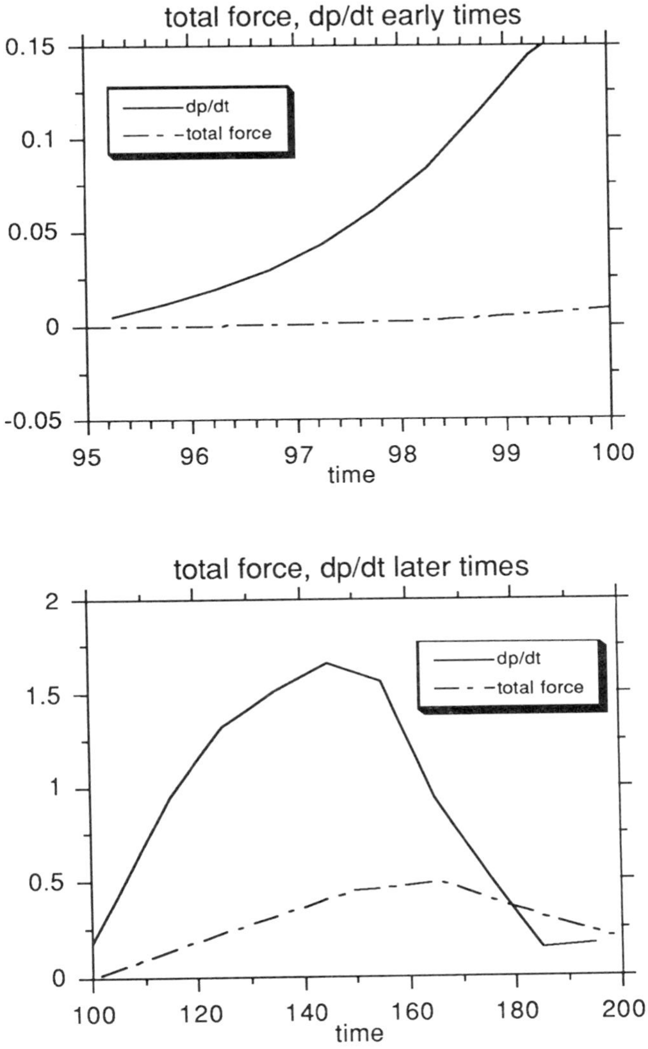


Fig. 7 Comparison between the total force and the time derivative of the total momentum for both time ranges considered.

forces inside the plasmoid region has to be responsible for the major part of the gain in plasmoid momentum. The answer to this puzzle is given by the fact that we ignored the inflow of momentum through the boundary, or, the growth of the volume causing an inclusion of accelerated flow into the plasmoid volume. Whereas the second term on the RHS represents the total internal force, the above features are accounted for by the first term on the RHS of the equation for the temporal change of the total plasmoid momentum.

$$\frac{d\mathbf{P}}{dt} = \int_{\partial V} (\mathbf{v}_s - \mathbf{v}) \cdot \mathbf{n}_s \rho \mathbf{v} \, ds + \int_V (\underline{j} \times \mathbf{B} - \nabla p) \, dV$$

Here $V$ denotes the volume of the plasmoid body, $\partial V$ the surface thereof, $\mathbf{n}_s$ the outward normal of this boundary, $\mathbf{v}_s$ the velocity of the surface, $\mathbf{v}$ the plasma flow velocity, and $\rho$ the plasma density.

The fact that the main momentum gain of the plasmoid stems from the accumulation of accelerated plasma implies that this plasma has been accelerated outside the plasmoid volume. Hesse and Birn [1990c] showed that the dominant momentum gain takes place at the Earthward end of the plasmoid just tailward of the near-Earth reconnection region ("neutral line"). A secondary minor contribution stems from the capturing of fast plasma in front of the plasmoid as field lines associated with that fast plasma become reconnected in the rear and thereby helical plasmoid field lines. Further Hesse and Birn [1990c] found that the acceleration outside the plasmoid is dominated by the Lorentz force. The region where the Lorentz force peaks has the qualitative properties of a slow shock or a slow mode compression. Similar shock layers are found at later times inside of the boundary magnetic surface connected with the near-Earth "neutral line" region. This stage is similar to the steady state models of reconnection which include a slow shock inside of the magnetic separatrix [e.g., Petschek, 1964].

Further evidence for the importance of acceleration outside the plasmoid can be found in Figure 8, which displays the average velocity $v_x$ of the plasmoid during its evolution. During the early times, the changes of $v_x$ are small compared to its initial value, indicating again that the most important change in momentum is due to previously accelerated plasma being added to the plasmoid. At later times, the average velocity decreases, due to the fact that the boundary of the plasmoid body progresses to regions which contain plasma that has not been accelerated.

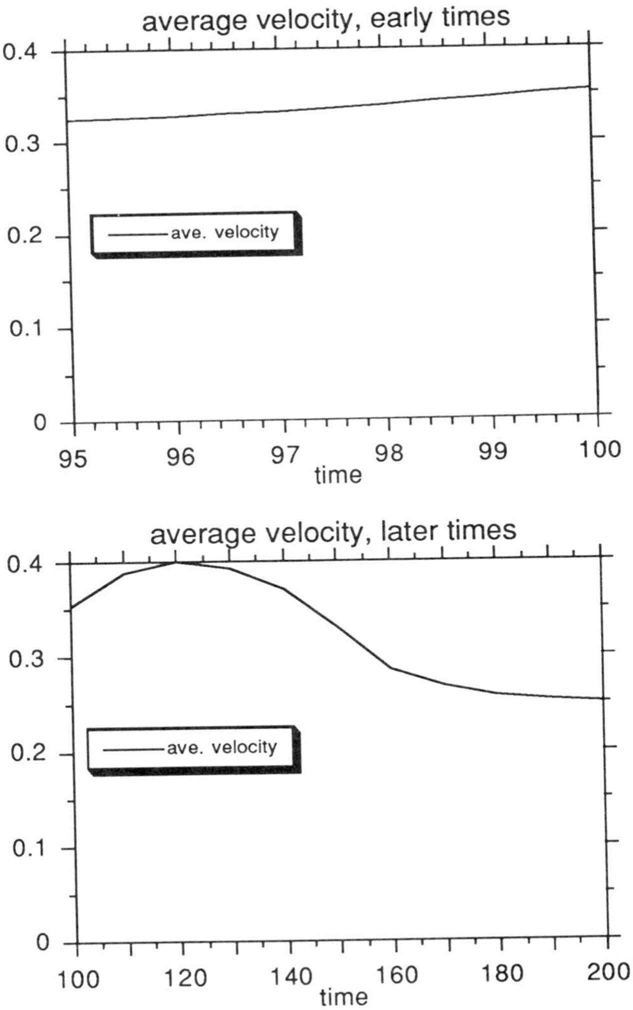


Fig. 8 Average plasmoid velocity for the early and late evolution.

The results of this section suggest that the initial stage of the substorm is dominated by the acceleration of plasma by the reconnection process. Initially, the formation of spiralling field lines seems to be a mere consequence, less important in the high $\beta$ plasma. Further, the magnetic field in the current sheet region is initially decoupled from the flow by the nonideal processes in the reconnection region. Only at later times, when the plasmoid becomes detached from the closed field line region and ultimately from the reconnection region, the plasmoid body is located in a region where the convective time scale, given by the fast tailward flow, is much shorter than the diffusive, given by the finite resistivity. This implies that the spiralling field lines become effectively frozen into the flow. The resulting system is then ejected tailward with a momentum that is determined by the reconnection process at the earlier stages.

At last we would like to comment on the apparent difference in velocity change between our three-dimensional model and the two-dimensional model of Otto et al. [1990]. The reason for this can be seen from Figure 17, which shows flow patterns in the equatorial plane for two simulations to be discussed later in more detail. The panels show a typical feature, namely a strong cross-tail variation of the tailward flow speeds. This implies that not only the fastest flow from the noon-midnight meridional region is added to the plasmoid, but also the slower flows from the flanks of the plasmoid region, whose extent is roughly indicated by the $B_z = 0$ lines in Figure 17. Clearly, the cross-tail flow variation then causes a lower average plasmoid velocity. Therefore a two-dimensional model has the tendency to overestimate the plasmoid velocity; in a more realistic three-dimensional configuration a lower speed results.

## 4. Three-Dimensional Magnetic Reconnection

As shown by Schindler et al. [1988] and Hesse and Schindler [1988], magnetic reconnection in general configurations without magnetic nulls is associated with a localized region ("diffusion zone") where an electric field component parallel to the magnetic field exists and ideal MHD is violated. Global effects, leading to new magnetic connections of plasma elements in ideal regions, require that the parallel electric field, integrated across the diffusion zone, leads to a finite value, equivalent to a potential difference

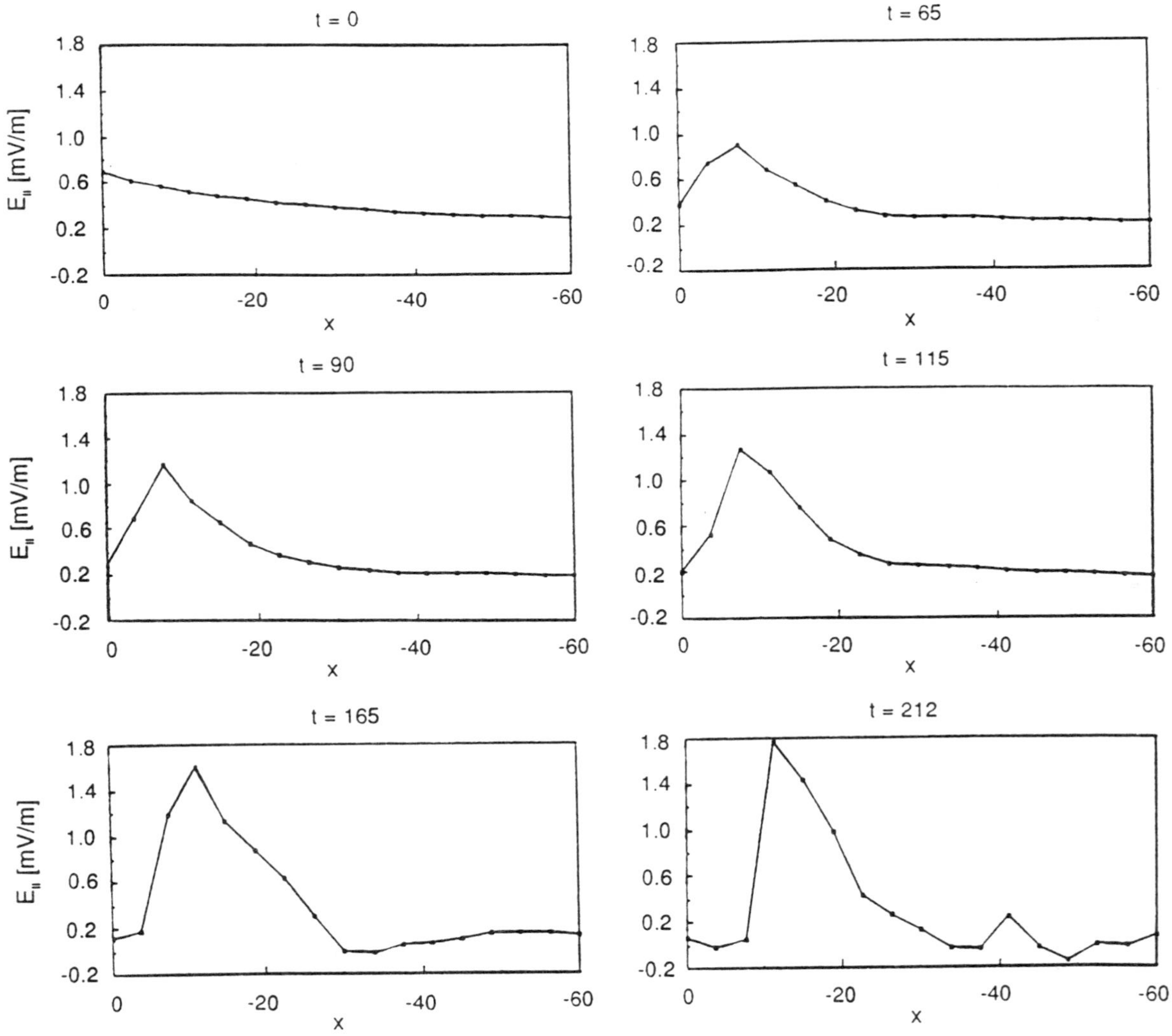


Fig. 9 Variation of $E_{\|}$ along the $x$-axis for the case $B_{yN} \neq 0$ for different times indicated in the different panels.

$$\Phi = \int_{\text{fieldline}} E_{\parallel}\, ds \neq 0 \qquad (12)$$

where the integration extends along a field line crossing the diffusion zone and $s$ is the arc length. We should emphasize here that the electric field need not be a potential field, so that the path along a field line is important in the definition of $\Phi$ by Equation (12). In the sense of general magnetic reconnection (GMR) [ Schindler et al., 1988], the formation and evolution of a plasmoid is a reconnection process. Note that, apart from the symmetric models without net cross-tail magnetic field (see below), all models presented in this paper do not exhibit zeroes of **B** anywhere in the simulation box. Furthermore, we point out that the frozen flux boundary condition employed in our simulations prohibits the formation of a helical flux rope like our plasmoid by an ideal ($\eta = 0$) process.

We investigated properties of three-dimensional magnetic reconnection in a shorter tail model with nonvanishing cross-tail magnetic field. In this case, the model extended up to $x = -140\mathrm{R}_E$ [ Birn and Hesse, 1990a,b; Hesse and Birn, 1990a]. Figures 9 and 10 demonstrate the localization of $E_{\parallel}$ in three different spatial representations. Figure 9 shows the variation of $E_{\parallel}$ along the $x$-axis at different times. The concentration of large $E_{\parallel}$ near $x = -10$ is quite obvious. This region may be identified as the diffusion zone.

To demonstrate the extent of the diffusion zone in the other space directions, Figures 10$a$ and $b$ show contour lines of constant $E_{\parallel}$ for $t = 165$ in the $x, y$ plane and the $x, z$ plane, respectively. The shaded areas in Figure 10$a$ and $b$ indicate the region where $E_{\parallel}$ exceeds 50 percent of its maximum value. It surrounds the near-Earth part of

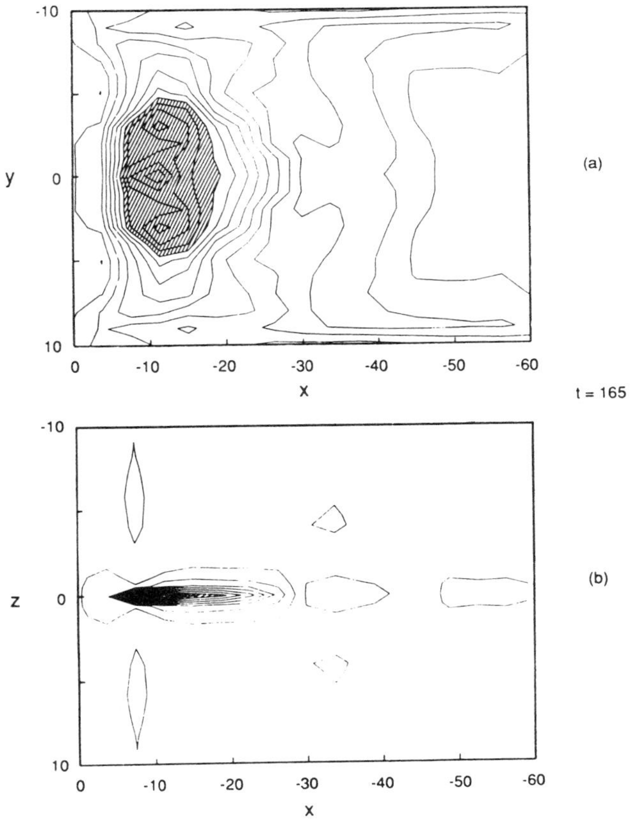


Fig. 10 Contour lines of constant $E_{\parallel}$ in the equatorial plane ($a$) and the midnight meridian plane ($b$) at $t = 165$ for the case $B_{yN} \neq 0$.

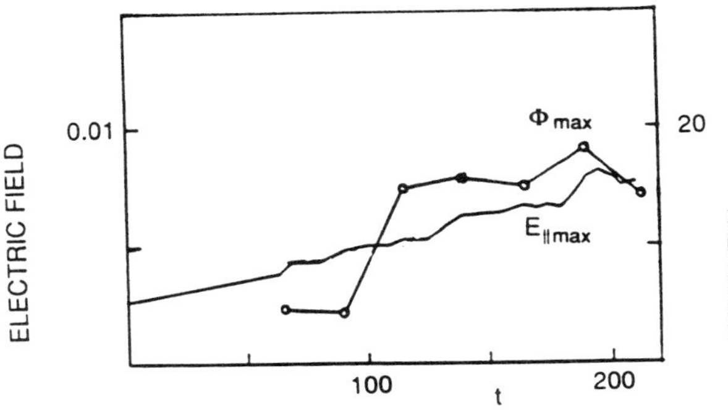


Fig. 11 Evolution of the maximum values of $E_{\parallel}$ (solid line) and the potential difference $\Phi$ (solid line with circles), integrated along field lines, for the case $B_{yN} \neq 0$.

the $B_z = 0$ line, which corresponds to the X-line in the symmetric case.

Figures 9 and 10 demonstrate that the extent of the diffusion zone and the magnitude of $E_{\parallel}$ do not vary very much in time. This might suggest that the effect of $E_{\parallel}$ on the acceleration of particles should not vary very much either. This, however, is not the case, as we will demonstrate. Predictions of the acceleration of non-adiabatic particles (that is, most of the ions) require the tracing of actual particle orbits and are outside the scope of the present paper. Electrons, however, can be expected to follow the field lines through nonvanishing magnetic fields more closely, so that the integrated potential difference, associated with the parallel electric field, should give a reasonable estimate of the expected electron acceleration. Accordingly we have integrated $E_{\parallel}$ along various field lines to find the field lines which carry the maximum value $\Phi_{max}$ of the potential difference, defined by (8). Figure 11 shows the temporal variation of $\Phi_{max}$ (solid line with circles) together with the evolution of the maximum values of $E_{\parallel}$ (solid line). Note the characteristic increase of $\Phi_{max}$ at about $t = 100$, which is the time when $B_z$ becomes negative in the equatorial plane, corresponding to the formation of the neutral line in the symmetric case. This increase of $\Phi_{max}$ is due to a characteristic change of the field configuration by reconnection. When $B_z$ becomes small in the equatorial plane, $B_y$ becomes the dominant component. Thus field lines are formed that stay for a longer distance near the equatorial plane in the region where the enhanced parallel electric fields are present. These field lines can therefore accumulate a larger amount of $E_{\parallel}$ and thus produce a larger total potential difference.

It is of particular interest to look for signatures ("foot prints") of these field regions at the near-Earth boundary. Figure 12 shows for different times contours of constant potential difference $\Phi$ at $x = 0$ for field lines crossing this boundary, where $\Phi$ is integrated along these field lines. The following features seem remarkable. The region of strong potential differences coincides closely with the boundary between closed and open field lines, i.e., with the plasma sheet boundary layer. This region evolves from a more diffuse to a rather narrow band in its extent in z. At the same time there is a clottiness, that becomes more and more pronounced in time. This is remarkable, because it is not related to a similar structure, that is a variation with $y$, of the fields in the tail. It is, however, related to the filamentary structure of the field connections, discussed in section 2, and to the fact that field lines that pass through the diffusion region and collect large parallel potential differences can either be connected with the near-Earth boundary directly or through some "excursion" around the plasmoid region in the more distant tail.

## 5. The Substorm Current Wedge

One of the major features of magnetotail dynamics, associated with geomagnetic substorms and providing a coupling mechanism between the tail and the ionosphere, is an alteration of the current

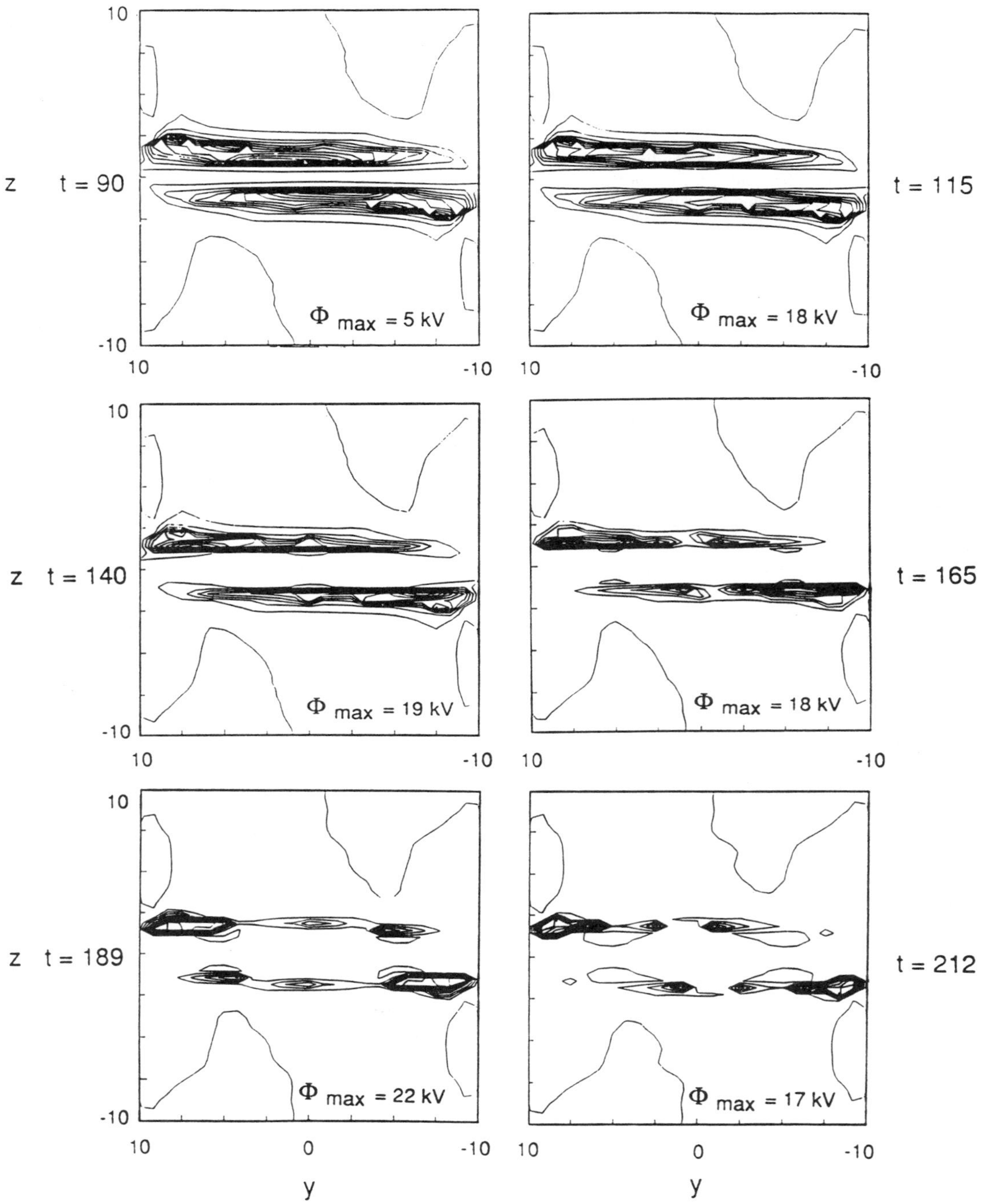


Fig. 12 Contours of constant potential difference $\Phi$, integrated along field lines, at $x = 0$ for the case $B_{yN} \neq 0$ and different times, indicated in the different panels.

system, inferred from observations, which is commonly called the substorm current wedge [ McPherron et al., 1973]. This current system consists of a deviation of the cross-tail current in the near tail toward the Earth on the dawn side and returning to the tail on the dusk side. While the diverted currents remain perpendicular to the magnetic field in the equatorial plane, which is identical with the neutral sheet in an idealized symmetric case, they are expected to become field-aligned away from the equatorial plane with their direction consistent with the region 1 field-aligned current system, observed nea the Earth [ Iijima and Potemra, 1976b]. The substorm current wedg thus provides an explanation for the observed increase of the regio 1 system with geomagnetic activity. The increase of the oppositel directed region 2 current system, observed at lower latitudes, is ofte considered as a response to the increase of the "primary" region system [e.g., Potemra et al., 1979].

As our experience with other initial equilibria shows, the stud of the generation and evolution of the substorm current wedge in a three-dimensional MHD simulation requires a suitable quiet tail model as initial condition. We therefore started one simulation from an equilibrium that exhibited the observed quiet time region 1 current distribution [ Birn and Hesse, 1990c; Birn, this volume]. In this case, the model is chosen symmetric, i.e., the net cross-tail magnetic field is zero, for simplicity.

The evolution of the field-aligned currents in the present case is demonstrated by Figure 13, which shows contour lines of constant $j_{\|}$ at the near-Earth boundary $x = 0$. The contours are spaced by differences in $j_{\|}$ of 0.025 (corresponding to about $3\times10^{-9}$ A/m$^2$). Single hatching represents tailward field-aligned currents, while cross-hatching corresponds to earthward field-aligned currents, exceeding 0.025 in magnitude in each case. The initial panel (upper left) clearly shows the equilibrium region 1 type system in the plasma sheet boundary region, peaking near $|y| = 7$, $|z| = 2$. In addition, there is a field-aligned current system of opposite polarity at higher latitudes, peaking at the corners $|y| = 10$, $|z| = 10$. This current system is due to the assumed increase of the scale length $L_z$ with $|y|$ toward the flanks of the tail, associated also with an increase of $B_z$ and thus a contribution to $j_x$ from $\partial B_z/\partial y$. Because they are located on open lobe field lines which map to the polar cap, these field-aligned currents may be associated with observed polar cap currents, for instance, cusp currents [ Iijima and Potemra, 1976*a*] or "NBZ" currents [e.g., Iijima et al., 1984], or the nightside polar cap currents inferred by Spence [1988] form the field model of Tsyganenko and Usmanov [1982], which all have the same polarity. They get reduced during the first tens of Alfvén times (see also Figure 14) and get also concentrated closer to the boundaries. There is a further, weaker, region 2 type system inside of the region 1 system, at smaller $|z|$ values, which becomes temporarily more pronounced at $t = 20$ (upper right panel), but disappears at later times. The region 1 system diminishes between $t = 0$ and $t = 20$ and stays roughly constant until about $t = 120$, when it develops a new strongly increasing component, which peaks near $|y| = 4$, $|z| = 1$, close to the boundary region of the now compressed current sheet.

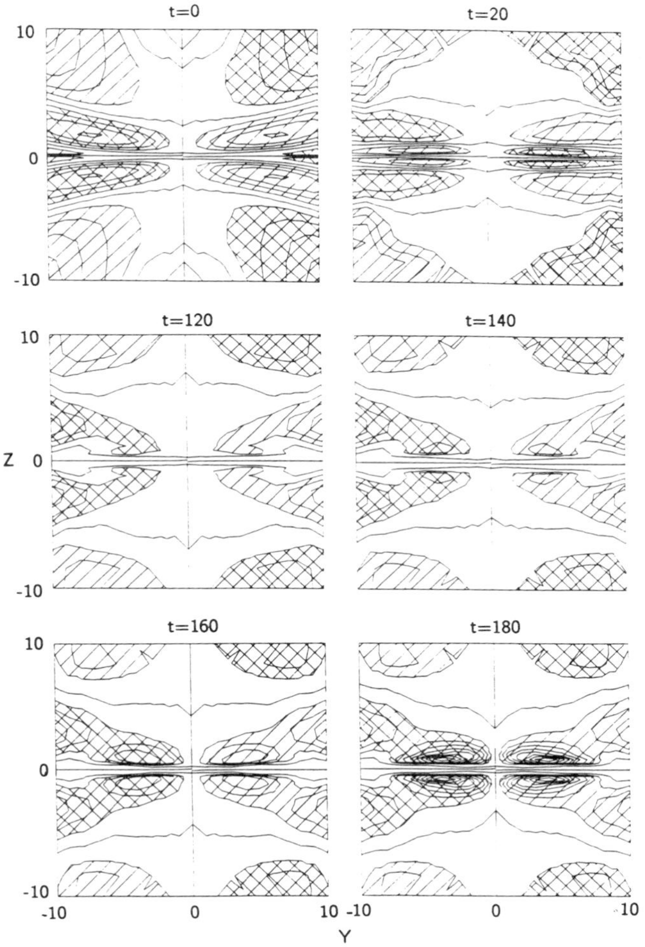


Fig. 13 Contours of constant field-aligned current density $j_{\|}$ at the near-Earth boundary $x = 0$ of the simulation box for different times as indicated in the panels. The increment of $j_{\|}$ between neighboring contours is 0.025 (corresponding to about $3 \times 10^{-9}$ A/m$^2$). Single hatching represents tailward field-aligned currents, while cross-hatching corresponds to earthward field-aligned currents, exceeding 0.025 in magnitude in each case.

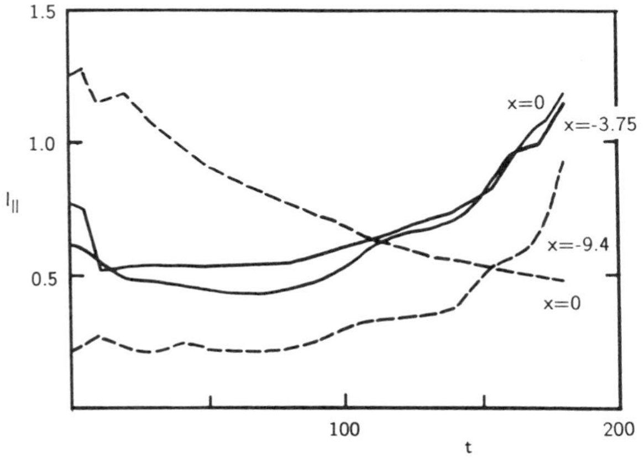


Fig. 14 Temporal evolution of total field-aligned currents, flowing earthward (solid lines) and tailward (dashed lines) on the dawn side of the tail through selected cross-sections at constant $x$ values as indicated in the figure. One dimensionless unit corresponds to about $0.3 \times 10^6$ A of current flowing into or away from each hemisphere.

The temporal evolution of the different field-aligned current systems, integrated over $y$ and $z$, is shown in Figure 14. The two solid curves represent the total region 1 type current in one sector, integrated at $x = 0$, and $x = -3.75$, as indicated. The two curves are quite similar, demonstrating the effective conservation of these currents as they approach the near-Earth boundary. They show the initial decrease, mentioned before, and a significant increase, starting

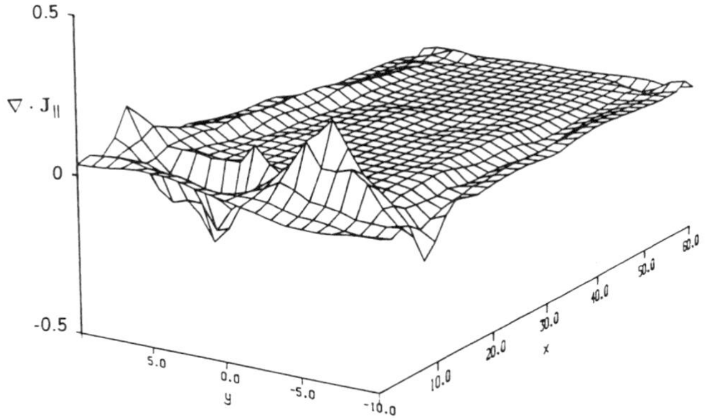


Fig. 15 Surface plot of $\nabla \cdot \mathbf{j}_{\|}$ as a function of $x$ and $y$ at $z = 0$ for $t = 150$.

at about $t = 100$, close to the time of neutral line formation. The two dashed curves represent the total current of opposite type, at $x = 0$, where it consists mostly of the previously discussed gradually decaying lobe current, and at $x = -9.4$, where additional region 2 type currents are found later in time inside of the region 1 system. These, however, do not propagate to the boundary $x = 0$ and are thus not present in Figure 13. This system is similar to the one found in our earlier simulations [ Birn, 1984].

To identify the source region of the field-aligned currents we have calculated the divergence of $j_{\|}$. This is represented in Figure 15 as a surface plot of $\nabla \cdot \mathbf{j}_{\|}$ as a function of $x$ and $y$ at the equatorial plane $z = 0$ for $t = 150$. The most prominent features are the two peaks of opposite signs near $x = -7.5$, $|y| = 4$. These peaks coincide closely with regions of strong velocity shear or vorticity. This becomes more obvious from Figure 16, which shows, in the same representation as Figure 15, the $z$ component of the vorticity $\mathbf{\Omega} = \nabla \times \mathbf{v}$ at the equatorial plane. This figure exhibits peaks of $\Omega_z$ close to those of $\nabla \cdot \mathbf{j}_{\|}$ in Figure 15. This association is quite plausible. The velocity shear (or vorticity) at the edges of the fast earthward flow in and near the equatorial plane enhances a pre-existing magnetic shear and causes a twisting of magnetic flux tubes connected with those regions, which directly produces the field-aligned currents. These currents flow mostly in the $x$ direction at some distance from the equatorial plane. We find that both terms in the (dimensionless) representation $j_x = \partial B_z/\partial y - \partial B_y/\partial z$ contribute to the enhanced current component. The first one is the consequence of the increase of $B_z$ in the central region near midnight, often called "dipolarization," which leads to a decrease of $B_z$ toward the flanks with increasing $|y|$. This is closely associated with the velocity shear contribution $\partial v_x/\partial y$. The second one corresponds to a rotation of the field direction, that is a change of $B_y/B_x$, with increasing $|z|$, associated with a velocity shear component $\partial v_x/\partial z$. This contribution to $j_x$ is qualitatively the same as the magnetic field shear between plasma sheet and lobes, already present in the initial equilibrium configuration. The shear, however, becomes enhanced as a consequence of the fast flow and the associated vorticity.

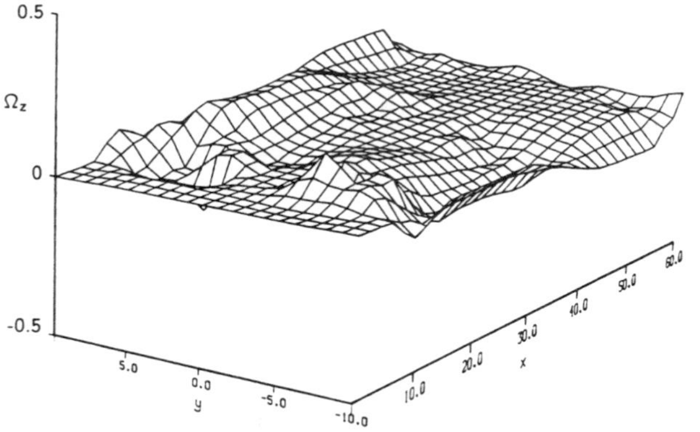


Fig. 16 Surface plot of the $z$ component, $\Omega_z$, of the vorticity $\mathbf{\Omega} = \nabla \times \mathbf{v}$ as a function of $x$ and $y$ at $z = 0$ for $t = 150$.

## 6. Magnetosphere-Ionosphere Coupling During Substorms

Apart from the question of the appropriate substorm model, one of the open problems in magnetospheric research is the question whether a substorm is significantly influenced by the ionosphere. A magnetosphere ionosphere coupling may either modify or even drive the substorm evolution due to previous changes in ionospheric properties. The coupling is brought about by field aligned currents that flow out of the ionosphere into the magnetosphere and vice versa, thereby establishing an information carrier between the two systems. Therefore an important aspect of this coupling is the change or generation of field-aligned current systems due the substorm current wedge [ McPherron et al., 1973] during the substorm development in the Earth's magnetotail.

The new aspect brought into this investigation is the inclusion of a model ionosphere, where the previously discussed (section 5) currents close, and its coupling to the magnetotail beyond the near-Earth region. In contrast to the investigations discussed in the previous sections, we here include a resistive closure of the field aligned currents generated during substorm evolution or pre-existing in the initial equilibrium. The aim here is to isolate the coupling between the magnetotail dynamics and the ionosphere; we therefore exclude an overlaid ionospheric potential (or magnetotail convection electric field) due to the influence of the solar wind, and the modification brought about by the ring current and the associated region 2 current system [ Iijima and Potemra, 1976a] in the region inward of $10R_E$. The basic logic behind our approach is the following (for details we refer to Hesse and Birn [1990c]): By the closure of the field aligned currents in a resistive medium representing the ionosphere, electric fields are generated that then induce a convection in the tail [e.g., Block, 1966]. This convection then influences the magnetotail evolution, either slowly, or even by means of causing the onset of some fast tail instability like the tearing mode. These changes in turn modify the field-aligned current system by some Alfvén waves that transports the information from the magnetotail into the ionosphere and bring about the magnetic field changes necessary for the field-aligned currents to flow. The self-consistency of the model therefore can be seen as a circle, where changes in the magnetotail introduce changes in the ionosphere which then change the magnetotail. This simulation will be referred to as "i run." For comparison, we also study a run with infinitely conducting earthward boundary ("c run").

In either simulation run we find that the dynamical evolution of the magnetotail was caused by the imposed finite resistivity in (3) leading to the formation of a plasmoid. Figure 17 shows the evolution of the plasma flow together with the magnetic neutral lines in the equatorial plane of the simulation box, again for both runs. In either case, we find the formation of a reversal region of the north-south magnetic field component $B_z$. This is accompanied by the appearance of initially fast tailward flows, which are later supplemented by fast earthward flows also. The upper two panels of Figure 17 show the flow pattern at a very early time ($t = 10$). As to be expected, the ionospheric coupling at these early times influences the flow in the region closer to the earthward end primarily, leading to a reduction of the tailward streaming caused by the slow resistive diffusion. A comparison of the two runs at later times again indicates that the i run involves a slightly faster evolution than the c run. This manifests itself in both the formation time of the neutral lines (around $t = 100$ for the i run, but between $t = 110$ and $t = 120$ for the c run), but also in the much more pronounced flow velocities in the i run.

Although the comparison between equal times seems to indicate that the two runs produce quite different results, a comparison of earlier times of the i run with later times of the c run suggests that the major effect is not a qualitative change in the evolution pattern but a time lag of the c run. Comparing the results of either run with previous simulations in a somewhat different configuration [ Birn and Hesse, 1990a] yields that the actual differences between the two evolutions, if properly aligned in time, turn out to be smaller than brought about by a different initial equilibrium.

The information transport between the magnetosphere and the ionosphere is provided by the field-aligned current density. Therefore, it is of interest to compare the field-aligned current densities and their distribution at the earthward end of the simulation box for both i and c runs. Figure 18 displays the maximum values of the current densities for both the i and the c run. Here we find that during the quiet phase for times $t \lesssim 100$ both maxima decay, with the one of the i run being slightly lower than in the c run. The overall decay can be attributed to the resistive diffusion given by the explicit resistivity in (3). The slight difference between the two runs, however, is due to the ionospheric coupling which should reduce current density maxima also. After $t \approx 100$ the two graphs differ considerably, due to the faster evolution in the i run. Note that a shape of the two graphs

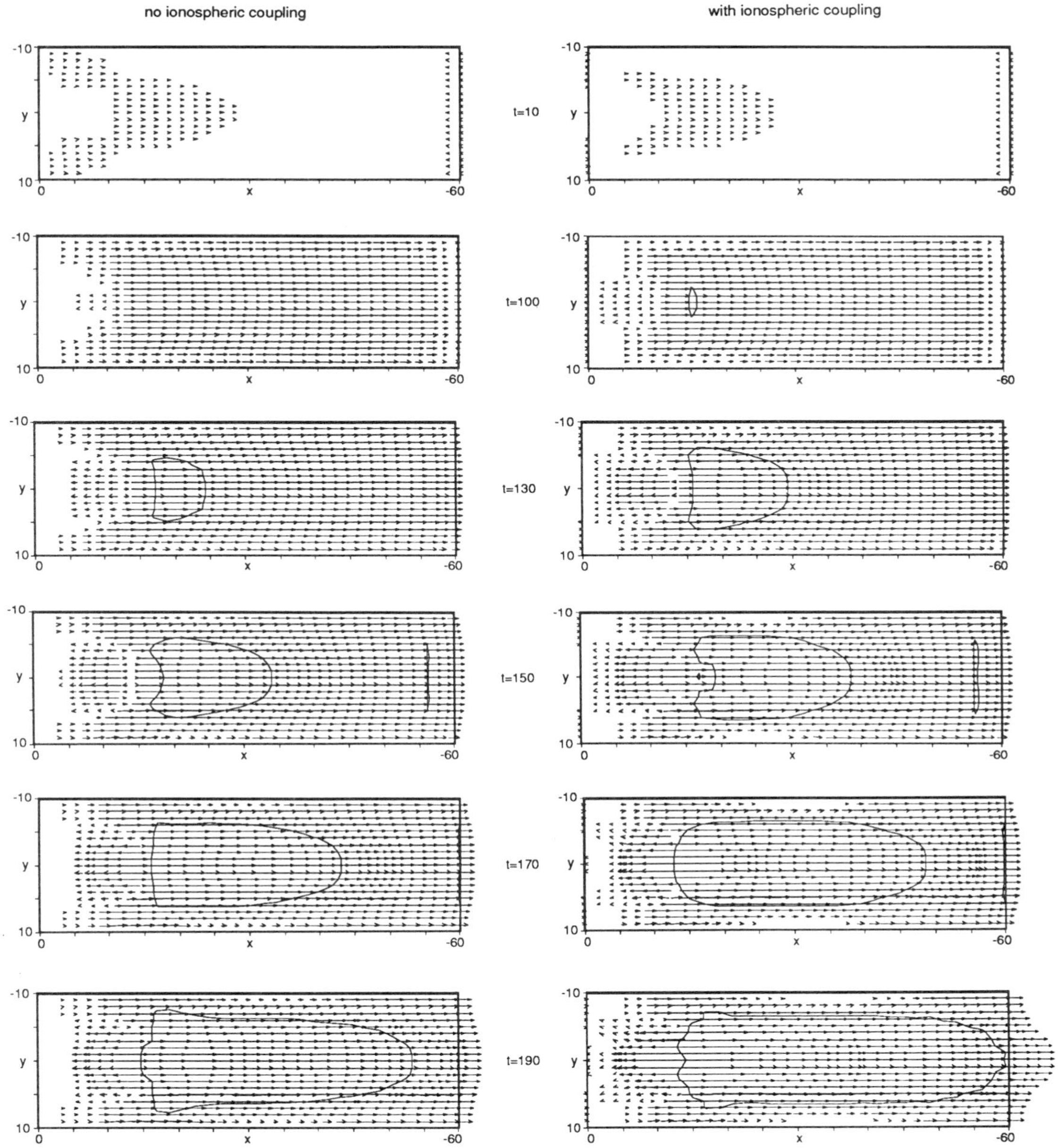


Fig. 17 Evolution of the plasma flow and magnetic neutral lines for the i and c simulations.

is very similar, despite their separation in time. Hence even in the evolution of the maxima of the field-aligned current density in the near Earth region we find the major difference to be a time shift.

The evolution of the maximum ionospheric potential difference in the i run is depicted in Figure 19. For comparison, Figure 19 also shows the evolution of the maximum ionospheric potential drop that would be incurred by a resistive closure of the parallel current systems of the c run, using the same ionospheric conductivity. We find that the evolution of the parallel current density distribution in the i run is influenced such as to lead to reduced maximum ionospheric potentials.

At last, the question remains as to what the cause for the apparent acceleration of the i run compared to the c run is. The explanation of the effect is given in Figure 20, which shows the flow patterns in the i run at the Earthward end of the simulation box for $t = 150$. The flow vectors (which have to be tangential to the potential contours at this time) show a global circulation pattern, which leads to inflow into the central plasma sheet region in the midnight meridional region of the magnetotail ($y \approx 0$), and outflow further to the flanks. The major effect is the inflow in the central region, since it is conceivable that the convective inflow acts as an enhancing mechanism for the inflow due to magnetic reconnection. This effect should be

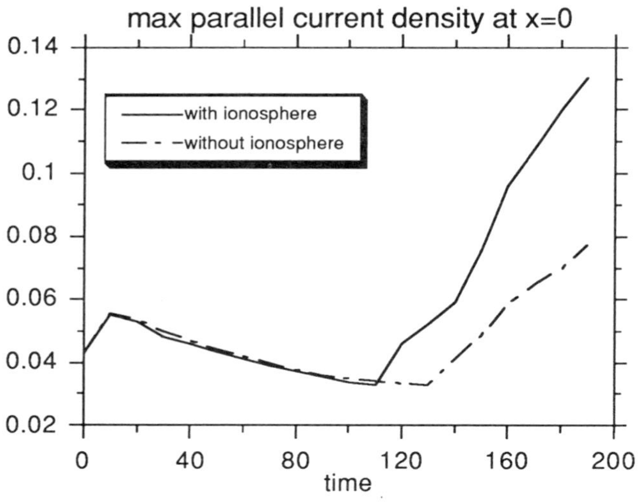


Fig. 18 Maximum values of the field-aligned current densities at $x = 0$ for both i and c runs vs. time.

even more pronounced as the evolution starts in the midnight region (cf. Figure 17) and persistently exhibits the fastest flows both into and out of the reconnection region there.

## Summary

In this paper we reviewed our present understanding of the substorm associated plasmoid evolution in three dimensions. We started out from three-dimensional magnetotail equilibria, most of which involved the presence of a cross-tail magnetic field component with about the average observed magnitude (3% of the typical lobe field strength).

The first investigation discussed here dealt with the specifics and changes of the magnetic topology of a three-dimensional plasmoid. We found, consistent with predictions based on a kinematical model [ Birn et al., 1989], that the plasmoid in generic systems becomes magnetically open, i.e., changes from the volume of magnetic loops envisaged in translationally invariant models to a flux rope like structure that is magnetically connected to its ambient field. Further, the time of severance from the initial state of the plasmoid, which involves a flux rope connected to the Earth, to the state of complete severance from the Earth is finite. The transition involves a gradual change of magnetic topology and exhibits tangled magnetic flux tubes of different magnetic topology at intermediate times. All of these features are intrinsically three-dimensional in nature and clearly different from two-dimensional investigations, which suggest an instantaneous severance from the closed field line region [e.g., Hones, 1979; Otto et al., 1990] and a magnetic topology consisting of closed loops only.

The same simulation was used to investigate the way the plasmoid gets accelerated up to its final speed after severance from the near Earth region. For this purpose, we found a generalization of a three-dimensional plasmoid volume that proved suitable for our studies although it was not topologically invariant. Another simulation involving a symmetric magnetotail model that enabled us to use the old concept of the plasmoid volume being the region of closed magnetic loops was performed for comparison; it showed no significant difference from the results we found for the model with net cross-tail magnetic field. Our results show that the plasmoid does not exhibit big changes in its velocity after formation. The initial speed is essentially given by the average velocity of the fast flow that pre-exists at the time of the formation of the magnetic helices (or the closed magnetic loops). Later, this velocity changes somewhat, but much less than seen in two-dimensional models [ Otto et al., 1990]. The reason for this was found to be the cross-tail variation of the velocity of the flow that was added to the plasmoid and therefore yielded a lower average velocity. Clearly, a two-dimensional model leads to an overestimation of the average velocity, since it describes only the fastest flow created in the noon-midnight meridional region (which is the onset region in the three-dimensional model) and lacks the ability to include the slower flows at $y \neq 0$. Consistent with Otto et al. [1990] our simulation exhibited a more pronounced pressure

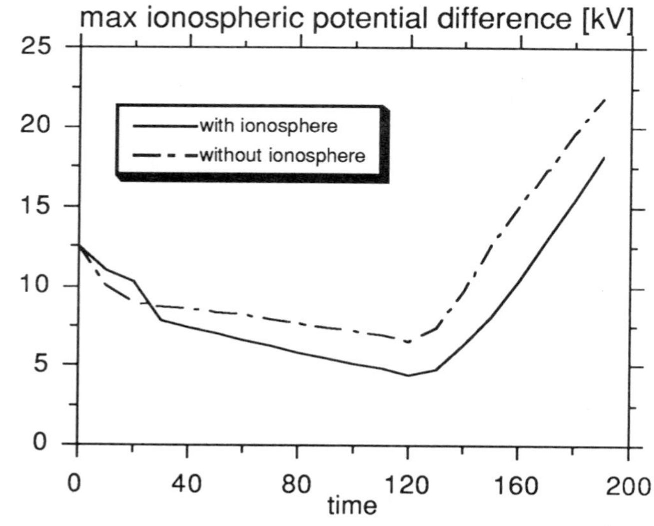


Fig. 19 Maximum ionospheric potential drop for the i and c runs vs. time. Note that the values for the c run are not computed self-consistently.

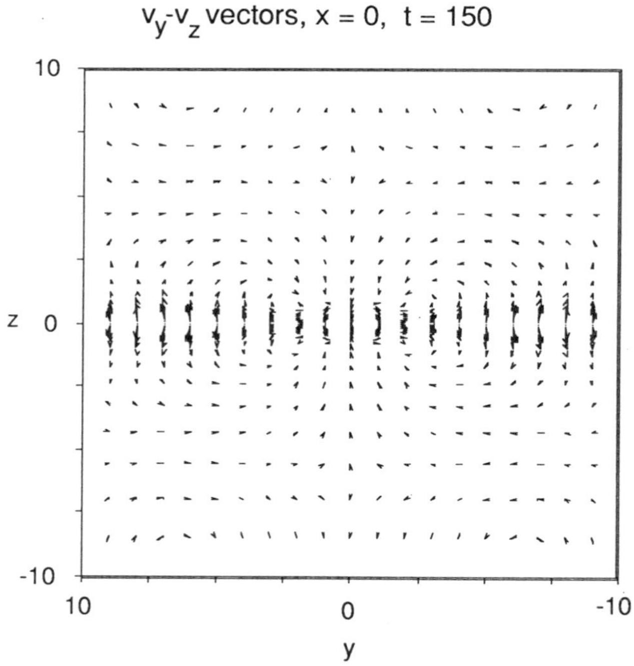


Fig. 20 Flow patterns as imposed by the ionospheric coupling at $x = 0$ for $t = 150$. Note the inflow toward the equatorial region around the noon-midnight meridian.

force contribution to the change of velocity of the plasmoid during the initial evolution. Only at later times the Lorentz force became more prominent. The total velocity, however, did not change appreciably due to these forces. Therefore we were led to conclude that the major plasma acceleration really takes place in regions with features reminiscent of slow shocks outside of the actual plasmoid volume.

Plasmoid evolution in three-dimensions as well as in symmetric models should be a general magnetic reconnection process [e.g., Schindler et al., 1988]. In order to investigate theoretical predictions about parallel electric fields [ Hesse and Schindler, 1988], we studied magnetic reconnection in a self-consistent MHD simulation. We found the formation of a region of enhanced $\mathbf{E}_{\|}$, which appeared to be highly localized in our simulation box. This lead us to identify this region as the diffusion region of our system. Further, an integration of the parallel electric field component along field lines showed that maximum values of the order of 20kV could be obtained from the self-consistent evolution. The foot points of the field lines carrying the high integrated values on the $x = 0$ plane exhibited an increasingly spotty structure as the evolution progressed. Further, the investigation showed that these electric fields can act as an accelerating mechanism for particles, particularly electrons, which could be guided by the magnetic field to the Earth after acceleration.

A question of recent interest in the space science community is the near Earth effects of magnetospheric substorms. An item of particular interest here is the disruption of the cross-tail current and its diversion to the ionosphere by the substorm current wedge [ McPherron et al., 1973]. Clearly, if the current wedge can be explained by the reconnection model at all, it has to be a three-dimensional effect. In our investigation, we found the generation of the substorm current wedge during the self-consistent plasmoid evolution. The generation of these field aligned currents was found to be closely associated to properties of the flow vorticity. In fact, in the regions of high vorticity the flow compresses and distorts the magnetic field thereby causing field aligned currents to flow corresponding to the field gradients. According to our results, theses regions are located earthward of the reconnection region at about the boundaries of the cross-tail extent of the neutral lines. The magnitude of the field aligned currents flowing to the Earth matched the observed ones [ Iijima and Potemra, 1976a,b].

A question associated with the generation and modification of field-aligned current systems is their closure in the ionosphere, which involves electric fields if the ionospheric conductance resembles the observed values. These electric fields will then drive a convection in the magnetotail, thereby inducing changes in the evolution, up to actually initiating a substorm evolution. We performed a first step of an investigation of the influence of magnetosphere-ionosphere coupling on substorm evolution. We assumed a current closure in an homogeneous ionosphere, and used the electric fields generated in the process to provide a boundary condition for the velocity. Compared to a simulation with frozen flux condition, we yielded a slightly accelerated evolution. Our explanation of this effect is that the electric fields drive a convection that enhances the plasma inflow into the reconnection region. Further, we found ionospheric potential drops of several 10kV resulting from the resistive closure of the substorm associated field-aligned current systems. Although the evolution with ionospheric coupling seemed to change appreciably, a closer inspection showed that the difference is essentially given by a time lag introduced by the ionospheric driving. We therefore came to the conclusion that an ionospheric model like the one investigated here does not significantly influence the evolution.

Acknowledgments. This work was supported by the U.S. Department of Energy through the Office of Basic Energy Science, and by NASA under contract W-7405-ENG-36. M. H. gratefully acknowledges support from the Los Alamos National Laboratory Director's postdoctoral program.

References

Birn, J., and E.W. Hones, Jr., Three-dimensional computer modelling of dynamic reconnection in the geomagnetic tail, *J. Geophys. Res.* *86*, 6802, 1981.

Birn, J., Computer studies of the dynamic evolution of the geomagnetic tail, *J. Geophys. Res., 85*, 1214, 1980.

Birn, J., Magnetotail equilibrium theory: the general three-dimensional solution, *J. Geophys. Res., 92*, 11101, 1987.

Birn, J., M. Hesse and K. Schindler, Filamentary structure of a three-dimensional plasmoid, *J. Geophys. Res., 94*, 241, 1989.

Birn, J., Three-dimensional computer modeling of dynamic reconnection in the magnetotail, in *Magnetic Reconnection in Space and Laboratory Plasmas, Geophys. Monogr. Ser.*, vol. 30, edited by E. W. Hones, Jr., p. 264, AGU, Washington, D. C., 1984.

Birn, J., and M. Hesse, MHD simulations of reconnection in a skewed three-dimensional tail configuration, *J. Geophys. Res.* in press, 1990a.

Birn, J., and M. Hesse, The magnetic topology of the plasmoid flux rope in a MHD simulation of magnetotail reconnection, in: *Physics of Magnetic Flux Ropes, Geophys. Monograph Series,* vol. 58, edited by C. T. Russell, E. R. Priest, and L. C. Lee, p. 655, AGU, Washington, D.C., 1990b.

Birn, J., and M. Hesse, The substorm current wedge and field-aligned currents in MHD simulations of magnetotail reconnection, *J. Geophys. Res.*, in press, 1990c.

Block, L. P., On the distribution of electric fields in the magnetosphere, *J. Geophys. Res., 71*, 855, 1966.

Fairfield, D. H. On the average configuration of the geomagnetic tail, *J. Geophys. Res., 84*, 1950, 1979.

R. Hautz and M. Scholer, Numerical simulations on the structure of plasmoids in the deep tail, *J. Geophys. Res., 14*, 969, 1987.

Hesse, M. and K. Schindler, A theoretical foundation of general magnetic reconnection, *J. Geophys. Res., 93*, 5559, 1988.

Hesse, M., and J. Birn, Magnetosphere-Ionosphere Coupling during Plasmoid Evolution: First Results, submitted to *J. Geophys. Res.*, 1990b.

Hesse, M., and J. Birn, Parallel electric fields in a simulation of magnetotail reconnection and plasmoid formation, in: *Physics of Magnetic Flux Ropes, Geophys. Monograph Series,* vol. 58, edited by C. T. Russell, E. R. Priest, and L. C. Lee, p. 679, AGU, Washington, D.C., 1990.

Hesse, M., and J. Birn, Plasmoid evolution in an extended magnetotail, submitted to *J. Geophys. Res.*, 1990c.

Hones, E. W., Jr., Plasma flow in the magnetotail and its implications for substorm theories, in *Dynamics of the Magnetosphere* (S. I. Akasofu, ed.), p. 545, D. Reidel Publ. Co., 1979.

Hones, E. W., S. J. Bame, and J. R. Asbridge, Proton flow measurements in the magnetotail plasma sheet made with Imp 6, *J. Geophys. Res., 81*, 227, 1976.

Hones, E. W., Jr., J. Birn, D. N. Baker, S. J. Bame, W. C. Feldman, D. J. McComas, R. D. Zwickl, J. A. Slavin, E. J. Smith, and B. T. Tsurutani, Detailed examination of a plasmoid in the distant magnetotail with ISEE 3, *Geophys. Res. Lett., 11*, 1046, 1984.

Hones, E. W., T. A. Fritz, J. Birn, J. Cooney, and S. J. Bame, Detailed observations of the plasma sheet during a substorm on April 24, 1989, *J. Geophys. Res., 91*, 6845, 1986.

Hones, E. W., Jr., and K. Schindler, Magnetotail plasma flow during substorms: a survey with IMP 6 and IMP 8, *J. Geophys. Res., 84*, 7155, 1979.

Iijima, T., T. A. Potemra, L. J. Zanetti, and P. F. Bythrow, Large-scale Birkeland currents in the dayside polar region during strongly northward IMF: A new Birkeland current system, *J. Geophys. Res.*, 7441, 1984.

Iijima, T., and T. A. Potemra, Field-aligned currents in the dayside cusp observed by Triad, *J. Geophys. Res., 81*, 2165, 1976*b*.

Iijima, T., and T. A. Potemra, The amplitude distribution of field-aligned currents at northern high latitudes observed by Triad, *J. Geophys. Res., 81*, 2165, 1976*a*.

McPherron, R. L., C. T. Russell, and M. A. Aubry, Satellite studies of magnetospheric substorms on August 15, 1968, 9. Phenomenological model for substorms, *J. Geophys. Res., 78*, 3131, 1973.

Otto, A., K. Schindler, and J. Birn, Quantitative study of the nonlinear formation and acceleration of plasmoids in the Earth's magnetotail, submitted to *J. Geophys. Res.*, 1989.

Petschek, H. E., Magnetic field annihilation, AAS-NASA Symposium on the physics of solar flares, N. F. Ness (ed.), *US Printing Office,* p.425, 1964.

Potemra, T. A., T. Iijima, and N. A. Saflekos, Large-scale characteristics of Birkeland currents, in *Dynamics of the Magnetosphere*, edited by S.-I. Akasofu, p. 165, D. Reidel, New York, 1979.

Schindler, K., A theory of the substorm mechanism, *J. Geophys. Res., 79*, 2803, 1974.

Schindler, K., M. Hesse and J. Birn, General magnetic reconnection, parallel electric fields and helicity, *J. Geophys. Res., 93*, 5547, 1988.

Slavin, J. A., D. N. Baker, J. D. Craven, R. C. Elphic, D. H. Fairfield, L. A. Frank, A. B. Galvin, W. J. Hughes, R. H. Manka, D. G. Mitchell, I. G. Richardson, T. R. Sanderson, D. J. Sibeck, E. J. Smith, and R. D. Zwickl, CDAW 8 observations of plasmoid signatures in the geomagnetic tail: an assessment, *J. Geophys. Res, 94,* 15153, 1989.

Spence, H. E., M. G. Kivelson, and R. J. Walker, Comparison of field-aligned currents at ionospheric and magnetospheric altitudes, *Adv. Space Res., 8,* 343, 1988.

Tsyganenko, N. A., and A. V. Usmanov, Determination of the magnetospheric current system parameters and development of experimental geomagnetic field models based on data from IMP and HEOS satellites, *Planet. Space Sci., 30,* 985, 1982.

---

M. Hesse and J. Birn, Los Alamos National Laboratory, MS D438, Los Alamos, N.M. 87545.

# PARTICLE ORBITS IN MAGNETOSPHERIC CURRENT SHEETS: ACCELERATED FLOWS, NEUTRAL LINE SIGNATURE, AND TRANSITIONS TO CHAOS

T. W. Speiser[1], P. B. Dusenbery[1,2], R. F. Martin, Jr.[3], and D. J. Williams[4]

Abstract. Theory of particle motion in current sheets is reviewed. For small, approximately constant normal magnetic field, $B_z$, particles oscillate about the current sheet and "live" within the sheet for one-half gyroperiod based on $B_z$. This lifetime replaces the mean collision time in the Lorentzian conductivity and thus gives rise to the concept of an inertial (or gyro-) conductivity. A substorm model by Coroniti [1985] utilizes this conductivity to allow reconnection to proceed without anomalous processes due to wave-particle interactions. Chaotic particle orbits may at times be important to the dynamics, depending on parameters such as particle energy, current sheet thickness, and field line curvature. A current sheet model with neutral line predicts a ridge structure and asymmetries in the distribution function. Observed ion distributions near the plasma sheet boundary layer, during the CDAW 6 interval, are consistent with the model predictions.

## 1. Introduction

A principal theme of solar-terrestrial research is to understand how particles and fields from the sun couple to and interact with plasma and fields in the geospace environment. Three regions in the magnetosphere appear to play crucial roles in this interaction: the magnetopause; the magnetotail plasma sheet; and the auroral region (Figure 1). Excellent reviews of solar wind interaction with the magnetosphere are given by Haerendel and Paschmann [1982] and Alexander et al. [1984].

The magnetopause and its associated boundary layers is a key region where mass, momentum and energy are transferred from the solar wind to the magnetosphere. This transfer is dominantly controlled by reconnection when the interplanetary magnetic field (IMF) is southward. Perreault and Akasofu [1978] introduced an energy coupling parameter proportional to $VB^2\sin^4(q/2)$, which correlates well with magnetospheric dissipation processes ($q$ is the polar angle of the IMF, $q = 0°$ for Northward, $q = 180°$ for Southward IMF, $V$ is the solar wind speed, and $B$ is the magnitude of the IMF).

A magnetospheric substorm is an event during which magnetospheric plasma, particles and fields change dramatically (e.g. Galeev, 1982; Baker et al., 1984). The magnetotail, including the plasma sheet and its boundary layer, plays a major role in energy storage and particle acceleration during substorms. Particle energization by magnetic reconnection, [Dungey, 1961] which takes place near a neutral point is the basis of many models of magnetospheric substorms, yet the fundamental physics of this process is still not well understood. For a recent review on reconnection, see Galeev [1982], Sonnerup et al. [1984] and Hones [1984].

In Figure 2, the topology for tail reconnection is illustrated. A diffusion region will exist around an $x$-type neutral point. A region with nearly constant normal field ($B_z$) within the current sheet may sometimes exist adjacent to the diffusion region. During reconnection, $\mathbf{E} = E_y\hat{e}_y$ should exist across the tail in the dawn-dusk direction. This electric field can accelerate particles in the current sheet and is a measure of the reconnection rate in the tail. $E_y(t)$ probably grows explosively around substorm onset, and may be roughly constant when reconnection saturates [Coroniti, 1985].

The magnetohydrodynamic (MHD) approximation breaks down in the diffusion region where small scale phenomena and nonadiabatic motion become important. In order to provide the diffusion of the magnetic field, the resistivity needs to be large in the diffusion region (if the resistivity were zero, or conductivity infinite, the electric field is shorted out and reconnection would not occur). However, we will see (section 5) that finite resistivity need not imply particle-particle or particle-wave interactions.

Because reconnection at the dayside magnetopause and in the magnetotail occur in regions which are essentially collisionless, a collisionless resistivity is required. A search for that resistivity involves either of two approaches, 1) an approach based on particle dynamics, or 2) one based on turbulence via plasma waves.

Plasma wave turbulence has been found to be associated with streaming particles and field aligned currents in boundary layers of the plasma sheet [Gurnett et al., 1976], the so-called plasma sheet boundary layers (PSBL). However, such turbulence is either in the wrong place (PSBL), or of the wrong magnitude to be responsible for the resistivity required for reconnection. As pointed out by LaBelle and Treumann [1988], the observed plasma wave amplitudes near the magnetopause resulted in diffusion coefficients which were always too small to explain reconnection. A similar result was found

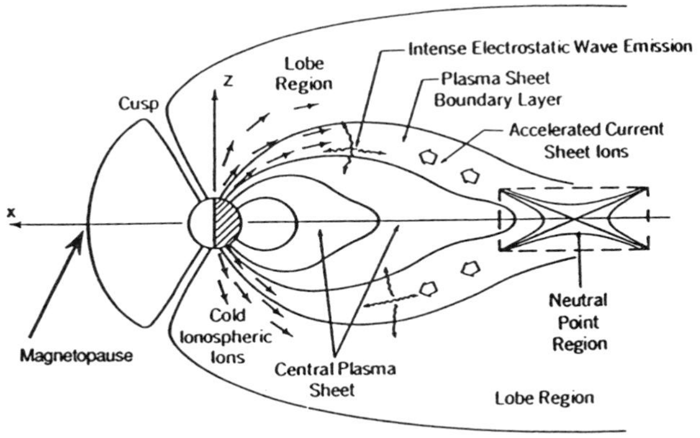


Fig. 1. Critical regions of the magnetosphere.

[1] APAS Dept., University of Colorado, Boulder, CO 80309-0391 and NOAA/SEL, Boulder, CO 80303.

[2] Magnetospheric Physics Branch, National Science Foundation, Washington, DC 20550.

[3] Physics Dept., Illinois State University, Normal, IL 61761-6901

[4] Applied Physics Laboratory, Johns Hopkins University, Johns Hopkins Road, Laurel, MD 20723-6099.

Modeling Magnetospheric Plasma Processes
Geophysical Monograph 62

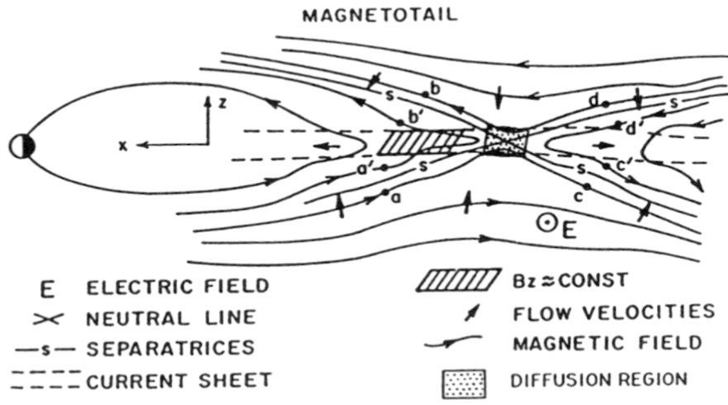


Fig. 2. Field topology in the magnetotail. Diffusion region highlighted around a neutral line. During reconnection, $a \rightarrow a'$.

by Anderson [1984] in the magnetotail, where wave intensities at the center of the current sheet decreased dramatically from the PSBL, making it difficult for turbulence to produce the required resistivity. In addition, Haerendel [1987] has noted that not once in eight AMPTE releases of heavy ions, has there been any evidence of anomalous resistivity processes (essentially, locally induced plasma wave turbulence), as diagnosed by in-situ waves. These releases were in the solar wind, magnetosheath and magnetotail. Therefore, because of the problems with anomalous resistivity, we will concentrate on the dynamics of single particles in the current sheet (CS) and near neutral lines, to provide a dynamical collisionless resistivity. For a review of particle, field and plasma observations and theories of the geomagnetic tail, see Speiser [1990].

## 2. Single Particle Dynamics

Figures 3 and 4, respectively, illustrate charged particle motion in a strictly neutral sheet, and in a current sheet with a small, approximately constant, $B_z$ [Speiser, 1965, 1967, 1968, 1990].

When $0 < B_z \ll B_0$ (where $B_0$ is the tail field outside the current sheet), the motion is a simple combination of a fast oscillation about the current sheet and a slower gyromotion about $B_z$, until the particle is ejected from the current sheet. This ejection is caused by a change in sign of the Lorentz force $z$-component (see section 3).

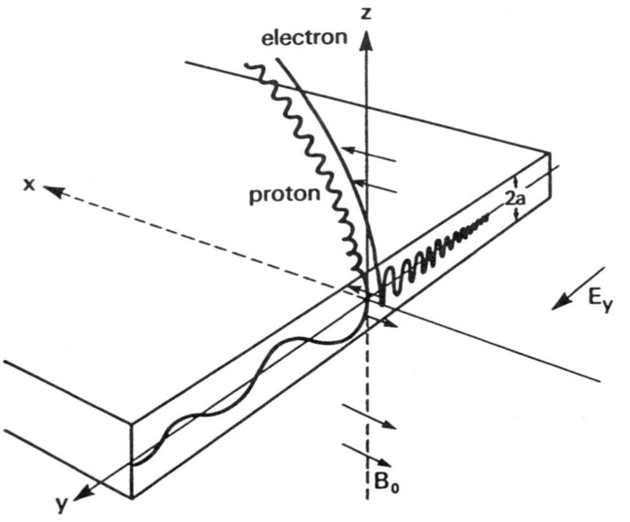


Fig. 3. Strictly "neutral" sheet, with constant $B_0$, and $E = E\hat{e}_y$.

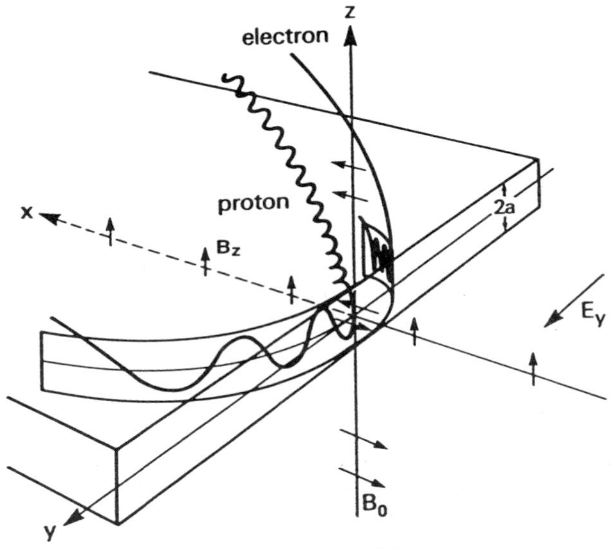


Fig. 4. CS with small $B_z$.

## 3. Transition to Chaos

The two fundamental motions of a charged particle in a current sheet can be described by the equations of two coupled one-dimensional oscillators. A nonlinear $z$-oscillation normal to the current sheet ($\omega_z$) and a gyromotion about $B_z$ ($\Omega_n$). When the frequencies of the two oscillators $\omega_z$ and $\Omega_n$ become commensurate, the particle motion is chaotic.

What is chaos? In the study of deterministic nonlinear dynamical systems, we often encounter chaotic behavior. This behavior arises from extreme sensitivity to initial conditions, or equivalently, from the exponential divergence of nearby orbits in certain domains of phase space. Chaotic dynamics implies stochastic behavior, at least over long times, and leads to diffusion in phase space from which irreversible behavior can arise.

Is there a chaos parameter which relates these frequencies and determines the degree of stochasticity? Let us begin with the Lorentz force equation:

$$ma = q(E + (\mathbf{v} \times B)) \tag{1}$$

The basic field model (see Figure 4) assumes that $B = B_x\hat{x} + B_z\hat{z}$, and $E = E_0\hat{y}$, where

$$B_x = \begin{cases} B_0(z/a), & -a < z < a \\ B_0 \quad , & |z| > a \end{cases}$$

$$B_z = \text{constant}$$

Because $E_y$ and $B_z$ are assumed constant, Speiser [1965] showed that $E_y$ could be transformed away. The force equation can then be written as

$$\begin{aligned} \ddot{x} &= C_2\dot{y} \\ \ddot{y} &= C_1 z\dot{z} - C_2\dot{x} \\ \ddot{z} &= -C_1\dot{y}z = -k(t)z \end{aligned} \tag{2}$$

where $C_1 = qB_0/ma$, and $C_2 = qB_z/m$.

Neglecting the $C_1 z\dot{z}$ term in (2), the $x$ and $y$ equations imply circular motion (Figure 5) about the $z$-axis with frequency $C_2 = \Omega_n$. That is, $x = -\rho_n \sin\Omega_n t$, and $y = -\rho_n\cos\Omega_n t$, where $\rho_n = v_\perp/\Omega_n$. The force in the $z$-direction is given by $\ddot{z} = -C_1\dot{y}z = -k(t)z = -\omega_z^2 z$. As $\dot{y} = v_\perp \sin\Omega_n t$, the $z$-motion is like a spring oscillator, but the spring constant, $k$, changes with time. If $k = C_1\dot{y} > 0$, the $z$-motion is oscillatory (so-called meandering orbit). (Thus, the $z\dot{z}$ term in (2) is bounded and may often, but not always, be neglected.) If $k = C_1\dot{y} < 0$, the $z$-motion results in acceleration away from the $z = 0$ plane (ejection). Now, consider the ratio of fast to slow oscillation frequency;

$$\frac{\langle \omega_z \rangle}{\Omega_n} = \frac{C_1^{1/2}}{C_2} \left\langle \dot{y}^{1/2} \right\rangle = \frac{2.4}{\pi} \left( \frac{\rho_n}{a b_0} \right)^{1/2}$$

where $\rho_n = v_\perp / (qB_z / m))$, $b_0 = B_z/B_0$, and define a parameter $\kappa_\perp$,

$$\kappa_\perp \equiv \left( \frac{a b_0}{\rho_n} \right)^{1/2}$$

$$= \left[ \frac{\text{minimum radius of curvature of a field line in the current sheet}}{\text{particle gyroradius at the current sheet center}} \right]^{1/2}$$

In this expression, $\langle \omega_z \rangle$ is the average of $\omega_z$ over the $z$-oscillation period. Therefore, we arrive at

$$\boxed{\frac{\langle \omega_z \rangle}{\Omega_n} = \frac{2.4}{\pi} \frac{1}{\kappa_\perp}} \tag{3}$$

The $\kappa_\perp$ parameter controls the stochasticity of charged particle orbits and is closely related to the $\kappa$ parameter of Büchner and Zelenyi [1986].

For particle motion far from the current sheet, the magnetic moment, $\mu$, is often a good adiabatic invariant. For particle motion inside the current sheet we can distinguish three cases:

1. $\langle \omega_z \rangle \ll \Omega_n$ ($\kappa_\perp \gg 1$)
   Guiding center motion is important and $\mu$ is a good adiabatic invariant.
2 $\langle \omega_z \rangle \gg \Omega_n$ ($\kappa_\perp \ll 1$)
   Current sheet motion is important and a current sheet invariant is defined.
3. $\langle \omega_z \rangle \approx \Omega_n$ ($\kappa_\perp \approx 1$)
   There is no adiabatic invariant and particle motion is chaotic.

How does $\kappa_\perp$ vary with the perpendicular energy, $\varepsilon_\perp$? Figure 6 is a plot

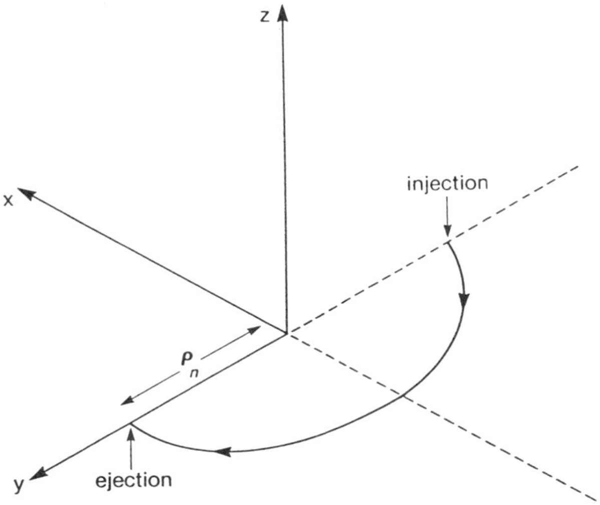


Fig. 5. $x$-$y$ motion of a positively charged particle.

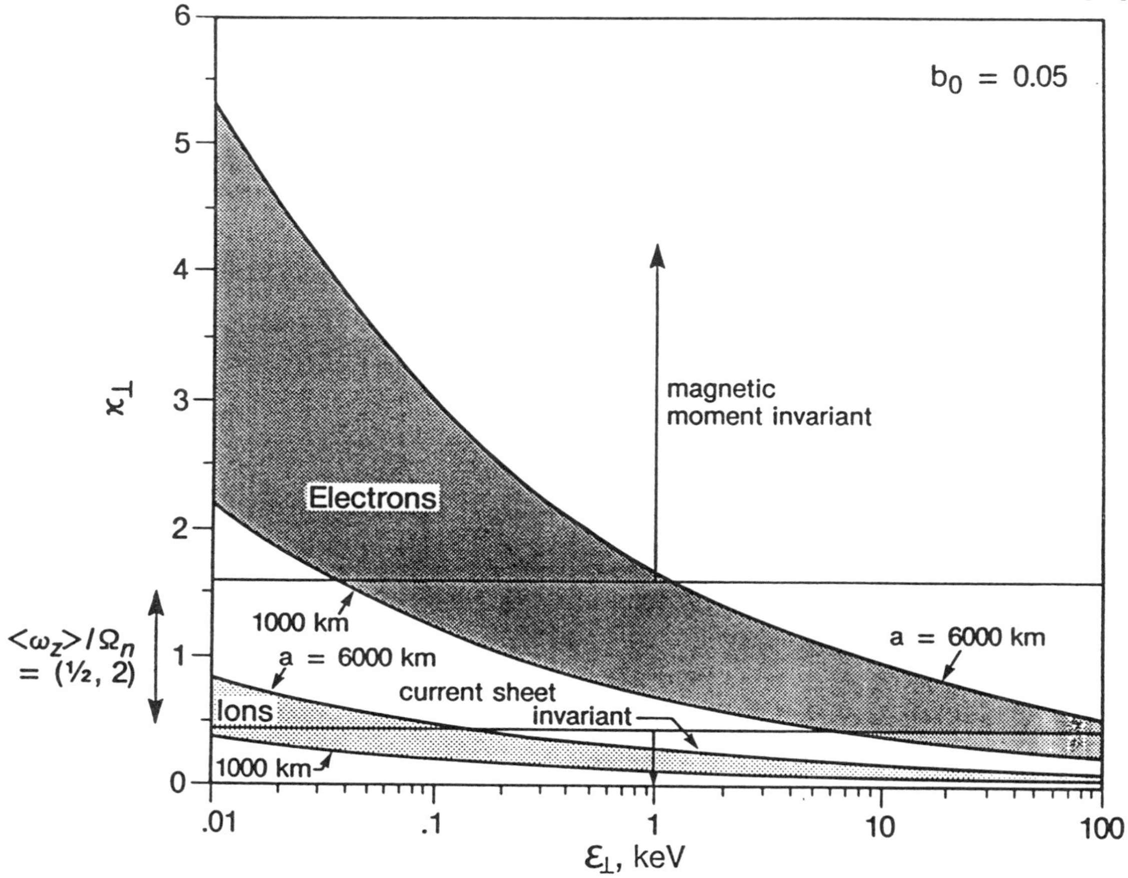


Fig. 6. $\kappa_\perp$ vs $\varepsilon_\perp$ for both ions and electrons and the range of $\kappa_\perp$ when $\langle \omega_z \rangle / \Omega_n$ varies between 1/2 and 2, which we arbitrarily define as the stochastic range.

of $\kappa_\perp$ versus perpendicular energy $\varepsilon_\perp = 1/2mV_\perp^2$, for ions (protons) and electrons, and for thin (1,000 *km*) and thick (6,000 *km*) current sheets. We assume that $B_z = 1\,nT$ and $B_0 = 20\,nT$. Regions are indicated for $\kappa_\perp \gg 1$, where $\mu$ is a valid adiabatic invariant and $\kappa_\perp \ll 1$, where because of the current sheet $z$-oscillation, a current sheet invariant [Speiser, 1970; Sonnerup, 1971] comes into play. In the intermediate range, where $1/2 < \langle \omega_z \rangle / \Omega_n < 2$ (or $1.6 \gtrsim \kappa_\perp \gtrsim 0.4$) both adiabatic invariants are violated and chaotic motion ensues.

## 4. Z-Motion in a Current Sheet

When the $z\dot{z}$ term in Equation (2) is negligible, the $x$-$y$ motion is a simple gyromotion about the weak $B_z$ field. We now consider the $z$-motion in more detail. The $z$-component equation of motion is given by

$$\ddot{z} + \omega_z^2 z = 0. \tag{4}$$

For non-constant $\omega_z$, Equation (4) is of the form of a nonlinear oscillator. Lichtenberg and Lieberman [1983] considered three such nonlinear systems. A simple (large amplitude) pendulum with one degree of freedom, a pendulum whose length changes on a slow time scale, and a pendulum resonating with a driver frequency are examples of these systems.

Since $\omega_z^2(t) \propto \dot{y}$, we can write $\omega_z(t)/\Omega_n = (\sin \Omega_n t)^{1/2} / \kappa_\perp$. Figure 7 is a plot of this ratio. For $\kappa_\perp \ll 1$, $\omega_z/\Omega_n \gg 1$ over a large part of the interval. Equation (4) can be solved analytically over the range where $\omega_z/\Omega_n \gg 1$, using the WKB method [Schiff, 1955]. The WKB solution is found by substituting $z = Ae^{i\phi(t)/\varepsilon}$ into our nonlinear oscillator equation, (4) ($\varepsilon$ is the smallness parameter $\dot{\omega}_z / 2\omega_z^2$). The result is [Speiser, 1968; Dusenbery et al., 1989]

$$z(t) = [A\sin\phi(t) + B\cos\phi(t)]/(\sin\Omega_n t)^{1/4} \tag{5}$$

where $\phi(t) = \int_0^t \omega_z(t')dt'$, and $A$ and $B$ are determined by initial conditions.

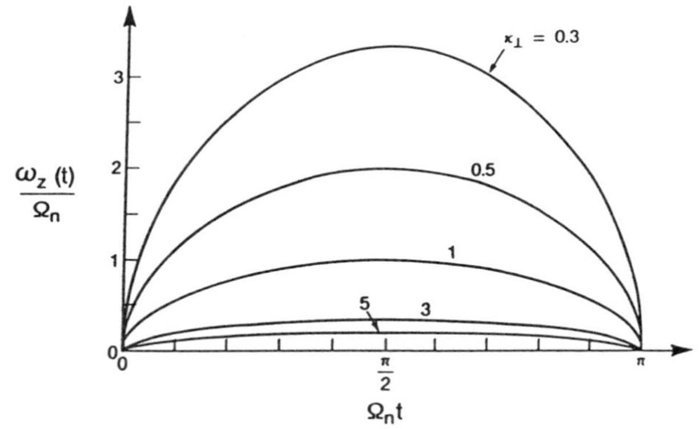


Fig. 7. Ratio of current sheet frequencies as a function of time.

Figure 8 shows a numerically calculated orbit for a current sheet with constant $B_z$, and Figure 9b shows the numerically calculated $z$-oscillation for a smaller value of $b_0$ and Figure 9a compares a plot of the analytic solution (Equation 5). In general, there is good agreement, but the analytic solution does not include the gradient drifts at the beginning and end of the orbit.

## 5. Inertial and Chaotic Conductivity

From the simple current sheet motion, a particle is trapped by the field reversal and accelerated by $E_y$, while it executes gyromotion about the weak, constant $B_z$. Thus, the acceleration time, lifetime of the particle in the system, or coherence time, replaces the mean collision time in an expression for the electrical conductivity [Speiser, 1970, 1990; Lyons and Speiser, 1985].

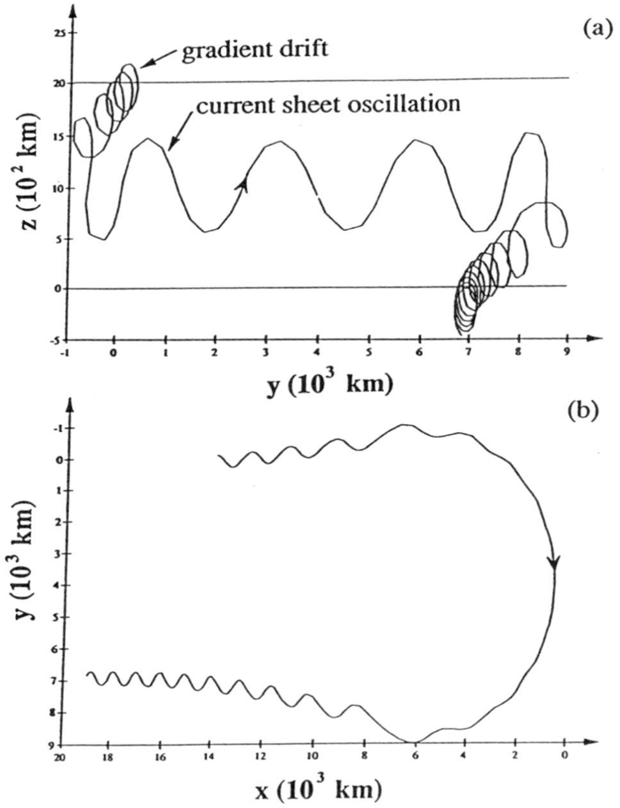


Fig. 8. Numerical current sheet orbit.

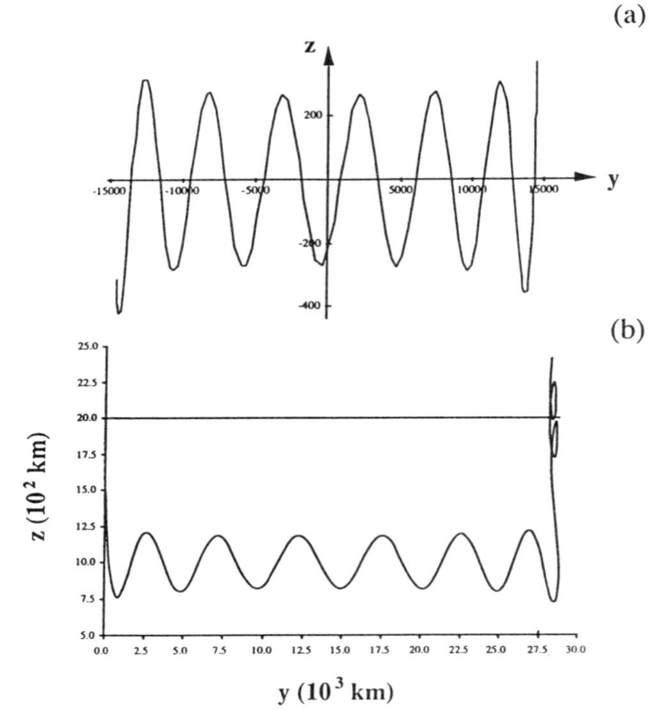


Fig. 9. Current sheet orbits; a) analytic theory, b) numerical integration.

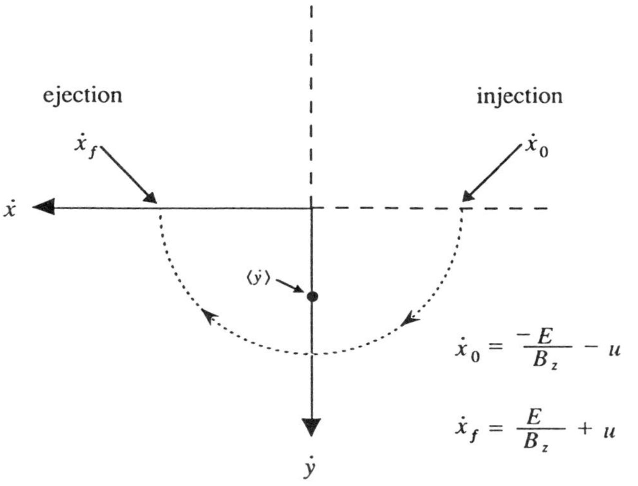


Fig. 10. Current sheet orbit in $\dot{x} - \dot{y}$ velocity space, in the frame where $E_y$ is transformed away.

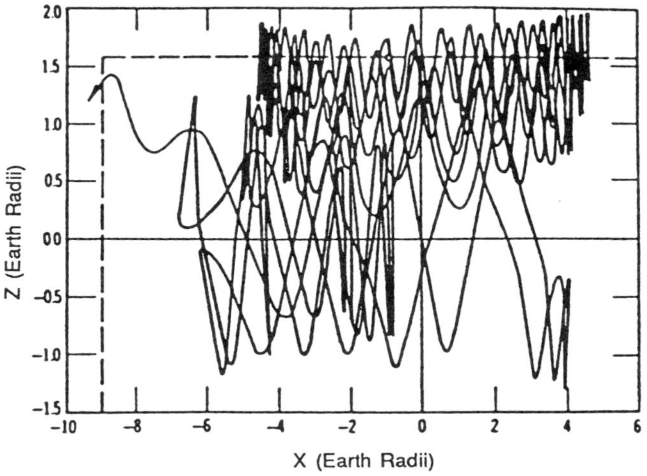


Fig. 11. Chaotic orbit for a thick current sheet ($2a = 3R_e$) [Martin, 1986].

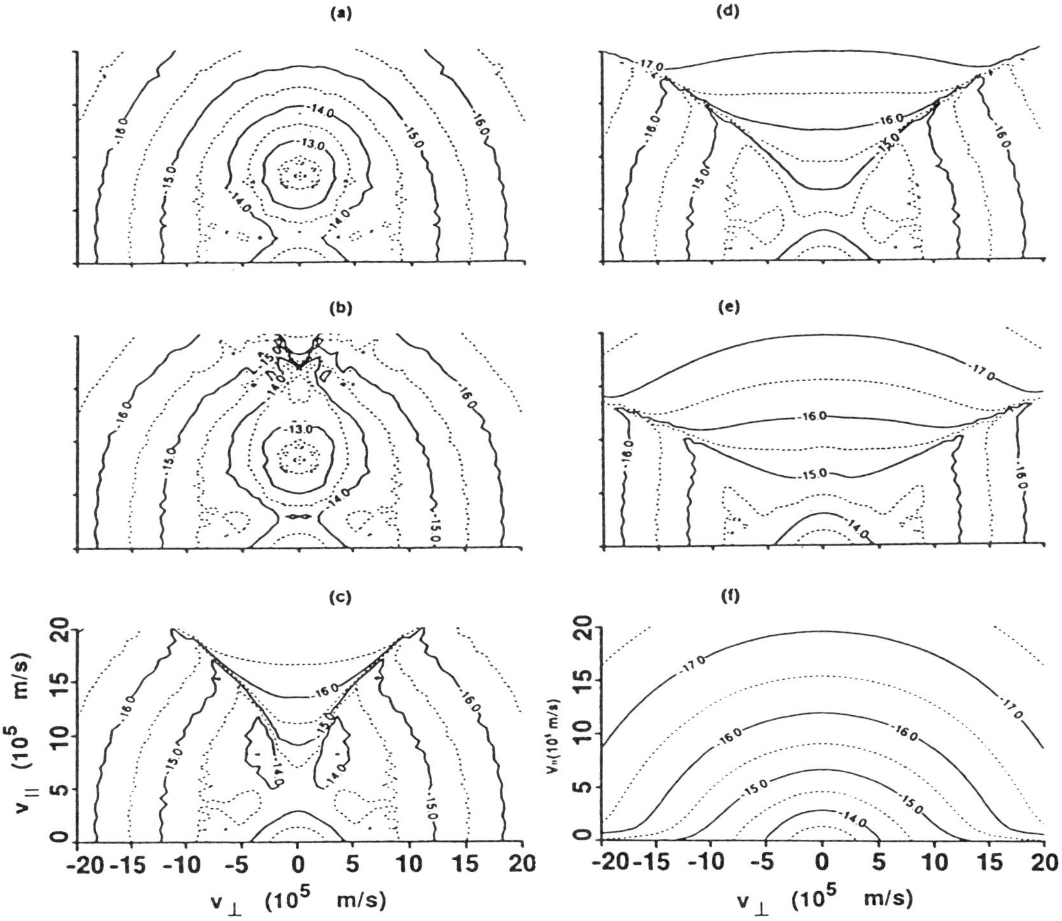


Fig. 12. Modelled distribution function. Panels a-f are at varying distances from the neutral line: a) $4L$; b) $3L$; d) $1.5L$; e) $L$; f) 0. $L$ = separatrix position = $10^4$ km.

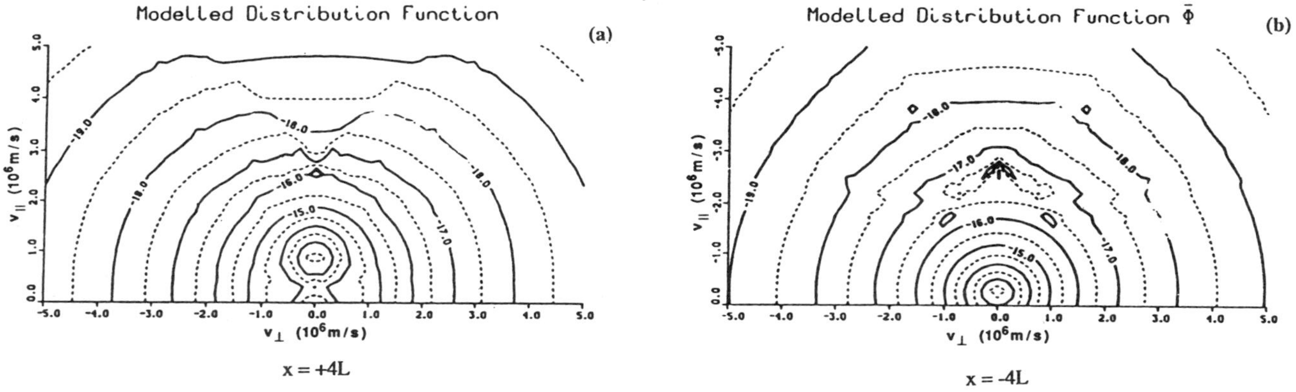


Fig. 13. a) modelled distribution function, $f$, at $x = +4L$, 4 separatrix distances earthward of the neutral line, and b) $f$ at $x = -4L$, tailward of the neutral line.

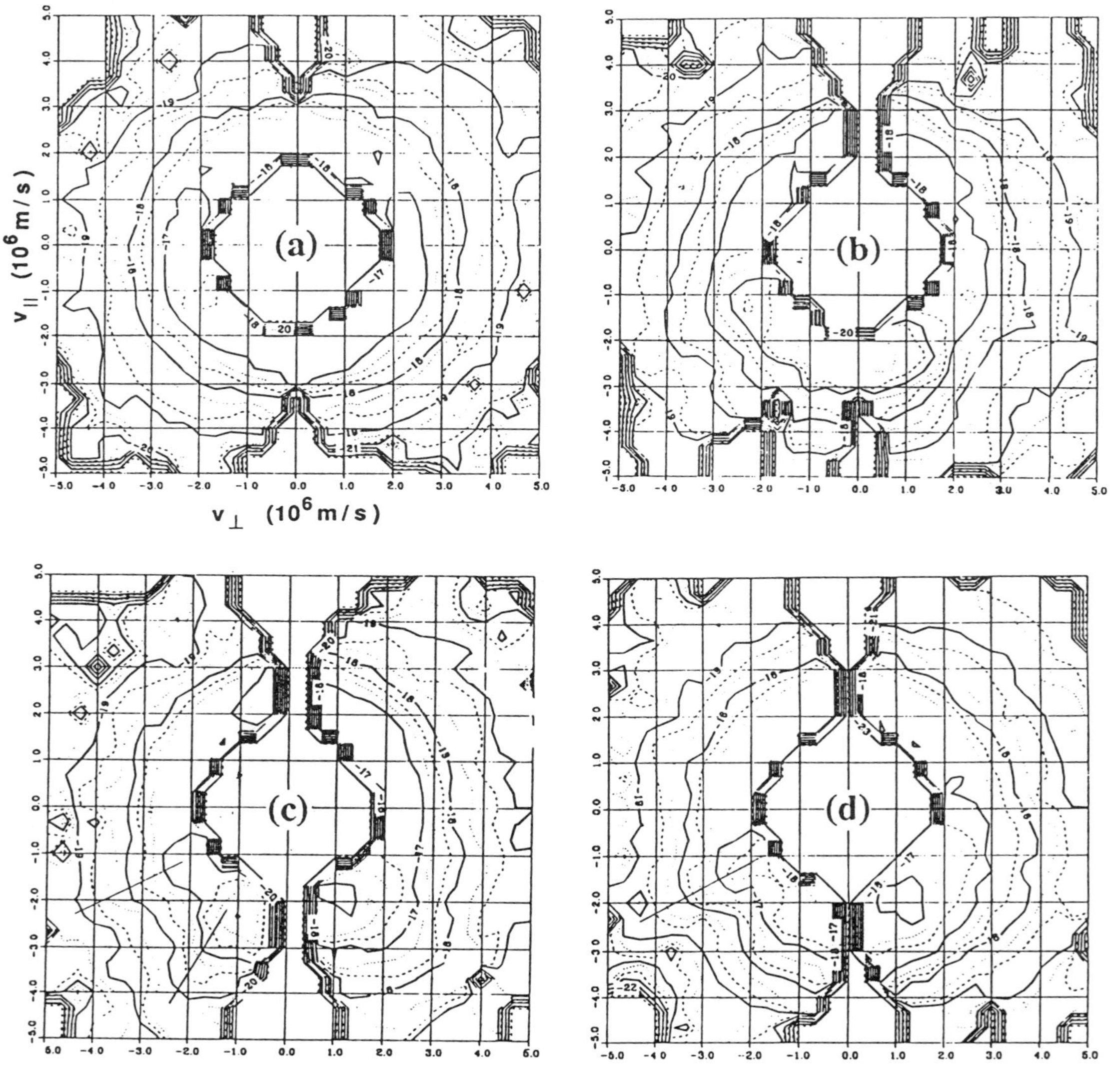

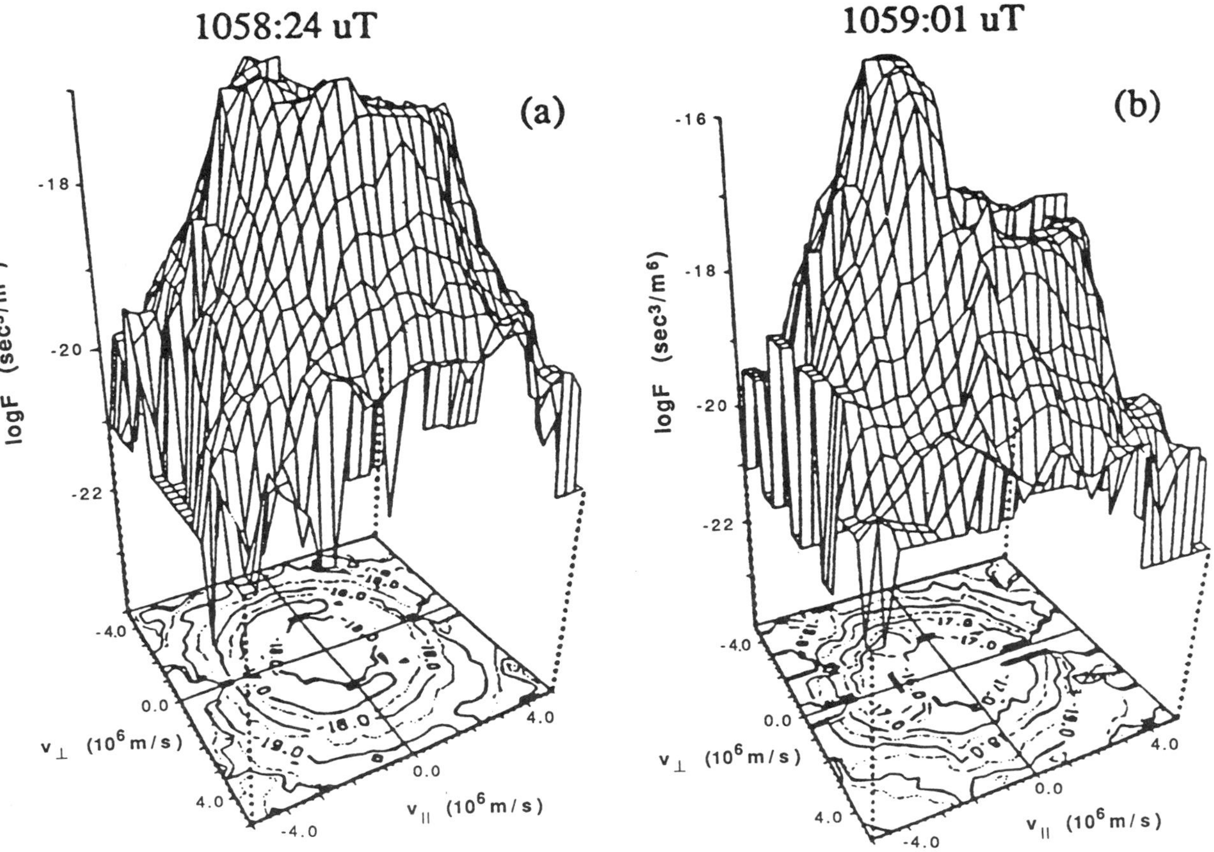


Fig. 15. Logarithmic perspective plots of the observed distribution function, $f$, corresponding to Figure 14 a and b.

In the transformed frame (low energy approximation) (Figure 10), $\langle \dot{y} \rangle \approx E / B_z$ and $j = ne\langle \dot{y} \rangle$, or $j = (ne/B_z)E = \sigma_g E$. $\sigma_g = \omega_z = ne/B_z = ne^2/m\Omega_n \approx 10^{-4}$ $mho/m$, taking typical values $n \approx 1\ cm^{-3}$, $B_z \approx 1\ nT$. $\sigma_g$ is thus a gyro or inertial conductivity (here, it is assumed $E = E\hat{e}_y$).

An alternate physical mechanism for a collisionless conductivity was suggested by Martin [1986]: since the particle dynamics in tail-like models is chaotic, one can use the timescale for decay of velocity correlations as an analog to a collision time, resulting in a "chaotic conductivity" due to the diffusion in phase space produced by chaotic dynamics. Martin used a model with an $X$-type neutral line, utilizing the Lyapunov characteristic exponent, $\lambda$, as a preliminary estimate for the chaotic timescale ($\lambda$ measures the timescale for exponential separation of chaotic orbits). With the same parameters as above, he obtained $\sigma_{ch} = ne^2/m\lambda \approx 10^{-4}$ $mho/m$, about the same as $\sigma_g$. Further calculations of the velocity autocorrelation function for this system [Martin, paper in progress] yield an exponential decay of correlations and an effective conductivity of $\sigma_{ch} = ne^2/m\nu \approx 4 \times 10^{-4}$ $mho/m$, where $\nu^{-1}$ is the correlation decay time. The correlation decay timescale is longer than the Lyapunov timescale, leading to a somewhat larger conductivity. Nonetheless, the chaotic and inertial conductivities are of a similar magnitude, implying that both effects may be relevant in the tail. As the CS thickness is increased, we find that chaotic motion is also increased, since particles spend more time mirroring back and forth in the CS. A chaotic orbit for a thick sheet is shown in Figure 11. These chaotic conductivity results are for a model with a neutral line, but as discussed in section 3, we expect chaos to be important even for the constant $B_z$ model when the CS parameters are such that $\kappa_\perp$ is close to one.

## 6. Motion in a Current Sheet with Neutral Line

Martin and Speiser [1988] showed that current sheet orbits become modified with the inclusion of a neutral line, and the signature of such a neutral line is a ridge in velocity space.

In Figure 12, from Martin and Speiser, the ion distribution function is modelled at the edge of the current sheet. Panels (a) to (f) show the distribution function as if a satellite were approaching a neutral line along the

---

Fig. 14. Observed MEPE energetic ion distribution function ($s^3m^{-6}$) for the CDAW 6 interval, Day 81, 1979: a) 10:58:24.4 $UT$, just outside plasma sheet boundary layer (PSBL); b) 10:59:1.4 $UT$, encounter with strong tailward beam; c) 10:59:37.4 $UT$; d) 11:00:50.4 $UT$. These diagrams cover the first four energy channels (34-120 $keV$) of the MEPE data. Figures a and b can be seen in perspective in Figure 15. Note the greater than one order of magnitude enhancement of the distribution function when the tailward beam appears in b. $v_\parallel < 0$ is tailward, $v_\parallel > 0$ is Earthward. The sign of $v_\perp$ is arbitrarily taken to be the negative of the sign of $v_z$. The white areas are indicative of lack of data, e.g. in the center of the diagrams, there is no data for energies below 34 $keV$. Similarly, for small pitch angles, i.e. along the $v_\perp = 0$ axis, there is often no data, and thus an artificial hole in the distribution function is seen. Ridge-like structures are highlighted by the straight line in c and d. For further discussion of this type of diagram, see Speiser, et al., 1981. See text.

plasma sheet boundary layer. The model parameters are: $E_y = 1/4\ mV/m$, $a = 1000\ km$, $B_0 = 20\ nT$, $B_z = 1\ nT$, $U_x$ (initial bulk flow speed) = -350 *km/sec*. In panel (a), the accelerated, Earthward directed, field-aligned beam is seen centered on $v_{\parallel} \sim 850\ km/sec$, which is just $2E_y/B_z + U_x$, as predicted by the analytic solutions (section 3).

As the neutral line is approached, the contours at large $v_{\parallel}$ start to be broken up (panel (b)), and then a ridge in velocity space appears which moves to larger pitch angles as the observing position gets close to the neutral line. In panel (f), the observing position is directly above the neutral line.

Figure 13 shows a comparison of simulated distribution functions earthward and tailward of a model neutral line [Speiser and Martin, 1989; Speiser et al., 1989]. Within the ridge, the distribution function appears asymmetric, comparing the earthward and tailward simulations. The depletion of $f$, along the $v_{\perp} = 0$ axis, is due to the assumed asymmetric initial bulk flow, $U_x$. That is, the particles within the ridge are not turned around by $B_z$, so they are initially earthward-going (Figure 13a), and thus come from a reduced part of the assumed initial tailward flowing distribution. For the tailward simulation (Figure 13b), the particles within the ridge are initially tailward-going, and thus come from an enhanced part of the assumed initial tailward flowing distribution. More details of these simulations, plus mappings of $f$ throughout the current sheet, are found in the study of Speiser and Martin [1990].

Figure 14 shows four consecutive 36-second ion observations from the MEPE (ISEE 1) instrument [Williams et al., 1978] for a plasma sheet boundary crossing, during the CDAW 6 (Coordinated Data Analysis Workshop) interval [Fritz et al., 1984]. Figure 15 shows perspective plots corresponding to 14a and b. Note the strong tailward beam seen in Figures 14b and 15b. These observations show qualitative agreement with our modelled ridge distributions (Figures 12 and 13), in that ridge-like distributions are clearly evident in Figures 14b-d, with some multiple ridge structure in Figures 14b and c (such multiple structure can also be seen in tailward models, e.g. Figure 13b). Of note is the outward displacement of the contours of $f$ at pitch angles smaller that the ridge, especially evident in Figure 14d. We believe this supports the model predictions of Figure 13, as discussed above, if the plasma sheet encounter is tailward of a neutral line. This interpretation (of a tailward encounter) would also agree with that of Fritz et al. [1984].

## 7. Summary

Anomalous resistivity processes seem to be incapable of producing the finite resistivity required for magnetospheric reconnection processes. Particle inertia plays this role in Coroniti's [1985] substorm model. Particle chaos may, at times, be important and can also give rise to an effective resistivity. Whether chaos or inertia is dominant may depend on current sheet parameters such as sheet thickness. A tail magnetic neutral line produces a ridge structure in modelled distributions. This structure is asymmetric on the earthward and tailward sides of the neutral line. For a few 36-second ion observations near the plasma sheet boundary layer, during the CDAW 6 interval, ridge-like structure is observed and the asymmetry is consistent with a distribution tailward of the neutral line, which also agrees with the direction of the observed beam.

*Acknowledgements.* We would like to thank D. Thaxton for her word-processing and editorial skills. We gratefully acknowledge the support of NASA (through grants NAGW-1176 and NAGW-1593), Johns Hopkins University (through JH-6018900), the National Science Foundation (through ATM-8803301 and ATM-9002447), the Research Corporation, and Illinois State University.

## References

Alexander, J.K., L. F. Bargatze, J. L. Burch, T. E. Eastman, J. G. Lyon, J. D. Scudder, T. W. Speiser, G. H. Voigt, and C. -C. Wu, Coupling of the solar wind to the magnetosphere, in *Solar Terrestrial Physics: Present and Future*, edited by D. M. Butler and K. Papadapoulos, *NASA Ref. Pub. 1120*, 1984.

Anderson, R. R., Plasma waves at and near the neutral sheet, in *Proceedings of the conference on Achievements of the IMS*, ESA SP–217, p. 199, 1984.

Baker, D. N., S. -I. Akasofu, W. Baumjohann, J. W. Bieber, D. H. Fairfield, E. W. Hones, Jr., B. Mauk, R. L. McPherron, and T. E. Moore, Substorms in the magnetosphere, in *Solar Terrestrial Physics: Present and Future*, edited by D. M. Butler and K. Papadapoulos, *NASA Ref. Pub. 1120*, 1984.

Büchner, J. and L. M. Zelenyi, Deterministic chaos in the dynamics of charged particles near a magnetic field reversal, *Phys. Lett. A, 118*, 395, 1986.

Coroniti, F. V., Explosive tail reconnection: the growth and expansion phases of magnetospheric substorms, *J. Geophys. Res., 90*, 7427, 1985.

Dungey J. W., Interplanetary magnetic field and the auroral zones, *Phys. Rev. Lett., 6*, 47, 1961.

Dusenbery, P. B., T. W. Speiser, I. Doxas, and R. F. Martin, Jr., Current sheet dynamics and particle chaos: Basic theory, *EOS, 70*, 1286, 1989.

Fritz, T. A., D. N. Baker, R. L. McPherron, and W. Lennartsson, Implications of the 1100 *uT* March 22, 1979 CDAW 6 substorm event for the role of magnetic reconnection in the geomagnetic tail, in *Magnetic Reconnection in Space and Laboratory Plasmas*, edited by E. W. Hones, *Geophys. Mono. No. 30*, American Geophysical Union, Washington, DC, 1984.

Galeev, A. A., Magnetospheric tail dynamics, in *Magnetospheric Plasma Physics*, Developments in Earth & Planet. Sciences Series 4, edited by A. Nishida, D. Reidel, Boston, Massachusetts, 1982.

Gurnett, D. A., L. A. Frank, R. P. Lepping, Plasma waves in the distant magnetotail, *J. Goephys. Res., 81*, 6059, 1976.

Haerendel, G., Tail exploration and tail formation with artificial plasma clouds, in *Magnetotail Physics*, edited by A. T. Y. Lui, The Johns Hopkins University Press, Baltimore, p. 337, 1987.

Haerendel, G. and G. Paschmann, Interaction of the solar wind with the dayside magnetosphere, in *Magnetospheric Plasma Physics*, Developments in Earth & Planet. Sciences Series 4, edited by A. Nishida, D. Reidel, Boston, Massachusetts, 1982.

Hones, E. W., Jr., *Magnetic Reconnection in Space and Laboratory Plasmas*, edited by E. W. Hones, *Geophys. Mono. No. 30*, American Geophysical Union, Washington, DC, 1984.

LaBelle, J. and R. A. Treumann, Plasma waves at the dayside magnetopause, *Space Sci. Rev., 47*, 175-202, 1988.

Lichtenberg, A. J. and M. A. Lieberman, *Regular and Stochastic Motion*, Springer-Verlag, New York, 1983.

Lyons, L. R. and T. W. Speiser, Ohm's law for a current sheet, *J. Geophys. Res., 90*, 8543, 1985.

Martin, R. F., Jr., Chaotic particle dynamics near a two-dimensional magnetic neutral point with application to the geomagnetic tail, *J. Geophys. Res., 91*, 11,985, 1986.

Martin, R. F., Jr. and T. W. Speiser, A predicted energetic ion signature of a neutral line in the geomagnetic tail, *J. Geophys. Res., 93*, 11,521, 1988.

Perreault, P. and S.-I. Akasofu, A study of geomagnetic storms, *Geophys. J. Roy. Astron. Soc., 54*, 547, 1978.

Schiff, L. I., *Quantum Mechanics*, McGraw-Hill, New York, 1955.

Sonnerup, B. U. Ö., Adiabatic particle orbits in a magnetic null sheet, *J. Geophys. Res., 76*, 8211, 1971.

Sonnerup, B. U. Ö., P. J. Baum, J. Birn, S. W. H. Cowley, T. G. Forbes, A. B. Hassam, S. W. Kahler, W. H. Matthaeus, W. Park, G. Paschmann, E. R. Priest, C. T. Russell, D. S. Spicer, and R. Stenzel, Reconnection of magnetic fields, in *Solar Terrestrial Physics: Present and Future*, edited by D. M. Butler and K. Papadapoulos, *NASA Ref. Pub. 1120*, 1984.

Speiser, T. W., Particle trajectories in model current sheets; 1. Analytical solutions, *J. Geophys. Res., 70*, 4219, 1965.

Speiser, T. W., Particle trajectories in model current sheets; 2. Applications to auroras using a geomagnetic tail model, *J. Geophys. Res.*, *72*, 3919, 1967.

Speiser, T. W., On the uncoupling of parallel and perpendicular motion in a neutral sheet, *J. Geophys. Res.*, *73*, 1112, 1968.

Speiser, T. W., Conductivity without collisions or noise, *Planet. Space Sci.*, *18*, 613, 1970.

Speiser, T. W., Geomagnetic tail, in *Geomagnetism 4*, edited by J. Jacobs, Academic Press Limited, London, in press, 1990.

Speiser, T. W., D. J. Williams and H. A. Garcia, Magnetospherically trapped ions as a source of magnetosheath energetic ions, *J. Geophys. Res.*, *86*, 723, 1981.

Speiser, T. W. and R. F. Martin, Jr., Energetic ion signature of a neutral line in the geomagnetic tail, *IAGA Bulletin*, *53C*, 423, 1989.

Speiser, T. W., R. F. Martin, Jr., and D. J. Williams, Discovery of the energetic ion ridge signature in the geomagnetic tail, *EOS*, *70*, 1286, 1989.

Speiser, T. W. and R. F. Martin, Jr., Energetic ions as remote probes of X-type neutral lines in the geomagnetic tail, in *progress*, 1990.

Williams, D. J., E. Keppler, T. A. Fritz, B. Wilken, and G. Wibberenz, The ISEE 1 and 2 medium energy particles experiment, *IEEE Trans. Geosci. Electron.*, *GE-16(3)*, 270, 1978.

# MERGING AND THE SINGLE PARTICLE

T. E. Moore
*NASA Marshall Space Flight Center, Space Science Laboratory, Mail Code ES53, Huntsville, AL 35812*

D. C. Delcourt
*ESA/ESTEC, Department of Space Physics, Noordwijk, The Netherlands*

Low-altitude dissipation of solar wind energy in the ionospheric plasma produces heavy ion outflows which are entrained into the global magnetospheric circulation, producing regions of enhanced plasma mass density. Outflows from the cusp region are particularly effective in providing ionospheric plasma to the near-Earth plasma sheet because of their favorable position upstream of the polar lobes and plasma sheet boundary layers, within this circulation. Steady state, adiabatic, single particle, guiding center description of this plasma flow in model fields appropriate to moderate activity levels, leads to the conclusion that typical ionospheric heavy ions are energized parallel to the magnetic field to plasma sheet energies on their first encounter with the neutral sheet region, forming earthward streaming plasma sheet boundary layer features. Estimates of the plasma energy density produced suggest that additional energization is required to support the magnetotail. Known dipolarization reconfigurations of the geomagnetic field in the near-Earth plasma sheet lead to temporary violation of the first adiabatic invariant, requiring a more general equation of motion. The result is much larger energization of ionospheric ions, both perpendicular and parallel to the magnetic field. While some characteristics of these phenomena appear to be contained in steady or time-dependent MHD descriptions of the plasma, the production of an extremely hot and isotropic plasma sheet seems to appear more naturally in the single particle description. On the other hand, the initiation and propagation of disturbances associated with dipolarization events cannot be described with a single particle model, nor have they been addressed by time-dependent MHD simulations. A qualitative two-dimensional analogy is suggested in which the solar wind interaction with the magnetotail region is represented by an offshore gale which produces ebb tide conditions at the shore line. An instability of this shear driven quasi-equilibrium then leads to a return to normal shoreline levels by means of a bore wave. A coupled MHD and single particle approach to substorm dynamics is suggested as a means of making further quantitative progress.

## Introduction

The injection of an energetic plasma into the inner magnetosphere appears as the most striking high-altitude aspect of substorm phenomena [*McIlwain*, 1974; *Smith and Hoffman*, 1974; *Konradi et al.*, 1975; *Williams*, 1981; *Mauk and Meng*, 1983]. During substorm expansion phase, the magnetic field relaxes from "tail-like" to a more dipolar configuration [*Cummings et al.*, 1968; *McPherron*, 1979; *Sauvaud and Winckler*, 1980], a front of hot plasma advances westward and toward the Earth in the evening to midnight sector [*Barfield et al.*, 1977; *Moore et al.*, 1981], and this is associated with the transient appearance of large (several mV/m) waves fields [*Shepard et al.*, 1980; *Aggson et al.*, 1983], presumably induced by and consistent with the magnetic field dipolarization.

Modeling Magnetospheric Plasma Processes
Geophysical Monograph 62


*Mauk* [1986] quantitatively investigated this dipolarization in terms of the "convection surge" mechanism of *Quinn and Southwood* [1982], using a guiding center approach to calculate single particle orbits in the near-Earth region. *Mauk* [1986] raised the question of what plasma populations exist in the region subject to substorm dipolarizations of the magnetic field and showed that low-energy ionospheric ion populations would be tremendously energized, primarily in the direction parallel to the magnetic field, by the curvature drifts associated with the stretched field in combination with the large induced electric field associated with the collapse.

Plasma heating and outflow [*Lockwood et al.*, 1985; *Moore et al.*, 1986] from the dayside magnetospheric cleft region has been shown to be an important source of plasma to the near-Earth plasma sheet, since it provides large fluxes of $O^+$ to the central plasma sheet during periods of moderate to high magnetospheric convection [*Cladis*, 1986; *Delcourt et al.*, 1989]. These authors have shown, using single particle guiding center calculations in model fields, that observed fluxes of ionospheric ions lead to plasma sheet boundary layer features very similar to those actually observed, namely energetic earthward moving ion beams flowing along the boundary layer magnetic field.

*Delcourt et al.* [1989] have shown in particular that the observed global distribution of ion outflux from the ionosphere leads in moderately active conditions to a "plume" of enhanced plasma density which feeds directly to the plasma sheet inside of 17 $R_E$. The result is earthward streams of ions of mean energy which is dependent upon the steady convection strength but is typically in the range of several keV. This energy is gained by the symmetric plumes of ions from each hemisphere's cleft on a single pass through the neutral sheet, forming likewise symmetric boundary layer streams. The result is illustrated schematically in Figure 1. Lighter ions from the polar and cleft regions tend to be lost from the magnetosphere or recirculated from much larger distances if they reach the neutral sheet earthward of the neutral line. During lower activity levels, the cleft heavy ion fountain is recirculated through the plasma sheet with much less energization. The energetic plasma sheet at low activity may then consist of a mixture of light ions of terrestrial and solar origin convecting earthward from the vicinity of the distant neutral (stagnation) line. To the extent that substorms are associated with a period of moderate convection activity [e.g., *Cattell and Mozer*, 1984], the state of the hot plasma sheet prior to expansion phase is expected to include ionospheric heavy ions.

In this paper, we consider the effects of plasma sheet dynamics on the preexisting ion population set up in a hypothetical substorm context of moderate magnetospheric convection. In particular, we employ a time-dependent single particle trajectory calculation, as developed by *Delcourt et al.* [1990], to integrate the trajectories of ionospheric cleft ion fountain ions in the electromagnetic field of the collapsing magnetotail. This simulation produces high-energy trapped particles in a time scale which is short compared with the evolution of a substorm, as well as substantial injections into the ring current region. We consider the strengths and shortcomings of the single particle description vis a vis the MHD

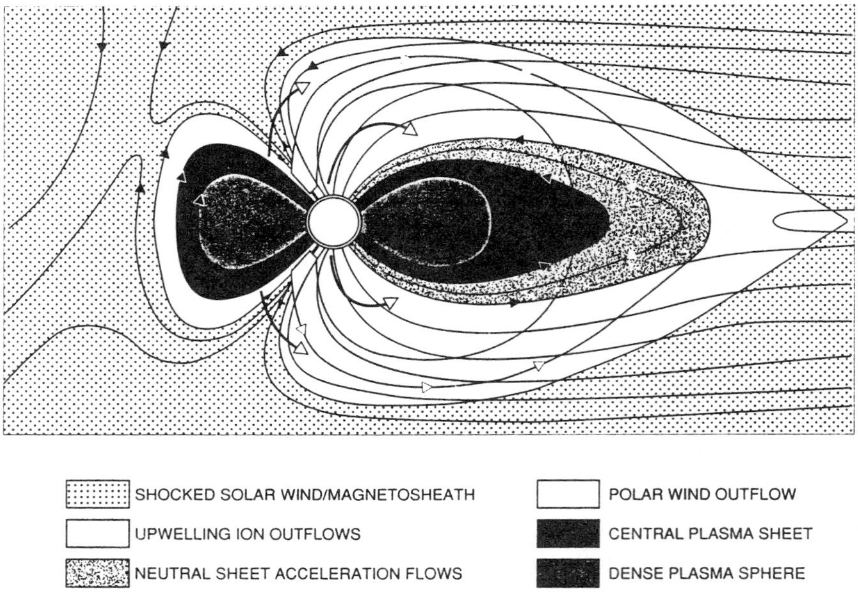


Figure 1. Schematic illustration of ionospheric outflow and acceleration, including the major regions as designated by shading.

description of magnetotail phenomena, and describe a simple hydrodynamic analog of substorm reconfigurations of the magnetotail. Possible means of merging the single particle and continuum descriptions are proposed.

## Single Particle Motion in Model Fields

The approach adopted here consists of computation of the trajectories of individual particles with specified initial position, velocity, and mass, in simple empirical models of the magnetospheric electric and magnetic fields. These trajectory calculations have been carried out using the magnetic field model of *Mead and Fairfield* [1975] (here after referred to as M/F). The magnetic field was thus considered to be the sum of a dipole field $\mathbf{B}_{dip}$ and an external contribution $\mathbf{B}_{ext}$ for which the M/F model provides an empirically-based polynomial description at four different levels of magnetic activity. Tilt angle effects have been neglected in these calculations. In order to simulate the relaxation of the geomagnetic field as in a "dipolarization" event, the field was evolved in time by a gradual variation of the perturbation $\mathbf{B}_{ext}$ from a more active M/F state to a less active one, using observations as a guide [*Aggson et al.*, 1983]. In order to infer the electric field induced by the time-varying magnetic field, the vector potential **A** was calculated. At any time t of the collapse, this vector potential is constructed as

$$\mathbf{A}(t) = \mathbf{A}(t_o) + F(t)[\mathbf{A}(t_f)-\mathbf{A}(t_o)] \qquad (1)$$

where $A(t_o)$ and $A(t_f)$ are the initial and final configurations and F(t) is a polynomial used to model the rate of change of the magnetic field. Accordingly, the instantaneous magnetic field and induced electric field in a position **r** are given by

$$\mathbf{B}(\mathbf{r},t) = \mathbf{B}_{dip}(\mathbf{r}) + \mathrm{curl}[\mathbf{A}(\mathbf{r},t)] \qquad (2)$$

$$\mathbf{E}_{ind}(\mathbf{r},t) = -\partial\mathbf{A}(\mathbf{r},t)/\partial t \qquad (3)$$

The electric field of (3) displays in general a substantial component parallel to the local magnetic field. The magnetospheric plasma was assumed to enforce the condition that $\mathbf{E} \cdot \mathbf{B} = 0$ [e.g., *Falthammar*, 1965; *Birmingham and Jones*, 1968; *Schulz and Eviatar*, 1969]. Free charges were thus assumed to cancel the induced parallel field, leading to a redistribution of the field in the perpendicular direction [*Heikkila and Pellinen*, 1977]. Details of the this technique may be found in *Delcourt et al.* [1990].

The choice of F(t) is somewhat arbitrary, but must be made consistent with observed substorm fields, as mentioned above. To insure realistic influences on the particles, a polynomial form was used in which the time scale for transition and the M/F initial and final states were chosen to produce peak-induced electric fields in qualitative agreement with those observed [*Aggson et al.*, 1983; *Shepard et al.*, 1980], namely a 2-minute magnetic transition time from the highest to lowest states of M/F.

This procedure may be considered questionable in some respects since the field model represents average states and not actual instantaneous states of the geomagnetic field. The four states of the M/F model were somewhat arbitrarily set to cover the full range of mean field variability, so that the transitions between them are somewhat arbitrary rather than physical. At present the dynamics of geotail collapse are not well documented or known from simulation studies, particularly with regard to initiation, propagation, and time delay effects within the plasma sheet [*Moore et al.*, 1981; *Lopez et al.*, 1989]. Nevertheless, it does seem clear that dipolarization is a fundamental feature of substorms. The collapse modeled here proceeds simultaneously at all positions within the plasma sheet without propagation effects, for lack of definitive information (however, see discussion below).

### *Guiding Center vs Exact Equation of Motion*

The previous calculations of ionospheric outflow trajectories in steady state field models have been made using a guiding center approximation. As noted by *Cladis* [1988] and *Delcourt et al.* [1989], the deviations from adiabaticity for these calculations are relatively small, the gyroradii obtained generally being limited to less than 10% of the field curvature radius in the neutral sheet. However, it is found that appreciable nonadiabatic behavior begins to be evident for gyroradii exceeding 5% of the local field curvature radius, so that these calculations are at the limits of what can be done accurately using a guiding center description of the trajectories.

The large induced electric field in the magnetotail, modeled through the changes in the magnetic field as described above, leads to clear nonadiabatic features in the particle orbits, i.e., breaking of the first adiabatic invariant due to finite gyroradius effects. Although these effects rapidly become negligible as particles convect to lower radial distances, they do require that trajectories be obtained from the full equations of motion, rather than from a guiding center approximation. Gravity has been included here, but is important only in the computation of trajectories for low-energy ions at low altitudes

$$\mathbf{r}'' = (q/m)[\mathbf{E} + \mathbf{r}' \times \mathbf{B}] + \mathbf{g} \tag{4}$$

Trajectories were obtained by integrating this equation using a fourth-order Runge-Kutta technique.

## Convective Transport and Energization

### *Source Population for Energization*

In order to establish an initial population of particles to be acted upon by the substorm dipolarization, a distribution of ions was developed by means of computing the $O^+$ flow morphology in steady state fields as described in *Delcourt et al.* [1989] (i.e., using a guiding center description). This was based upon the cleft ion outflow of $O^+$ originating in the 1000 to 1400 local time sector and between 74° to 82° invariant latitude (extending over 4° of invariant latitude centered on the M/F cusp field lines). The average properties of the particles are taken from *Moore et al.* [1986] and the flux spatial distribution from *Yau et al.* [1985], as displayed in Tables 1 and 2 of *Delcourt et al.* [1989]. The resulting distribution of plasma bulk parameters such as density and mean energy was then estimated from the computed trajectories of representative particles using the flux weighting and residence time technique of *Delcourt et al.* [1989].

### *Initial Density/Energy Distribution*

The result of this procedure for $O^+$ density is shown in Figure 2, where an approximately noon-midnight cut through the three-dimensional grid of 1 $R_E$ resolution is plotted using a shaded contouring of density values. As demonstrated by the numerical study of *Cladis* [1986], a velocity analysis of the trajectories reveals gradual acceleration as the flow is picked up by the anti-sunward convection, followed by very large parallel acceleration (several keV) by curvature drift effects on passage through the neutral sheet, and finally precipitation into the conjugate atmosphere, at invariant latitudes of approximately 67°. As expected from this steady acceleration, combined with the diverging field line geometry, the result is a steadily decreasing density along the trajectories. This exercise establishes a population of heavy ionospheric ions in the mid-geotail, which will be acted upon by substorm reconfigurations of the magnetic field.

## Particle Acceleration in Dipolarization Events

### *Stability of Neutral Sheet Acceleration*

While the calculations described here do not compute self-consistent fields, and therefore cannot truly address issues of stability of the situations studied, it is possible to examine the estimated energy density of the particle populations formed in comparison with the field energy density of the model magnetic field. As shown in Figure 3, the acceleration of the cleft particles leads to a particle energy density in the mid-tail which is comparable to or larger than the magnetic field energy density, which contains a deep minimum in the same region of space, particularly for the initial (superdisturbed) version of the M/F field.

While this enhanced plasma pressure is qualitatively consistent with the pronounced stretching of the magnetotail, quantitative comparison can be made with the plasma pressure requirements of the average magnetotail [*Spence et al.*, 1989]. The energy density shown in Figure 3 near 12 $R_E$ at the equator is approximately $3 \times 10^2$ eVcm$^{-3}$ ($5 \times 10^{-10}$ dyne cm$^{-2}$) whereas *Spence et al.* [1989] find that approximately an order of magnitude larger pressure to be consistent with the magnetic field model. This indicates that the plasma pressure produced by our steady state calculation is lower than that required to support the average steady state plasma sheet. While this indicates a deficiency within the simple assumptions of this test particle approach, we believe it to be the best available basis for initiating a study of substorm dipolarization effects.

To simulate a substorm expansion phase, test particles were assigned to equi-distant steps along the steady state drift paths, initialized with the corresponding energies and pitch angles throughout a 3-hour range of local times centered at local midnight. The fields were subsequently allowed to evolve as specified above, during a 2-minute long dipolarization event. The results are displayed in a number of different ways in Figures 4-6.

### *Results*

Several features of interest result from the magnetotail collapse. First, as noted by *Mauk* [1986], a significant latitudinal dependence is evident, since the particles residing above ~30° or below ~-30° latitude remain relatively unaffected. In contrast, a progressive scattering of the well-defined ion streams is evident at low latitudes. In these regions, a prominent earthward displacement (a few $R_E$) results, in conjunction with intense accelerations reaching hundreds of keV. The pitch angle evolution is likewise of interest, since the acceleration is mainly perpendicular to the local magnetic field, in contrast with the steady state curvature drift accelerations. This perpendicular acceleration is closely related to

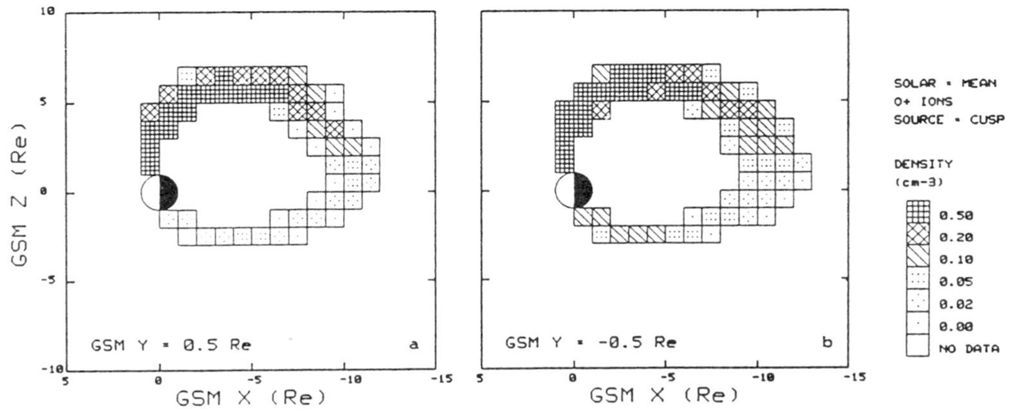


Figure 2. Model results for the cleft ion fountain density distribution of $O^+$ resulting from the cusp upwelling ion region in the dayside cleft, under steady but active conditions.

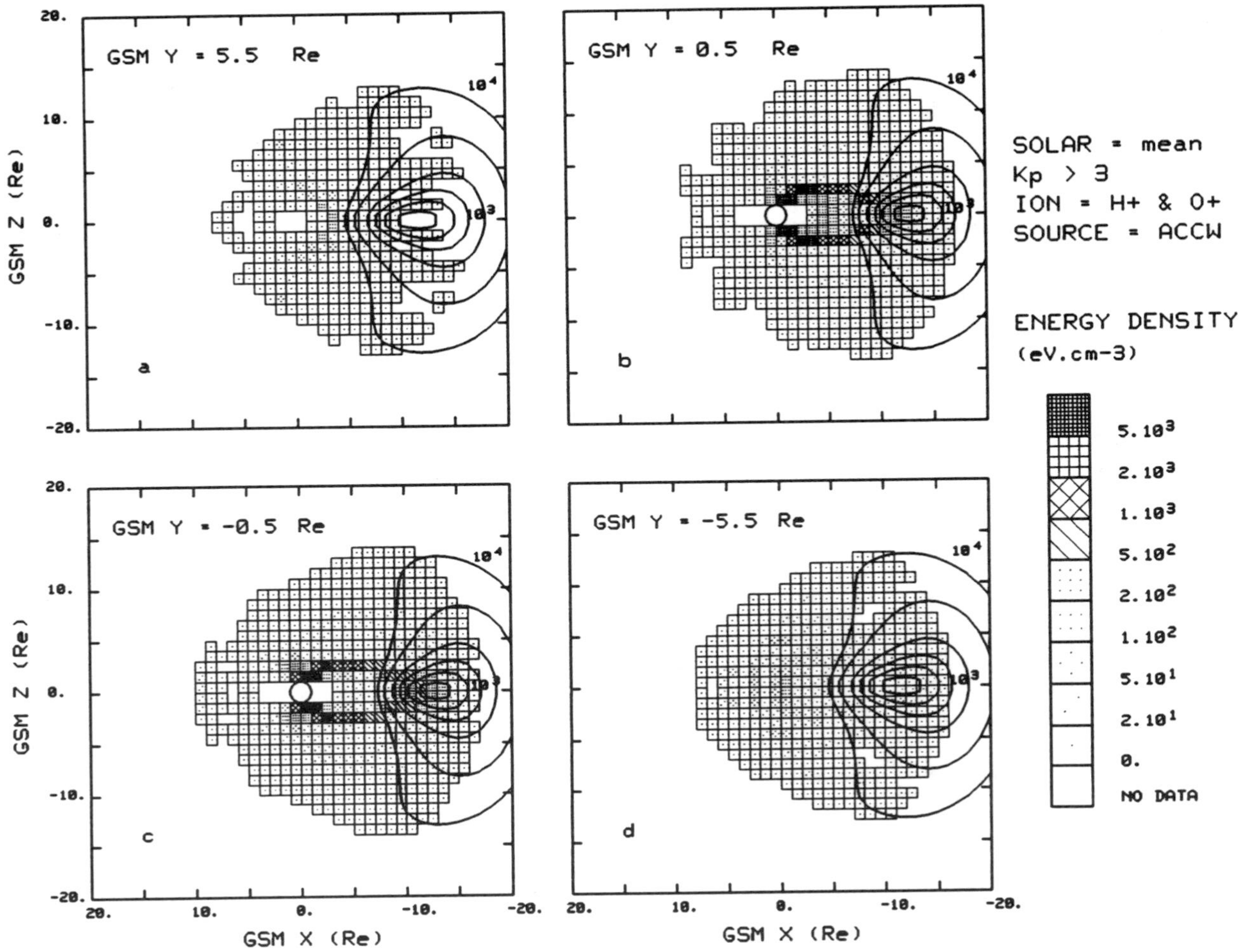


Figure 3. Comparison of magnetic field (line contours) and particle energy density (shading contours) in the Mead-Fairfield magnetotail produced by cleft ion fountain $O^+$ under disturbed conditions. The innermost field contour corresponds to an energy density of 2 x $10^2$ eV $cm^{-3}$ while the particle energy density exceeds this value within that contour.

temporary violations of the first adiabatic invariants for the particles so affected [*Delcourt et al.*, 1990].

As the particles are rapidly displaced earthward, they simultaneously travel westward under the increasing influence of gradient and curvature drifts. There results the build up of a plasma cloud in the vicinity of the equatorial plane, which extends over the evening sector on the time scale of the dipolarization. Due in part to differential drift rates, a pitch angle anisotropy favoring energetic perpendicular fluxes develops in the westward most part of the cloud. In view of the source intensity of the ionosphere, these ion streams, which have been strongly accelerated and thermalized, should contribute substantially to the inner ring current populations.

When bulk plasma parameters are estimated using the technique of *Delcourt et al.* [1989], it is possible to derive, at each step of the geotail collapse, a density and average energy in each spatial bin. Figure 4 shows the time evolution of the density, while Figure 5 shows the corresponding evolution of average energy as realized in the equatorial plane. These quantities are displayed at distinct steps of the dipolarization process, from onset through completion. Prior to the geotail collapse, both figures indicate the initial substantial supply of ionospheric $O^+$ to the mid-tail at times of moderate activity, originating from the dayside cleft region and transported to the tail by anti-sunward convection over the poles. As discussed above, this enhanced supply of $O^+$ to the mid-tail in active conditions may contribute substantially to the plasma sheet pressure.

Subsequently, the geomagnetic field relaxes toward a more dipolar configuration, according to the procedure also described earlier. Figures 4 and 5 reveal the formation of a westward and earthward propagating front of energized particles, similar to that reported by *Moore et al.* [1981] and *Arnoldy et al.* [1982]. The results are highly suggestive of the local creation of an embryonic ring-like distribution in the vicinity of geosynchronous orbit. The speed of propagation is essentially that of the field line collapse speed (with drift effects superposed) yielding displacements of 2 to 4 $R_E$ per minute (300 km $s^{-1}$), with duskward drift effects superposed.

With regard to the adequacy of the plasma pressure developed in terms of ring current populations, it may be noted from Figures 4 and 5 that the advancing westward front of plasma contains pressures on the order of 300 keV x 0.3 $cm^{-3}$ ~ $10^4$ eV $cm^{-3}$ ~ 2 x $10^{-8}$ dynes $cm^{-2}$. Such pressures

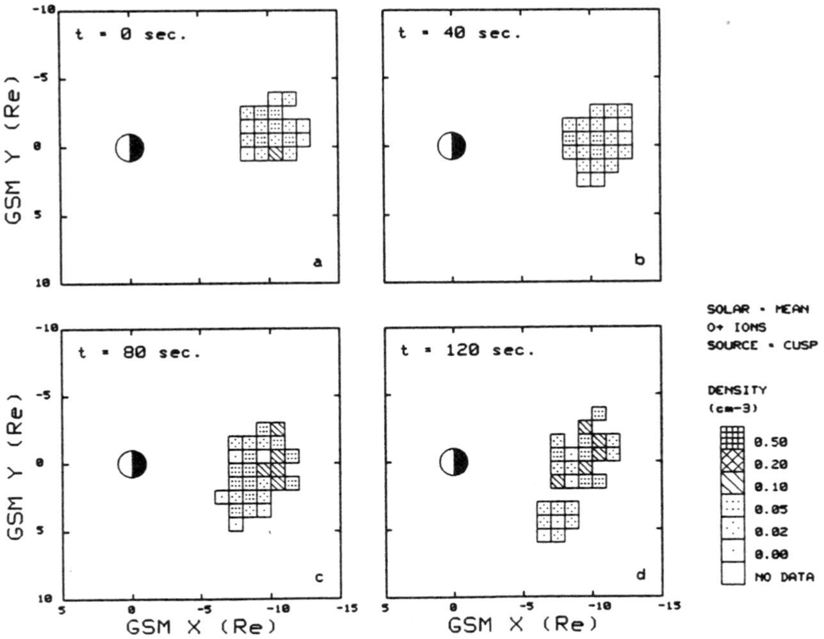


Figure 4. Modeled distribution of $O^+$ density in the equatorial plane resulting from the induced motions of the initial distribution of $O^+$ shown in Figure 2, at four times during a magnetotail collapse.

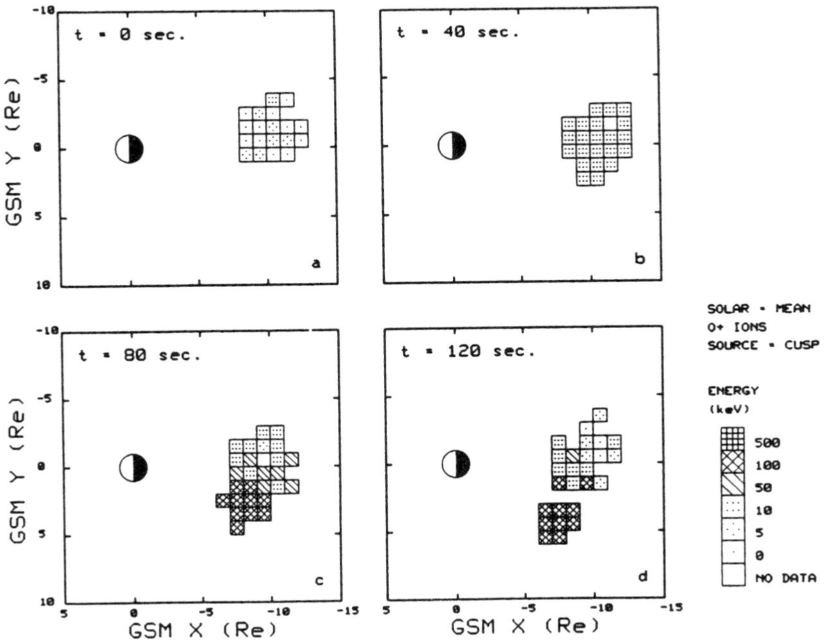


Figure 5. Modeled distribution of $O^+$ mean energy in the equatorial plane resulting from the induced motions of the intitial cleft ion fountain $O^+$ distribution as shown in Figure 2, at four times during a magnetotail collapse.

are in good agreement with plasma pressures at 5-7 $R_E$ as reported by *Spence et al.* [1989]. Evidently the powerful induced electric fields associated with magnetotail dynamics are better able to generate typical pressures in the plasma sheet than the steady state convection fields which are thought to exist there.

The radial plasma transport can be better appreciated in Figure 6 which presents the computed energies as a function of geocentric distance for the individual particles lying within 3 $R_E$ of the equatorial plane. The convection surge of *Quinn and Southwood* [1982] is clearly illustrated, but may be better described as an "induction surge" since it is clearly driven by the change in the magnetic field geometry. These results corroborate those of previous studies [e.g., *Lyons and Williams*, 1980; *Lyons and Speiser*, 1982; *Lui et al.*, 1986; and *Lyons and Schulz*, 1989], confirming that dipolarization of the tail is an effective mechanism to populate the storm-time ring current.

It is interesting that these simulations generate propagation effects on very short time scales similar to that of the field collapse speeds. Of course, this reflects simply that the induced electric field is just that required to maintain the plasma frozen to the evolving magnetic field lines. Equally interesting, though not addressed by this single particle modeling effort, is the redistribution of currents within the magnetotail, which should be accomplished by waves traveling away from the initiation site at magnetohydrodynamic speeds on the order of 10 $R_E$ per minute, fast enough to traverse the length of the near-Earth plasma sheet during a collapse event as modeled here. While the direction of the propagation of such waves is probably of little consequence for the effects noted here, their potential for localized dissipation in the plasmapause region has been noted by *Moore et al.* [1987].

In simulating the creation of energetic $O^+$ at the inner edge of the plasma sheet, these results are consistent with the AMPTE study of *Kremser et al.* [1987] which revealed high abundances of $O^+$ in this location and a gradual decrease with radial distance. Moreover, the *Kremser et al.* [1987] study presents evidence for a distinct behavior of $O^{6+}$ (namely, increasing abundance with radial distance), which is of course not addressed by the present work. Indeed, lighter or higher charge state particles will undergo weaker field variations within a cyclotron period, and consequently display non-adiabatic features only in more remote regions. In the near-Earth region, such ions will experience nearly adiabatic transport, yielding betatron or Fermi energization which cannot produce such high energy gains as found here. For these populations, distinct energization processes in the more distant tail must be invoked.

### Discussion: Tidal Basin Analogy for Substorms

The convection electric field assumed to exist for the purpose of computing particle trajectories under steady conditions (and establishing initial conditions for substorm tail collapse) requires for its existence an interaction with the solar wind whereby plasma flows anti-sunward over the polar caps and returns sunward within the plasma sheet proper.

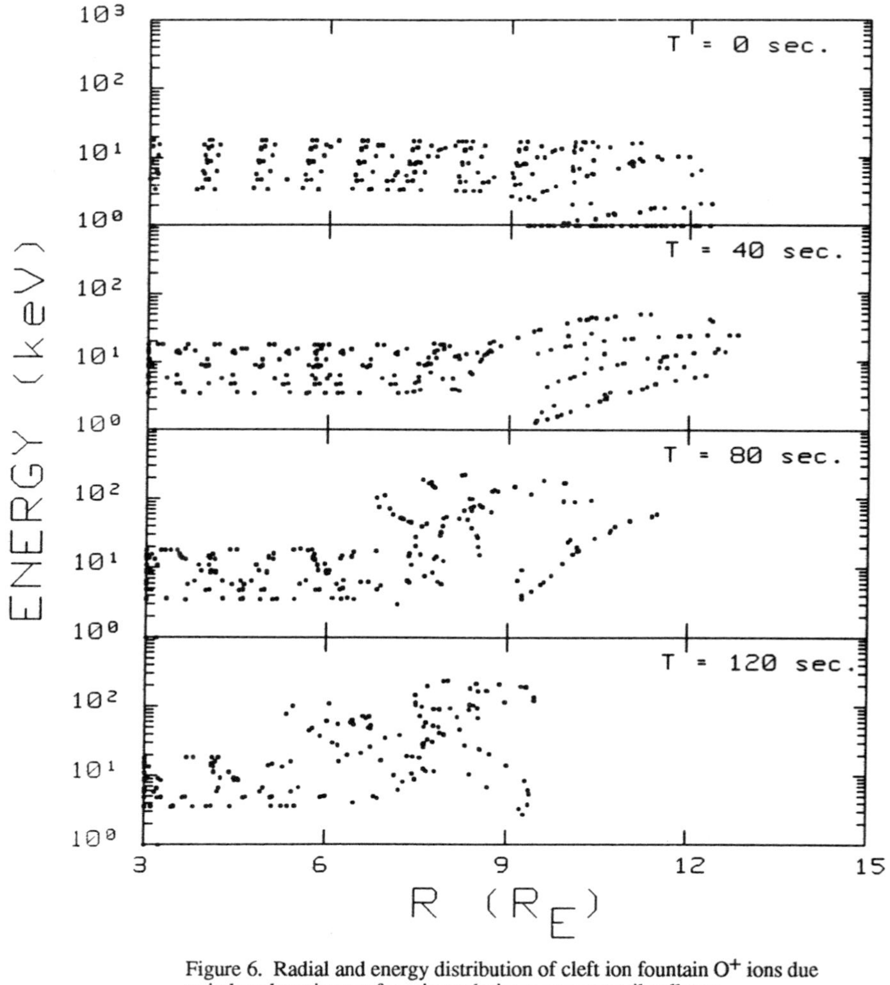


Figure 6. Radial and energy distribution of cleft ion fountain $O^+$ ions due to induced motions at four times during a magnetotail collapse.

Whether this process is driven by direct linking of the polar field lines to the solar wind or by anomalous shear stress transport across magnetic field lines at the magnetopause, or a combination of both, the result must be the same in the plasma sheet. Ionospheric plasma flows through the lobes and along the boundary layers toward the neutral sheet as it flows anti-sunward, decelerating and turning sunward in the near-Earth part of the neutral sheet or continuing off downstream in the solar wind in the distant neutral sheet. The location of the stagnation point marking the boundary between the return flow and the downstream escape is apparently determined by the balance between the stress required to reverse the flow and the limiting stress which can be supported by field line tension.

The X-line merging model of the neutral sheet begins with plasma moving laterally toward the neutral sheet from both sides, then turning sunward and earthward to form the return and escape flows, with total symmetry about the plane through the neutral line and normal to the neutral sheet. In some formulations, the return flow concentrates into localized streams along or near the separatrices of the X-line, very reminiscent of the earthward streams formed in the single particle steady state picture illustrated in Figure 1. However, this simple X-line merging picture misses completely the stress balance implicit in the creation of the plasma sheet. The anti-sunward flowing plasma is decelerated, turned around and then accelerated earthward by its passage into the neutral sheet, and the strong symmetry about the neutral sheet is not present with respect to the plane normal to the neutral sheet at the neutral line. Observationally, thermalization of this acceleration is required to account for a hot isotropic plasma sheet with pressure sufficient to support the inflated tail magnetic field. However, this is not a part of the MHD X-line model. Three-dimensional MHD simulations are able to produce a stretched magnetotail, and therefore must capture the essence of the required physics. However, they do not incorporate ionospheric plasma except as a resistive medium for field-aligned current closure, nor do they describe acceleration or heating effects beyond the bulk flows involved. Inasmuch as observations and single particle calculations indicate that ionospheric plasma with substantial pressure develops (as normalized by the field energy density), the MHD descriptions are likely to suffer from quantitative errors as well.

The single particle approach, on the other hand, is completely unable to address the dynamic behavior of the plasma sheet in dipolarization events. Given the dipolarization description, it leads to the creation of an earthward traveling front in the energetic particles, traveling at the field line collapse (induced electric field) speed. However, it is unlikely that dipolarization proceeds uniformly throughout the magnetotail. A recognized problem in substorm studies is the identification of the location of onset of the collapse, and the nature of the propagation of the collapse by a redistribution of the neutral sheet currents, which should be accomplished by magnetohydrodynamic waves. This problem can perhaps be best approached through a hybrid of MHD and single particle approaches.

As a basis for qualitative discussion of these concepts, we propose the following two-dimensional hydrodynamic analogy for substorm behavior:

Consider the situation illustrated in Figure 7 in which a tidal basin is formed by the establishment of a sand bar which partially entraps the near shore ocean. In the quiescent state, the sand bar is slightly submerged by the level water surface. Now a strong offshore wind blows which is coupled to the ocean by shear stresses at the air/sea boundary, and transports surface water away from shore with the wind. A new quasi-equilibrium could be established in which a region of elevated water level forms just inside the sand bar. Part of the flow would then be trapped by the sand bar, returning along the bottom to the inner parts of the basin and forming a closed circulation pattern. Part of the flow would escape beyond the sand bar and be carried off by the wind.

Consider further that the sand bar is established by the action of the circulation flow by slow deposition of sand which is put in suspension at the shore line and settles out on the bar where the flow stagnates. The sand would then accumulate to greater depth and slope until it exceeded its angle of repose and slumped back into the basin, or grew to the water level and cut off all escape. Depending upon the physical characteristics of the sandy basin bottom, it is conceivable that an extra mass of water swept against the sand bar peak by the action of the wind could contribute to this slumping by placing sufficient pressure upon the sand bar to trigger its relaxation back into the basin and/or outward toward the open sea. If the sand were to have a static viscosity much larger than its dynamic viscosity, the slumping process could be very rapid in comparison with the "growth" process, so that the excess water built up in the quasi-equilibrium would rather suddenly find itself without a confining barrier and would proceed to spread both into the basin and out to sea. The inward moving slope would encounter the shore, steepen, and break. The outward moving slope would encounter greater depths and spread until imperceptible.

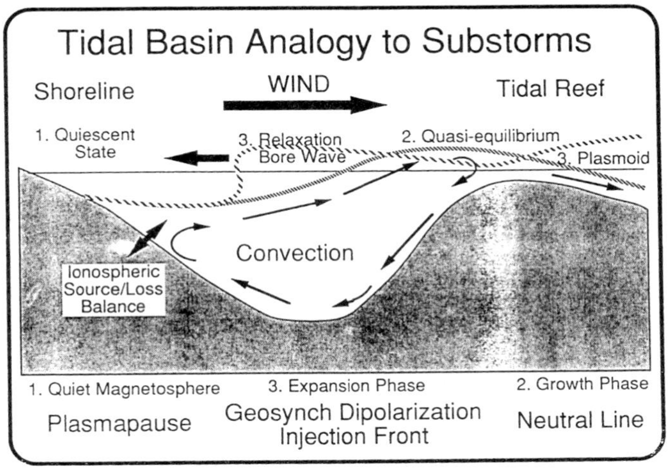


Figure 7. Schematic illustration of the "tidal basin" analogy for substorm behavior.

It should be apparent by now how the general features of this analogy relate to the phenomenology of substorm behavior. The sand plays the role of the magnetic field, which is in some sense convected with the flow, and piles up in the magnetotail due to the action of the solar wind interaction. The sand bar slope plays the role of the magnetic field tension in confining the water to the basin. As for the sand bar and water, the field and hot plasma attain a quasi-equilibrium which is consistent with enhanced convection but which is inherently unstable and subject to disruption. Disruption begins near the point of maximum countervailing stresses and propagates both upstream into the closed convection cell and downstream with the wind flow.

Potential energy is clearly converted into flow energy but not nearly as clearly into thermal energy. Like the MHD models of substorms, this hydrodynamic analogy does not seem to lead naturally to any pronounced heating or dissipation effects. A possible exception might lie in the turbulence associated with the wave breaking, though the effect is clearly weak in the hydrodynamic analogy. One significant difference lies in the magnitude of the stored potential energy per involved particle. Another significant difference lies in the breaking of the adiabatic invariant for particles in the magnetotail. This virtually guarantees that available free energy will be dissipated at least as much into gyrational energy of the particles as into directed flow energy.

## Conclusions

We have argued here that single particle trajectory calculations have enhanced our understanding of many substorm phenomena, particularly the acceleration and heating of terrestrial plasmas to form very energetic particles, which are not clearly predicted by MHD theories of magnetotail dynamics. Very simple single particle models of the dynamic magnetotail lead to very large energizations of ionospheric outflows. In the steady state these effects lead to pronounced plasma sheet boundary layers consisting of earthward streaming ion flows. During dipolarization events with demonstrably realistic characteristics, the boundary layer flows of ionospheric heavy ions are accelerated and isotropized to ring current energies, forming an earthward and westward propagating front of energetic particles which will subsequently disperse in the vicinity of synchronous orbit, qualitatively in agreement with well-established observations.

While suggestive, these results are not based upon fields which are self-consistent or derivable from fundamental principles, and further work is needed to repeat such single particle modeling in fields derived from three-dimensional simulations of the solar wind interaction with the magnetosphere. On the other hand, such simulations must for the present be based upon MHD equations, even though it is clear from the single particle results that the current systems which support the fields are carried not by resistive flows of electrons in a cold ion gas, but by complex drifts and nonadiabatic motions of ions and electrons.

Such conclusions give rise to worries about the basic applicability of the MHD equations to cases where a simple Ohm's law is inadequate. Yet the complex wave-like phenomena so important to substorm physics require a description which self-consistently couples the plasma with the fields. Perhaps the key lies in the use of results from single particle trajectory computations to gain better insight into the physics embodied in the generalized Ohm's law, so that shrewd (possibly nonlocal) choices can be made in relating currents to plasma parameters.

*Acknowledgments*. This work was supported in part by the NASA Space Physics Division Supporting Research and Analysis Program. Support for D.C.D. came from the ESA Research Fellowship Program.

References

Aggson, T. L., J. P. Heppner, and N. C. Maynard, Observations of large magnetospheric electric fields during the onset phase of a substorm, J. Geophys. Res. , 88, 3981, 1983.

Arnoldy, R. L., T. E. Moore, and S.-I. Akasofu, Longitudinal structure of substorm injections at geosynchronous orbit, J. Geophys. Res., 87, 77, 1982.

Barfield, J. N., S. E. DeForest, and D. J. Williams, Simultaneous observations of substorm electrons: Explorer 45 and ATS6, J. Geophys. Res., 82, 531, 1977.

Birmingham, T. J., and F. C. Jones, Identification of moving magnetic field lines, J. Geophys. Res., 73, 5505, 1968.

Cattell, C. A., and F. S. Mozer, Substorm electric fields in the Earth's magnetotail, in Magnetic Reconnection in Space and Laboratory Plasmas, Geophysical Monograph #30, edited by E. W. Hones, Am. Geophys. Un., Washington, DC, p. 208, 1984.

Cladis, J. B., Parallel acceleration and transport of ions from polar ionosphere plasma sheet, Geophys. Res. Lett., 13, 893, 1986.

Cladis, J. B., Transport of ionospheric ions in the magnetosphere: Theory and observations, Adv. Space Res., 8, 165, 1988.

Cummings, W. D., J. N. Barfield, and P. J. Coleman, Jr., Magnetospheric substorms observed at synchronous orbit, J. Geophys. Res., 73, 6687, 1968.

Delcourt, D. C., C. R. Chappell, T. E. Moore, and J. H. Waite, Jr., A three-dimensional numerical model of ionospheric plasma in the magnetosphere, J. Geophys. Res., 94, 11,893, 1989.

Delcourt, D. C., A. Pedersen, and J. A. Sauvaud, Dynamical acceleration of ions by magnetotail dipolarization, J. Geophys. Res., 95, in press, 1990.

Falthammar, C.-G., Effects of time dependent electric fields on geomagnetically trapped radiation, J. Geophys. Res., 70, 2503, 1965.

Heikkila, W. J., and R. J. Pellinen, Localized induced electric field within the magnetotail, J. Geophys. Res., 82, 1610, 1977.

Konradi, A., C. L. Semar, and T. A. Fritz, Substorm-injected protons and electrons and the injection boundary model, J. Geophys. Res., 80, 543, 1975.

Kremser, G., W. Studemann, B. Wilken, G. Gloeckler, D. C. Hamilton, and F. M. Ipavich, Average spatial distributions of energetic $O^+$, $O^{2+}$, $O^{6+}$, and $C^{6+}$ ions in the magnetosphere observed by AMPTE-CCE, J. Geophys. Res., 92, 4459, 1987.

Lockwood, M., M. O. Chandler, J. L. Horwitz, J. H. Waite, Jr., T. E. Moore, and C. R. Chappell, The cleft ion fountain, J. Geophys. Res., 90, 9736, 1985.

Lopez, R. E., A.T.Y. Lui, D. G. Sibeck, K. Takahashi, R. W. McEntire, L. J. Zanetti, and S. M. Krimigis, 1989, On the relationship between the energetic particle flux morphology and the magnetic field during substorms, J. Geophys. Res., 94, 17,105, 1989.

Lui, A.T.Y., R. W. McEntire, S. M. Krimigis, and E. P. Keath, Acceleration of energetic oxygen (E>137 keV) in the storm time ring current, in Ion Acceleration in the Magnetosphere and Ionosphere, Geophysical Monograph #30, Am. Geophys. Un., Washington, DC, p. 149, 1986.

Lyons, L. R., and M. Schulz, Access of energetic particles to storm time ring current through enhanced radial "diffusion," J. Geophys. Res., 94, 5491, 1989.

Lyons, L. R., and T. W. Speiser, Evidence for current sheet acceleration in the geomagnetic tail, J. Geophys. Res., 87, 2276, 1982.

Lyons, L. R., and D. J. Williams, A source for the geomagnetic storm main phase ring current, J. Geophys. Res., 85, 523, 1980.

Mauk, B. H., and C.-I. Meng, Dynamical injections as the source of near geostationary quiet time particle spatial boundaries, J. Geophys. Res., 88, 10,111, 1983.

Mauk, B. H., 1986, Quantitative modeling of the convection surge mechanism of ion acceleration, J. Geophys. Res., 91, 13,423.

McIlwain, C. E., Substorm injection boundaries, in Magnetospheric Physics, edited by B. M. McCormac, Reidel, Hingham, MA, 1974.

McPherron, R. L., Magnetospheric substorms, Rev. Geophys., 17, 657, 1979.

Mead, G. D., and D. H. Fairfield, A quantitative magnetospheric model derived from spacecraft magnetometer data, J. Geophys. Res., 80, 523, 1975.

Moore, T. E., R. L. Arnoldy, J. Feynman, and D. A. Hardy, Propagating substorm injection fronts, J. Geophys. Res., 86, 6713, 1981.

Moore, T. E., M. Lockwood, M. O. Chandler, J. H. Waite, Jr., C. R. Chappell, A. Persoon, and M. Sugiura, Upwelling $O^+$ ion source characteristics, J. Geophys. Res., 91, 7019, 1986.

Moore, T. E., D. L. Gallagher, J. L. Horwitz, and R. H. Comfort, 1987, MHD wave breaking in the outer plasmasphere, Geophys. Res. Lett., 14, 1007, 1987.

Quinn, J. M., and D. J. Southwood, Observations of parallel ion energization in the equatorial region, J. Geophys. Res., 87, 10,536, 1982.

Sauvaud, J.A., and J. R. Winckler, Dynamics of plasma energetic particles and fields near synchronous orbit in the nighttime sector during magnetospheric substorms, J. Geophys. Res., 85, 2043, 1980.

Schulz, M., and A. Eviatar, Diffusion of equatorial particles in the outer radiation zone, J. Geophys. Res., 74, 2182, 1969.

Shepard, G. G., R. Bostrom, H. Derblom, C.-G. Falthammar, R. Gendrin, K. Kaila, A. Korth, A. Pedersen, R. Pellinen, and G. Wrenn, Plasma and field signatures of poleward propagating auroral precipitation, J. Geophys. Res., 85, 4587, 1980.

Smith, P. H., and R. A. Hoffman, Direct observations in the dusk hours of the characteristics of the storm time ring current during the beginning of magnetic storms, J. Geophys. Res., 79, 966, 1974.

Spence, H. E., M. G. Kivelson, R. J. Walker, and D. J. McComas, Magnetospheric plasma pressures in the midnight meridian: Observations from 2.5 to 35 $R_E$, J. Geophys. Res., 94, 4264, 1989.

Williams, D. J., Phase space variations of near equatorially mirroring ring current ions, J. Geophys. Res., 86, 189, 1981.

Yau, A. W., E. G. Shelley, W. K. Peterson, and L. Lenchyshyn, Energetic auroral and polar ion outflow at DE 1 altitudes: Magnitude, composition, and magnetic activity dependence and long term variations, J. Geophys. Res., 90, 8417, 1985.

T. E. Moore, NASA Marshall Space Flight Center, Space Science Laboratory, Mail Code ES53, Huntsville, AL 35812.

D. C. Delcourt, ESA/ESTEC, Department of Space Physics, Noordwijk, The Netherlands.

# The Auroral Zone

# EIC WAVES, DOUBLE LAYERS, AND SOLITARY WAVES IN THE AURORAL ACCELERATION REGION

David Tetreault

Massachusetts Institute of Technology

*Abstract.* Electric field fluctuations identified as electrostatic ion cyclotron waves, double layers, and near solitary wave potential structures have been observed in the auroral particle acceleration region by the S3-3 and Viking satellites. We propose here that these electric fields can be explained in terms of a three dimensional model of nonlinear ion hole/clump instability.

## Introduction

The formation of ion acoustic double layers has been invoked as the cause of parallel electric field observations by the S3–3 and Viking satellites [Sato and Okuda, 1981; Hudson et al., 1983; Hasegawa and Sato, 1982]. The double layer potentials are characteristic of accelerating ion holes formed when ions become trapped in the potentials wells of growing ion acoustic waves. However, ion holes can self–organize and grow nonlinearly in a linearly stable plasma where, for weak currents and comparable temperatures, ion acoustic waves cannot grow [Berman et al., 1985; Berman et al., 1986; Dupree, 1983; Tetreault, 1983]. Because of its much lower instability threshold, nonlinear hole/clump instability has also been proposed as the cause of the observed double layers [Tetreault, 1988].

An ion hole can self–organize from a localized depletion in the ion phase space density. Such depletions are everpresent, even at thermal equilibrium, and provide the "seeds" for hole self–organization. A local, virialized (BGK equilibrium) hole forms when, because of the neutralizing charge density of the plasma electrons, ions become trapped in a depletion's negative potential. In a drifting electron/ion plasma, an ion hole can become nonlinearly unstable and grow in amplitude as electrons are resonantly reflected by the ion hole's negative potential. If $f_{0e}'(u)$ denotes the velocity gradient of the mean electron distribution function, hole growth occurs when, for $f_{0e}'(u)\geq 0$, a charge imbalance forms across the hole as a net number of electrons are reflected by the hole's potential. From Poisson's equation, this charge imbalance implies a potential drop across the hole of, approximately,

$$\Delta\phi/\phi_0 \sim -4v_e^2 f_{0e}'(u) \tag{1}$$

where $\phi_0$ is the hole potential depth. This potential drop (electric field) gives the hole its double layer potential structure and accelerates the hole through the ion distribution function. As the hole accelerates, analytical calculations show that it grows in amplitude at the approximate rate

$$\gamma_h^i \sim -8(\Delta v_{\parallel}/\Delta x_{\parallel})v_e^2 f_{0e}'(u)\, v_i^2 f_{0i}'(u) \tag{2}$$

where $\Delta v_{\parallel}=(2e\,\phi_0/m_i)^{1/2}$ is the velocity trapping width of the hole.

Numerous parameter studies of hole/clump dynamics (self–organization, growth rates, thresholds, sensitivities to mass ratio and system length etc.) have been done with computer simulations and compared with the analytical models [Berman et al., 1985; Berman et al., 1986; Dupree, 1983; Tetreault, 1983]. While the range of parameters studied in the simulations were limited by computer resources, the analytical models are not constrained. For example, the theoretical models (being valid for any mass ratio) agree in detail with the $m_i/m_e=4$ simulations of Ref. 4 and can then be used to predict the threshold for hole/clump formation in a realistic plasma where $m_i/m_e=1836$. Such analysis and comparisons predict that hole/clump instability can occur in real plasma [Tetreault, 1983].

Though this hole/clump model is one dimensional, its predictions agree well with double layer (parallel electric field) observations from the S3-3 and Viking satellites. This agreement results from the fact that, in a strongly magnetized plasma, the magnetic field restricts the particle trapping to one dimension (the component of velocity along the magnetic field).

However, the satellite data indicates that the double layers

Modeling Magnetospheric Plasma Processes
Geophysical Monograph 62

also have perpendicular structure, the scale perpendicular to the geomagnetic field ($\Delta x_\perp$) being somewhat larger than the field aligned dimension ($\Delta x_\parallel$) [Temerin et al., 1982; Bostrom et al., 1988; Koskinen et al., 1987; Bostrom et al., 1989]. In addition, the data reveals the presense of electrostatic ion cyclotron (EIC) waves and near solitary wave (SW) potential structures. We sketch here how, when the one dimensional hole/clump model is extended to three dimensions, it can explain these various fluctuations in a selfconsistent and unified way. A more detailed theory will appear elsewhere.

## Three Dimensional Model

We invoke the disparity between the ion dynamics parallel and perpendicular to the magnetic field. The characteristic time for the parallel ion motion is the trapping time ($\Delta x_\parallel/\Delta v_\parallel$) of an ion that is trapped in the negative potential ($\phi$) of an ion phase space hole. The ion gyro frequency ($\Omega_i$) characterizes the ion perpendicular dynamics. For typical parameters inferred from the satellite data ($e\phi/T_e \lesssim 1$, $\rho_i \sim 10\lambda_D$ and $\Delta x_\parallel \sim 20\text{–}50\lambda_D$),

$$\frac{\Delta v_\parallel/\Delta x_\parallel}{\Omega_i} \sim \left[\frac{\Delta v_\parallel}{v_i}\right]\left[\frac{\rho_i}{\Delta x_\parallel}\right] \sim \left[\frac{e\phi}{T_e}\right]^{1/2}\left[\frac{\rho_i}{\Delta x_\parallel}\right] < 1 \tag{3}$$

where $v_i$ and $\rho_i$ are the ion thermal velocity and gyro radius, and we have assumed that $T_e = T_i$. With this disparity in time scales, the Vlasov equation splits up into a response, $\delta f_\parallel^i$, for parallel ion motion, and a response, $\delta f_\perp^i$, for perpendicular ion motion. The ion hole can thus be thought of forming in a background plasma of gyrating ions.

Assuming that the perpendicular ion dynamics are linear, the Fourier transform of $\delta f_\perp^i$ is given by the usual linear response,

$$\delta f_\perp^i(\mathbf{k},\omega) = \frac{1}{4\pi e}\frac{\omega_{pi}{}^2 k_\perp{}^2}{\omega^2-\Omega_i{}^2} f_{0i}\,\phi(\mathbf{k},\omega) \tag{4}$$

Poission's equation is

$$\frac{1}{4\pi e}\nabla^2\phi = \int dv_\parallel \frac{e}{T} f_{0i}\,\phi - \int dv_\parallel \delta f_\parallel^i - \int dv_\perp \delta f_\perp^i \tag{5}$$

where the first term on the RHS is the electron response. With (4), the Fourier transform of (5) can be written as

$$\phi(\mathbf{k},\omega) = \frac{\tilde\phi(\mathbf{k},\omega)}{\varepsilon(\mathbf{k},\omega)} \tag{6}$$

where

$$\tilde\phi(\mathbf{k},\omega) = \frac{4\pi e}{k^2}\int dv_\parallel \delta f_\parallel^i \tag{7}$$

is the ion hole potential and, with $k^2 = k_\parallel{}^2 + k_\perp{}^2$,

$$\varepsilon(\mathbf{k},\omega) = 1 + \frac{1}{k^2\lambda_D{}^2} - \frac{1}{k^2}\frac{\omega_{pi}{}^2 k_\perp{}^2}{\omega^2-\Omega_i{}^2} \tag{8}$$

is the linear dielectric function for EIC waves. Eq. (6) describes the shielding of the hole/double layer potential by the background plasma. Neglecting any nonlinear correlations ("clumping") of the ions in gyro–angle, the perpendicular spatial scale of the hole is given by the ion gyro radius. We can thus think of the hole potential in the numerator of (6) as arising from a group ("macroparticle") of ions. Note that (6) is in the form of a shielded potential as in the discrete test particle model [Rosenbluth and Rostoker, 1962; Dupree, 1972]. The hole/double layer potential in the numerator of (6) can thus excite EIC waves within the hole's resonant velocity range, as in the standard discrete test particle model. The parallel (trapping) dynamics of the double layer is as described in the one dimensional model , and enters the 3D model here via (7) in the numerator of (6). The perpendicular (EIC) dynamics enters through the dielectric function in the denominator of (6).

From (6), we expect to see a spectrum due to both holes and EIC waves. If we sum (6) over $\mathbf{k}$ and $\omega$, we have

$$\phi(\mathbf{x},t) = \int d\mathbf{k}\int d\omega\, \frac{\tilde\phi(\mathbf{k},\omega)}{\varepsilon(\mathbf{k},\omega)} \exp i(\mathbf{k}.\mathbf{x}-\omega t) \tag{9}$$

The portion of the sum due to the holes comes from that part of the sum in (9) where $\omega \sim k_\parallel v_\parallel \lesssim k_\parallel v_i < \Omega_i$ so that the denominator of (9) is large, *i.e.*, $\varepsilon(\mathbf{k},\omega) \sim 1$ (recall (8)). These low frequencies are far from the ion cyclotron frequency and the linear stability boundary for EIC waves. This is the region of ion ballistic frequencies characteristic of the deepest ion holes and, therefore, $\tilde\phi(\mathbf{k},\omega)$ is also large. In the frequency range $\omega \sim \Omega_i$, we are near the EIC wave resonance so that $\varepsilon(\mathbf{k},\omega) \sim 0$ for $k\lambda_D \ll 1$ (see (8)). There are fewer ion holes in this region. Moreover, they have just formed out of the background thermal fluctuations and have not had time to grow to significant amplitudes. Therefore, though $\varepsilon(\mathbf{k},\omega)$ is small in this region, $\tilde\phi(\mathbf{k},\omega)$ is also small. While a precise evaluation must await a detailed mathematical model of $\tilde\phi$ for EIC holes, these physical considerations indicate that $\tilde\phi/\varepsilon$ in (9) appears to be finite for both the EIC and hole portions of the spectrum.

## Comparison With Data

The S3-3 and Viking data have indicated the presence of both weak double layers as well as "near" solitary wave structures in the auroral acceleration region. A particular example of these measurements is shown in Fig. 4 of Mozer and Temerin [1983], where several individual parallel electric field structures are identified as either DL or SW [Mozer and Temerin, 1983]. We have used the qualifier "near" here because the magnitudes of the measured electric fields identified as SW are so small that they may just be very weak double layers (Temerin and Kintner, private communication). Indeed, it has been suggested that the "near" SW and weak DL may be "different phases of evolution" of the same nonlinear structure [Bostrom et al.,

1988]. These observations are consistent with predictions of the nonlinear hole model. Consider an ion hole with speed $u=0$ in the rest frame of the ion distribution function, where the depth of the hole has just reached its maximum. As the ion hole continues to accelerate through the ion distribution, its depth will begin to decrease (for $f_{0i}'(u) \geq 0$ in (2), the hole growth rate is negative). Since, from Poisson's equation the hole depth is proportional to the hole potential, the depth of the potential well will also decrease. Also, since the number of electrons reflected by the hole potential will be larger the larger the potential, the double layer potential drop $\Delta\phi$ will also decrease. Therefore, as the hole accelerates further and further into the ion ditribution, its potential drop will decrease precipitously as it moves down the steep velocity gradient into regions of smaller $f_{0i}$. The final stage of the hole/double layer structure is thus a "near" SW structure, i.e., a potential structure with $\Delta\phi \sim 0$.

In the S3–3 data, the frequencies of the electric fields measured perpendicular to the magnetic field, $\omega \sim \Omega_i$, are characteristic of electrostatic ion cyclotron waves. This ion cyclotron emission is consistent with the shielded hole model (6) and (9). Both the Viking and S3–3 satellite data indicate that the double layers are found in the same regions where EIC waves have been observed. This observation is also consistent with the Cerenkov emission model (6) where the holes excite, or drive EIC waves in the cold background plasma. Of course, the EIC waves could coincidently also be excited by some other dynamical mechanism unrelated to the DL's, *e.g.*, EIC waves may be lineary unstable [Kindel and Kennel, 1971]. However, the 3D model presented here shows that the hole/double layers can form and grow in a background plasma containing EIC waves, *i.e.*, they can coexist.

The hole/double layers are described by the low frequency ($\omega \approx k_\parallel \Delta v_i < \Omega_i$) portion of (6). For this frequency range, we revert back to perpendicular position and, using (7) and (8), write (6) as

$$\left[ -\left(1 + \frac{\omega_{pi}^2}{\Omega_i^2}\right)\frac{\partial^2}{\partial x_\perp^2} + k_\parallel^2 + \frac{1}{\lambda_D^2} \right] \phi(x_\perp, k_\parallel, \omega) = 4\pi e \int dv_\parallel \delta f_\parallel^i (x_\perp, k_\parallel, \omega) \tag{10}$$

From (10), we can get an estimate of the relative size of the perpendicular and parallel spatial scales for a 3D hole/double layer:

$$\left[\frac{\Delta x_\parallel}{\Delta x_\perp}\right]^2 \sim \frac{1+(k_\parallel \lambda_D)^{-2}}{1+(\rho_i/\lambda_D)^2} \tag{11}$$

For typical parameters inferred from the satellite observations ($\rho_i \sim 10\lambda_D$ and $\Delta x_\parallel = 2\pi/k_\parallel \sim 20\text{–}50\lambda_D$), this ratio is on the order of unity or less and is consistent with the satellite data [Berman et al., 1985; Berman et al., 1986; Dupree, 1983; Tetreault, 1983].

*Acknowledgements.* The author would like to thank T. Chang and T. Dupree for discussions related to this work, and to P. Kintner, H. Koskinen, and M. Temerin for clarification of satellite data. Thanks to G. Crew for help in preparing the camera-ready copy. This work was supported by AFOSR(AFSC) under contract F49620-86-C-0128.

## References

Berman, R. H., D. J. Tetreault, and T. H. Dupree, Simulation of Phase Space Density Hole Growth and the Development of Intermittent Plasma Turbulence, *Phys. Fluids, 28*, 1055, 1985.

Berman, R. H., T. H. Dupree, and D. J. Tetreault, Growth of Nonlinear, Intermittent Fluctuations in Linearly Stable and Unstable Plasma, *Phys. Fluids, 29*, 2860, 1986.

Bostrom, R., G. Gustafsson, B. Holback, G. Holmgren, H. E. J. Koskinen, and P. Kintner, *Phys. Rev. Lett., 61*, 82, 1988.

Boström, R., B. Holback, G. Holmgren, and H. E. J. Koskinen, Solitary structures in the magnetospheric plasma observed by Viking, *Physica Scripta, 39*, 782–786, 1989.

Dupree, T., Theory of Phase Space Density Granulation in Plasma, *Phys. Fluids, 15*, 334, 1972.

Dupree, T., *Phys. Fluids, 26*, 2460, 1983.

Hasegawa, A. and T. Sato, Existence of a Negative Potential Solitary–wave Structure and Formation of a Double Layer, *Phys. Fluids, 25*, 632, 1982.

Hudson, M. K., W. Lotko, I. Roth, and E. Witt, Solitary Waves and Double Layers on Auroral Field Lines, *J.G.R., 88*, 916, 1983.

Kindel, J. and C. Kennel, Topside Current Instabilities, *J.G.R., 76*, 3055, 1971.

Koskinen, H. E. J., R. Bostrom, and B. Holback, in *Physics of Space Plasmas (1987), SPI Conference Proceedings and Reprint Series, Number 7*, edited by T. Chang, G. B. Crew, and J. R. Jasperse, p. 147, Scientific Publishers, Cambridge, Mass., 1987.

Mozer, F. S. and M. Temerin, *High-Latitude Space Plasma Physics*, Plenum, 1983.

Rosenbluth, M. and N. Rostoker, Scattering of Electromagnetic Waves by a Nonequilibrium Plasma, *Phys. Fluids, 5*, 776, 1962.

Sato, T. and H. Okuda, Numerical Simulation on Ion Acoustic Double Layers, *J. Geophys. Res., 86*, 3357, 1981.

Temerin, M., K. Cerny, W. Lotko, and F. S. Mozer, *Phys. Rev. Lett., 48*, 1175, 1982.

Tetreault, D., *Phys. Fluids, 26*, 3247, 1983.

Tetreault, D., *Geophysical Research Letters, 15*, 164, 1988.

# PARTICLE SIMULATION OF THE INTERACTION BETWEEN KINETIC ALFVÉN WAVES AND DOUBLE LAYERS

Niels F. Otani

School of Electrical Engineering, Cornell University, Ithaca, NY 14853

Perry C. Gray and Mary K. Hudson

Wilder Physics Lab, Dartmouth College, Hanover, NH 03755

*Abstract.* A new simulation method is used to study the interaction of kinetic Alfvén waves and double layers in the auroral acceleration region. The simulation model is designed to clarify the confusion over the source of free energy, which was effectively imposed by boundary conditions in previous double layer simulations. The validity and characteristics of the model are discussed. Early results from the simulation show the presence of "current-driven" double layers. When an infinitely wide current channel is modeled, a parallel electric field is found to be generated to maintain a constant current through the double layers. For a finite width channel, the parallel current is observed to decay away due to an effective resistance presented by the double layers via a magnetic diffusion effect. This effective resistance is apparently reduced in the finite width channel case, possibly due to a charge-neutralizing effect of the ion polarization drift.

## 1. Introduction

Localized parallel electric field structures have been observed in the acceleration region along auroral field lines [Temerin et al., 1982; Boström et al., 1988]. It has been speculated that these structures are *double layers*, the same structures which have appeared in electrostatic computer particle simulations [e.g., Sato and Okuda, 1981; Hudson et al., 1987; Barnes et al., 1985; Otani, 1984; Kim, 1987]. One point of confusion concerns the boundary conditions employed by the simulation models in which double layers are observed. When a fixed voltage is applied across the simulation system, "voltage-driven" double layers are observed. The potential drop across this type of double layer can be much larger than $T_e/e$. A different type of double layer appears when a constant current is injected from one end of the simulation system and removed from the other. If the electrons carrying the current have drift velocity of order the electron thermal speed, the ion acoustic instability develops. The nonlinear evolution of this instability is characterized by the formation of "current-driven" double layers with potential drop of order $T_e/e$. The behavior of the double layers also depends on whether the simulation is bounded or periodic and, in the latter case, on whether a non-self-consistent mean electric field is imposed on the system.

These various models produce double layers with differing characteristics because the nature of the free energy responsible for forming the double layers is in each case different. A natural solution to this dilemma is to include in the model the source of the free energy and then let the simulation determine the nature of the free energy and resulting nonlinear structure self-consistently. A prominent candidate for this free energy source is *kinetic Alfvén waves.* We implement this solution by including the necessary physics for the existence of the kinetic Alfvén wave into a basic electrostatic one-dimensional particle simulation code. The model includes the effects of the Alfvén wave ion polarization drift on the particle simulation Poisson equation and the wave's overall parallel electric field on double layers. As a side benefit, the model should also allow investigation of how the presence of double layers modifies kinetic Alfvén wave propagation.

## 2. The Simulation Model

It is convenient to think of the simulation as a model for a specific region of a specific structure in the magnetosphere. This definitive approach facilitates our discussion

Modeling Magnetospheric Plasma Processes
Geophysical Monograph 62

and understanding of the model, but we hasten to mention that the results should apply to a larger class of magnetospheric structures, regions, and situations in which the same general physics is operative. In particular, the transverse fields of the kinetic Alfvén wave described below may be thought of as a specific model for general global electromagnetic induction.

We consider the simulation as a model of a portion of a V-shaped potential structure in the auroral acceleration region of the Earth's magnetosphere. The simulated region is the shaded region in Fig. 1. The region is 1000 or more Debye lengths long along the background magnetic field $\mathbf{B}_0$, or about 20 km in the auroral acceleration zone, and $20\rho_s$ or about 3 km wide across the field. Here $\rho_s$ is the ion gyroradius at the electron temperature.

When the equations governing linear inertial (i.e., $v_{the}/v_A < 1$) kinetic Alfvén waves [Goertz and Boswell, 1979] are applied to a V-shaped potential structure, the fields and currents are found to be arranged as shown in Fig. 2 [cf. Lysak and Dum, 1983]. The central region of the V-shaped structure carries the bulk of the parallel current in the form of downward precipating electrons. In the linear picture, the parallel current is caused by an earlier passage of a parallel electric field associated with the leading edge of the downward-propagating kinetic Alfvén wave. Flanking the parallel current channel are the transverse electric and magnetic fields associated with the kinetic Alfvén wave.

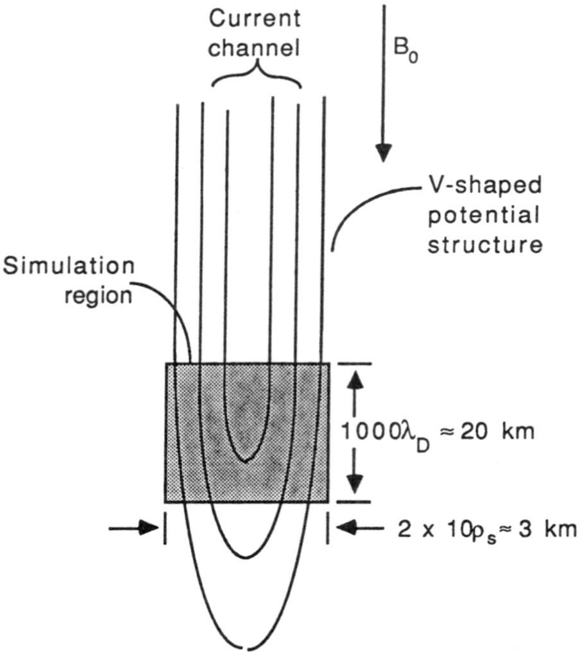


Fig. 1. Typical location of the simulation region, in grey, in a V-shaped potential structure.

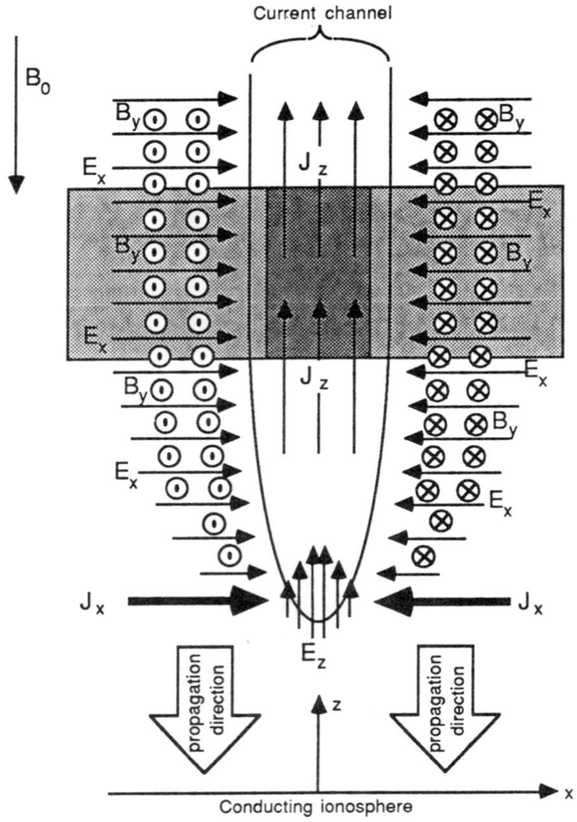


Fig. 2. Location of the simulation region, in grey, in a downward-propagating kinetic Alfvén wave. The dark grey area represents the region in which the particle dynamics of the code is considered to occur.

The simulation is also capable of modeling the current channel left behind after the kinetic Alfvén wave reflects off the ionosphere, a closely related situation. This scenario is depicted in Fig. 3 for the case of a perfectly conducting ionosphere. In this case, as shown in the Figure, the transverse electric fields of the downward- and upward-propagating waves cancel, leaving just the current channel and its associated magnetic field.

The simulation equations are motivated by these models. We assume fields of the following form:

$$E_x = \hat{E}_x(z,t)\sin k_x x, \tag{1a}$$

$$B_y = \hat{B}_y(z,t)\sin k_x x, \tag{1b}$$

$$J_z = \hat{J}_z(z,t)\cos k_x x, \tag{1c}$$

$$E_z = \hat{E}_z(z,t)\cos k_x x, \tag{1d}$$

for a *single*, fixed value of $k_x$. The perpendicular ($x$-

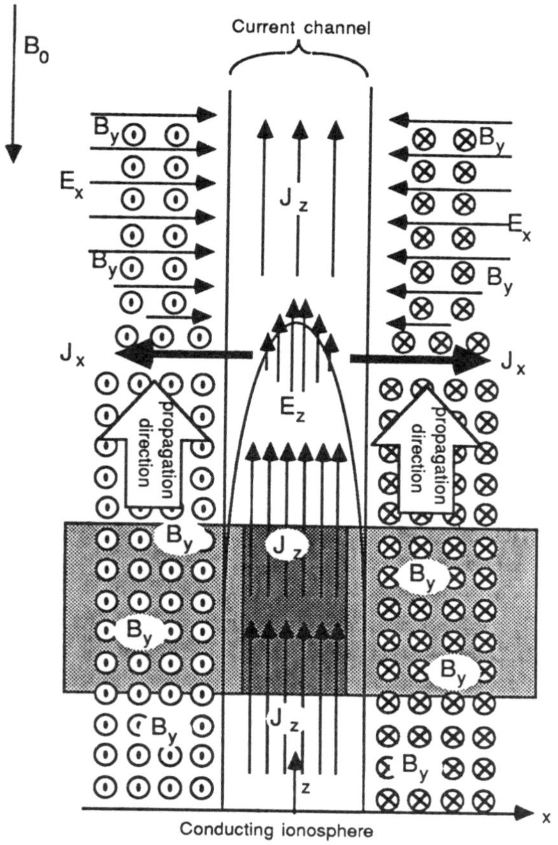


Fig. 3. Location of the simulation region following the reflection of a kinetic Alfvén wave front off the ionosphere.

direction) dependence for the fields is meant to model the perpendicular structure depicted in Figs. 2 and 3, with $\pi/k_x$ chosen equal to the V-structure width. As illustrated, $x = 0$ corresponds to the center of the current channel. Since $E_z$ and $J_z$ are largest in the current channel, these quantities are assumed to depend on $x$ as $\cos k_x x$. Similarly, the transverse fields $E_x$ and $B_y$ are maximal and oppositely directed on either side of the current channel and therefore are assumed to vary as $\sin k_x x$. The transverse current $J_x$ is assumed by both linear kinetic Alfvén wave theory and the simulation to be carried by the ion polarization drift, and thus appears in the simulation as a quantity proportional to $\partial E_x/\partial t$. These assumptions can in fact be shown to hold rigorously for a theoretical model based on the linear equations described by Goertz and Boswell [1979].

This assumed form of the $x$-dependence essentially reduces the model to a time-dependent, one-dimensional system. The system is presently assumed to be periodic in the $z$ direction, implying that the $z$-averaged fields appearing in the simulation are meant to represent the amplitudes of the fields whose wavelengths are much longer than the length of the simulation in the $z$ direction. The simulation follows the evolution of the $z$-Fourier transforms of the quantities $\hat{E}_x(z,t)$, $\hat{B}_y(z,t)$, etc., by applying a leapfrog method to the equations. The simulation equations primarily associated with the Alfvén wave are:

$$\frac{\partial}{\partial t}\langle \hat{B}_y \rangle = -ck_x \langle \hat{E}_z \rangle, \tag{2}$$

$$\frac{\partial}{\partial t}\langle \hat{E}_z \rangle = ck_x \langle \hat{B}_y \rangle - 4\pi \langle \hat{J}_z \rangle, \tag{3}$$

$$\langle \hat{E}_x \rangle = 0, \tag{4}$$

for $k_z = 0$, and

$$\frac{\partial}{\partial t}\hat{B}_y(k_z) = -c[ik_z \hat{E}_x(k_z) + k_x \hat{E}_z(k_z) W(k_z)], \tag{5}$$

$$\frac{\partial}{\partial t}\hat{E}_x(k_z) = -ik_z \frac{v_{A0}^2}{c} \hat{B}_y(k_z), \tag{6}$$

for $k_z \neq 0$. In these equations, the brackets ($\langle\rangle$) refer to an average over $z$, $W(k_z)$ is a fixed weighting function to be explained later, and $v_{A0}$ is the zero-order Alfvén speed $B_0/(4\pi m_i n_0)^{1/2}$. For computational convenience, the simulation actually uses quantities proportional to $k_x \hat{E}_x$ and $k_x \hat{B}_y$ instead of $\hat{E}_x$ and $\hat{B}_y$ so that the $k_x \to 0$ case can be handled by simply setting $k_x = 0$. $\hat{B}_y$ and $\hat{E}_x$ are infinite in this case, but $k_x \hat{B}_y$ and $k_x \hat{E}_x$ remain finite, and are proportional to the total parallel current density, $\hat{J}_z + (1/4\pi)\partial \hat{E}_z/\partial t$, and the charge density due to ion polarization drift, respectively. If a linear electron response of the form $\partial J_z/\partial t = (n_0 e^2/m_e)E_z$ is combined with Poisson's equation and Eqs. (1) through (6) with $W(k_z) \equiv 1$, the waves present are simply (1) linear kinetic Alfvén waves of the type described by Goertz and Boswell [1979], and (2) linear, perpendicularly-propagating ordinary light waves.

To include double layers in the simulation, we simply replace the linear electron dynamics described above with, essentially, a one-dimensional particle simulation capable of producing double layers. The particle simulation may be thought of as occurring in the dark grey region in Figs. 2 and 3. Both electrons and ions are represented as particles, as is typically required for the simulation of double layers. The particle simulation portion of the model consists of the following equations:

$$k_z^2 \hat{\phi}(k_z) = \hat{\rho}_{\text{ptcl}}(k_z) + \hat{\rho}_\perp(k_z) W(k_z), \tag{7}$$

$$\hat{E}_z(k_z) = -ik_z \hat{\phi}(k_z), \tag{8}$$

$$\hat{E}_z(k_z\!=\!0) = \langle \hat{E}_z \rangle, \tag{9}$$

$$\hat{\rho}_{\text{ptcl}}(z) = n_0 \sum_s q_s \frac{N_g}{N_{ps}} \sum_i S(z - z_{si}), \qquad (10)$$

$$\frac{dv_{zsi}}{dt} = \frac{q_s}{m_s} \hat{E}_z(z_{si}), \qquad (11)$$

$$\frac{dz_{si}}{dt} = v_{zsi}, \qquad (12)$$

$$\langle \hat{J}_z \rangle = n_0 \sum_s \frac{q_s}{N_{ps}} \sum_i v_{zsi}, \qquad (13)$$

where $\phi$ is an effective electrostatic potential for $E_z$, $\rho_{\text{ptcl}}$ is the charge density due to the simulation particles, $\rho_\perp = -(c^2/4\pi v_{A0}^2)\partial E_x/\partial x$ is the charge density due to the polarization drift of ions into particle simulation region, the subscript $s$ refers to the two species, electrons and ions, the subscript $i$ refers to the particle index within each species, $n_0$ is the mean number density, $N_g$ is the number of simulation grids in the $z$ direction, $N_{ps}$ is the number of simulation particles of species $s$, and $S(z)$ is the standard particle-in-cell shape function [Birdsall and Langdon, 1985].

We note that it is quite a fortuitous situation that all the field quantities required by the particle simulation portion of the model depend on $x$ as $\cos k_x x$. This means these quantities are maximized in the dark grey region in Figs. 2 and 3 and are relatively slowly varying in $x$ there. We can therefore ignore their $x$-dependence and treat them as one-dimensional quantities in the $z$ direction in the particle equations. This tremendously simplifies the equations and permits the particle simulation portion of the model to be one-dimensional, a huge savings in computer time.

The model equations may be simply summarized. Equations (2) through (6) with $W(k_z) = 1$ would describe the propagation of light waves in the $x$ direction and linear Alfvén waves in the $z$ direction if the $\langle \hat{J}_z \rangle$ term in Eq. (3) and the $\hat{E}_z(k_z)$ term in Eq. (5) were not present. These two terms couple these equations to the particle equations. If the two terms were calculated using a linear parallel electron inertial response and Poisson's equation Eq. (7), the light waves become ordinary light waves in a plasma, and the linear Alfvén waves become linear kinetic Alfvén waves. Instead, however, we have used full electron and ion parallel dynamics to calculate these two terms. The equations used, Eqs. (7) through (12), are, in fact, identical to conventional one-dimensional unmagnetized electrostatic particle simulation equations, except for Eq. (9) and the $\hat{\rho}_\perp(k_z)$ term in Eq. (7). These two contributions couple the particle equations to the Alfvén wave equations. It is clear, then, that two important topics can be investigated with this simulation: (1) the effect of kinetic Alfvén wave dynamics on the production and subsequent behavior of double layers, and (2) the effect of double layers on kinetic Alfvén wave propagation.

## 3. Validity of the Model

Since this model was intuitively rather than analytically formulated, it must stand up to careful scrutiny to be accepted as a viable model. Here are presented some potential criticisms of the model together with their resolutions.

1. The wavelength of the kinetic Alfvén wave along $\mathbf{B}_0$ is much, much longer than the simulation system length. How can both be represented in the simulation?

Resolution: Our system is not long enough to fit in the hundreds or thousands of double layers that typically would form in an Alfvén wavelength. There are two ways to proceed. If we choose the initial Alfvén wave to have a wavelength equal to the system length, our computer resources still allow us to have typically four, and as many as forty double layers per wavelength. Our statistics will not be as good as if we had had thousands of double layers, but it seems reasonable to expect that much could still be learned. Alternatively, we can also use an infinite-wavelength Alfvén wave ($k_z = 0$). Again, the statistics could be better, but interesting things could happen, and in fact, already have (see results).

2. Doesn't the wide separation of frequencies between the electron plasma frequency (the highest frequency followed by the model) and the Alfvén frequency necessitate the use of an unacceptably large number of timesteps?

Resolution: We can use Alfvén waves which have artificially high frequencies by taking advantage of the fact that the ion gyro-resonance does not exist in this model. It does not hurt, for example, to use Alfvén wave frequencies above what would otherwise be the ion gyrofrequency. We can further narrow the gap between the two frequencies by using a reduced mass ratio, a standard particle simulation trick. There is one danger. If the Alfvén frequency is too high, the corresponding wavelengths will become unacceptably short compared to the various length scales associated with the double layers. We require, for example, that the wavelengths of all Alfvén waves be long compared to the distance between double layers. This is enforced in the model by means of the weighting function $W(k_z)$. This function is chosen to be close to 1 for long wavelengths and near 0 for unacceptably short Alfvén wavelengths.

3. Doesn't the presence of light waves require the simulation to use very small timesteps?

Resolution: Because the light waves propagate exclusively in the $x$-direction, the only restriction is that $(k_x^2 c^2 + \omega_{pe}^2)\Delta t^2 < 1$, where $\omega_{pe}$ is the electron plasma frequency and $\Delta t$ is the timestep size. This is only slightly more restrictive than $\omega_{pe}\Delta t < 1$, a condition we have anyway for the particle simulation portion of the code.

4. Since $E_z$, $J_z$, and $\rho$ go like $\cos k_x x$, aren't the parti-

cle dynamics ridiculous a distance of $\pi/k_x$ away from the main particle simulation channel?

Resolution: We are attempting to model only the main current channel and the Alfvén waves on either side. A diffuse, featureless return current is assumed to exist outside of that. This is not the description of a perfectly sinusoidal system in the $x$-direction, but the sinusoidal model we are using should be a reasonable approximation.

5. Doesn't this model assume that double layers and kinetic Alfvén waves have the same scale length in the $x$ direction?

Resolution: It does, so it is very fortuitous that the evidence so far is that the two do have approximately the same scale length. In the acceleration region, $k_x c/\omega_{pe} \approx 1$ for the Alfvén wave, while results from a two-dimensional electrostatic particle simulation by Barnes, Hudson, and Lotko [1985] suggest that $10 k_x \rho_s \approx 1$ for double layers. The latter is consistent with Viking data [Boström, et al., 1988], which sets the lower limit on the double layer perpendicular scale length at 80 meters. It is therefore indeed fortunate that $c/\omega_{pe}$ and $10\rho_s$ are about the same in the acceleration region, both being about 1 km.

6. What about the perpendicular ion dynamics at or above the ion cyclotron frequency?

Resolution: In fact, this model does not include these effects, an important limitation. The interaction of the waves in this system with electrostatic ion-cyclotron (EIC) waves in particular is very possibly very important. We do not claim, therefore, that this model does everything; we only claim it is useful for the study of the interaction of Alfvén waves with parallel kinetic effects such as double layers.

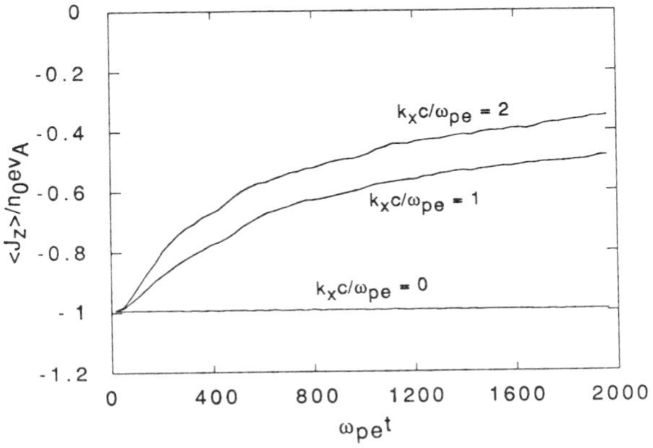


Fig. 4. The mean parallel current $\langle \hat{J}_z \rangle$ vs. time for three simulations runs with different values of $k_x$. All simulations are initialized with an electron parallel drift equal to the electron thermal velocity and $T_e/T_i = 100$. A reduced mass ratio is used: $m_i/m_e = 100$. The data has been time-averaged over the plasma period.

## 4. Early Results

Thus far, we have only examined the cases started with a constant current $\hat{J}_z(z, t = 0) =$ const. The current is initialized in the form of a drifting Maxwellian electron distribution. When the simulation is run with an infinite channel width ($k_x = 0$), we find that, apart from superimposed plasma oscillations, the parallel-averaged parallel current $\langle \hat{J}_z \rangle$ remains constant at its initial value throughout the development, evolution, and subsequent decay of double layers (Fig. 4). We emphasize that the current is not being forced artificially to be constant, but rather, the simulation is generating *self-consistently* the necessary mean electric field $\langle \hat{E}_z \rangle$ required to maintain the current. Since, for $k_x = 0$, there are no other interactions with the transverse fields, we observe standard current-driven double layers. We observe, as shown in Fig. 5, that the magnitude of the electric field increases with time as the double layers form from the nonlinear evolution of the ion acoustic instability. The mean electric field maximizes at about $t = 35\omega_{pi}^{-1} = 350\omega_{pe}^{-1}$, when double layer activity is greatest. The double layers decay beyond this time, as does the electric field, as the ions heat, stabilizing the instability. The double layers appear to behave like an anomalous resistance to the parallel current.

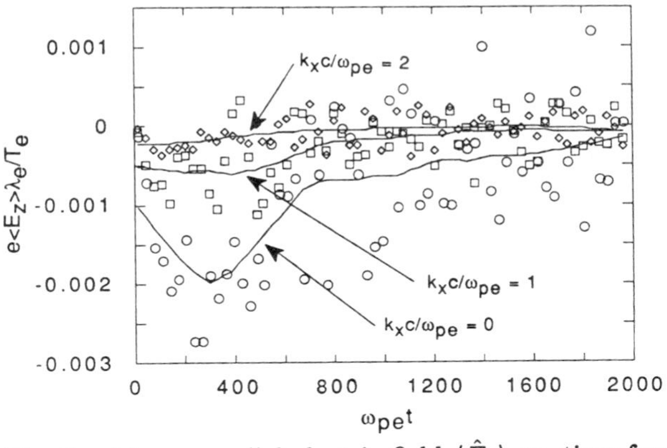


Fig. 5. Mean parallel electric field $\langle \hat{E}_z \rangle$ vs. time for the same three simulations as Fig. 4. The scatterplot points represent averages over roughly the plasma period: the circles denote data from the $k_x c/\omega_{pe} = 0$ simulation, the squares, the $k_x c/\omega_{pe} = 1$ simulation, and the diamonds, the $k_x c/\omega_{pe} = 2$ simulation. The line plots are calculated time-averaged fits to these data.

This implies that, if there is no mean ion drift in the auroral acceleration region, double layers should only be seen for a brief period of time. On the other hand, if the ions are drifting upward, for which there is some evidence, fresh cold ions will constantly be streaming in from below,

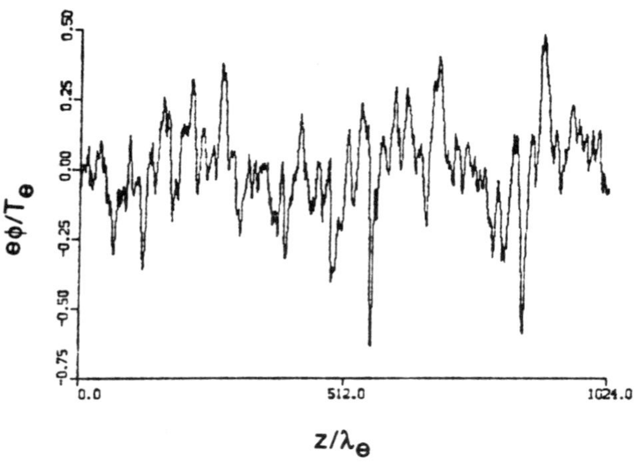


Fig. 6. The potential vs. z at $\omega_{pe}t = 500$ with $k_x c/\omega_{pe} = 2$, time-averaged over $100\omega_{pe}^{-1}$. The electron drift direction is left to right.

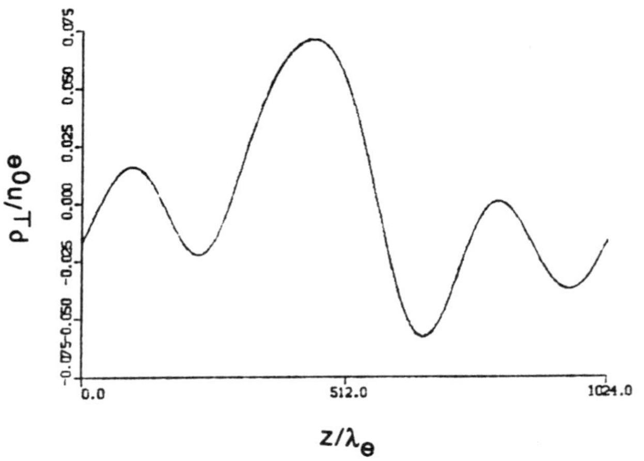


Fig. 7. The charge density due to ion polarization drift vs. z at $\omega_{pe}t = 500$ for the $k_x c/\omega_{pe} = 2$ simulation. Short wavelengths do not appear in this quantity because of the presence of the weighting function $W(k_z)$.

and then double layers should be observed within some fixed altitude range.

When the simulation is run with $k_x \neq 0$, we observe that the current is no longer constant (Fig. 4) but instead decays away. To the extent that the double layers behave as a resistivity to the parallel current ($\langle \hat{E}_z \rangle = \eta \langle \hat{J}_z \rangle$), the model equations imply

$$\frac{\partial}{\partial t}\langle \hat{B}_y \rangle = -\frac{\eta c^2 k_x^2}{4\pi}\langle \hat{B}_y \rangle; \qquad (14)$$

that is, the decay is due to magnetic diffusion. We also find that the mean parallel electric field associated with the double layers is considerably smaller in magnitude (Fig. 6).

The smaller electric field might be explained by the behavior of the ion polarization drift present in these $k_x \neq 0$ simulations. It is observed that there is a tendency for the charge $\rho_\perp$ due to ion polarization drift to be net positive just upstream of each of the double layers (cf. Figs. 6 and 7). There is a corresponding tendency for $\rho_{\mathrm{ptcls}}$ to be net negative in the same areas (not shown), but apparently the two effects do not quite cancel. The residual charge apparently cancels much of the net jump which usually accompanies double layers. In short, there appears to be a tendency for ion polarization current arriving from either side of the current channel to neutralize charge imbalances arising from the presence of nonlinear structures within the current channel. This effect is at present plausible, but unproven, because the requirement that the Alfvén wavelength be long compared to the spacing between double layers (see the previous section) is being violated in this case. Should these results hold up, the prediction is that wide V-shaped potential structures should show a net potential drop where double layers appear, while narrow structures should show a smaller potential drop due to the reduced effective resistivity presented by the double layers.

## 5. Conclusions

A novel computer simulation code has been devised to study the interaction of double layers and kinetic Alfvén waves. The simulation effectively replaces the linear electron response in the equations for linear kinetic Alfvén waves with the full electron and ion parallel dynamics capable of modeling double layers. Early results from the simulation show that the double layers are current-driven and that, in fact, the parallel current is constant in the limit of an infinitely wide current channel. A mean parallel electric field sufficient to maintain this current is simultaneously observed. When a channel of finite width is simulated, the current is seen to decay, apparently because of the effective resistivity presented in the form of double layers. The mean electric field and therefore the effective resistivity is also reduced, an effect we tentatively attribute to a charge-neutralizing mechanism associated with the polarization drift of ions into the current channel.

*Acknowledgments.* We gratefully acknowledge useful discussions with C. E. Seyler, P. Kintner, Y. Matsuda, W. M. Nevins, C. K. Birdsall, and W. W. Lee. This work was supported by Cornell University, at Lawrence Livermore National Laboratory through the Plasma Physics Research Institute, and at Dartmouth College through NASA Grant NAGW-1652.

References

Barnes, C., M. K. Hudson, and W. Lotko, Weak double layers in ion-acoustic turbulence, *Phys. Fluids, 28*, 1055, 1985.

Birdsall, C. K., and A. B. Langdon, *Plasma Physics Via Computer Simulation*, McGraw-Hill, New York, 1985.

Boström, R., G. Gustafsson, B. Holback, G. Holmgren, H. Koskinen, and P. Kintner, Characteristics of solitary waves and weak double layers in the magnetospheric plasma, *Phys. Rev. Lett., 61*, 82, 1988.

Goertz, C. K., and R. W. Boswell, Magnetosphere-ionosphere coupling, *J. Geophys. Res., 84*, 7239, 1979.

Hudson, M. K., T. L. Crystal, W. Lotko, C. Barnes, Weak double layers in the auroral ionosphere, *Laser and Particle Beams, 5*, 295, 1987.

Kim, K. Y., Theory of nonmonotonic double layers, *Phys. Fluids, 30*, 3686, 1987.

Lysak, R. L. and C. T. Dum, Dynamics of magnetosphere-ionosphere coupling including turbulent transport, *J. Geophys. Res., 88*, 365, 1983.

Otani, N. F., in *Proceedings, Second Symposium on Plasma Double Layers and Related Topics, Innsbruck, Austria, 1984*, p. 290, edited by R. Schrittwieser and G. Eder, Medienenhaber, Innsbruck, 1984.

Sato, T., and H. Okuda, Numerical simulations on ion acoustic double layers, *J. Geophys. Res., 86*, 3357, 1981.

Temerin, M., K. Cerny, W. Lotko, and F. S. Mozer, Observations of double layers and solitary waves in the auroral plasmas, *Phys. Rev. Lett., 48*, 1175, 1982.

# The Polar Cap/CUSP

# EFFECT OF A SUDDEN IMPULSE IN ELECTRON TEMPERATURE ON THE POLAR WIND: A TIME-DEPENDENT SEMI-KINETIC MODEL

C.W. Ho, J. L. Horwitz and G. R. Wilson

Department of Physics and Center for Space Plasma and Aeronomic Research,
The University of Alabama in Huntsville

*Abstract.* The time-dependent semi-kinetic model that Wilson et al. [1990] developed is used to investigate the effects on the polar wind outflow subjected to an abrupt electron temperature increase and then decrease. Immediate effects which can be seen are the formation of $H^+$ density depletions and $O^+$ density enhancements at the lower boundary of the simulation region which subsequently travelling up the flux tube. 'Kinetic phenomena' such as ion dispersion and multiple streams are observed. Steep density gradients are formed but they are smoothed out by dispersion as they propagate along the flux tube. Such effects are indications of the advantages of the kinetic model in simulating plasma outflows when details of the distribution function are desired.

## Introduction

Numerous models have been developed to study the polar outflow of ionospheric plasma along open geomagnetic field lines. There are hydrodynamic models [Banks and Holzer, 1968; Gombosi and Nagy, 1988], hydromagnetic models [Holzer et al. 1971], generalized transport models [Ganguli and Palmadesso, 1988], semi-kinetic models [Barakat and Schunk, 1983; Li et al., 1988; Wilson et al., 1990] and kinetic models [Lemaire and Scherer, 1973]. A list of different models can be found in Raitt and Schunk [1983] and Schunk [1988]. Hydrodynamic models are generally applicable to collision-dominated regions. Generalized transport equations provide a continuous transition from the collision-dominated to the collisionless regimes. Kinetic models are obtained by directly integrating the collisionless Boltzmann equation (Vlasov equation) or following individual particle trajectory using the particle-in-cell technique.

It is now generally accepted that the ionosphere is an important supplier of plasma to the higher-altitude magnetosphere [Horwitz, 1982; Chappell et al., 1987], and it is essential to understand the various ways in which the ionospheric plasma is accelerated and transported out into the magnetosphere. The vast region above the exobase ($\sim$ altitude of 4000 km) is basically a collisionless region; therefore, to study the transport of plasma in the magnetosphere, a kinetic model seems appropriate. The kinetic model also has the advantage of giving the distribution functions of the ions.

Much work has been done in studying polar plasma outflow from a kinetic point of view. The kinetic approach was first outlined by Dessler and Cloutier [1969] and developed by Lemaire and Scherer [1970, 1971]. Donahue [1971] pointed out that both the hydrodynamic and kinetic approaches are not contradictory but must be considered as complementary. Holzer et al. [1971] showed that the density and bulk velocity profiles obtained in both the hydrodynamic and kinetic models are not much different if the same boundary conditions were chosen at an appropriate exospheric level. Lemaire [1972] showed that low altitude hydrodynamic solutions to the polar wind could be fitted to high altitude kinetic solutions applicable in the collisionless regime. Barakat and Schunk [1984] have examined the steady-state outflow of $O^+$ when hot isothermal electrons are present in a semi-kinetic approach in which the electrons are taken to obey the Boltzmann relation. Li et al. [1988] have followed a similar approach in investigating the effect of significant ion heating at the ionospheric base in such a semi-kinetic model, but have improved the model so as to eliminate potential jumps and allow anisotropic ion temperatures at the lower boundary. They have shown that raising the ion temperatures results in a large $O^+$ flux, which is not very sensitive to the electron temperature at very large ion temperatures. A small-scale time-dependent kinetic model was developed [Singh and Schunk 1982; 1983] to study transient expansion in the polar wind. Two- and three-dimensional non self-consistent kinetic models of ion transport have also been developed [Horwitz, 1984, 1986a, 1986b, 1987; Horwitz and Lockwood, 1985; Horwitz et al., 1985, 1986] to examine the population of the polar cap magnetosphere and tail lobe by $O^+$ ions from the cleft ion fountain.

Recently, Wilson et al. [1990] have developed a semi-kinetic model which represents the first time-dependent, self-consistent, kinetic plasma outflow model. Such a model incorporates the contribution of higher velocity moments without the equation closure problem of the time-dependent generalized transport models. It can also provide a knowledge of the ion distribution function at points along the flux tube at different times and would allow us to study the development of the distribution of ion velocities seen in the outflowing polar plasma.

In this paper, we will use the same model as Wilson et al. [1990] to study the effect on outflowing polar plasma subjected to an abrupt increase and decrease in electron temperature. This situation may occur as a convecting flux tube passes through the cleft region where the electrons presumably have a higher temperature. We will make full use of the advantages of the kinetic model by plotting the ion velocity distribution functions.

## Time-Dependent Self-Consistent Kinetic Model

The model used in this paper is self-consistent in the sense that the ambipolar electric field is calculated at each time-step by assuming a Boltzmann relation between the electron density and the ambipolar electrostatic potential

Modeling Magnetospheric Plasma Processes
Geophysical Monograph 62

$$n_e(r) = n_{eo}\exp(e\phi/kT_e) \qquad (1)$$

Here $n_e(r)$ and $n_{eo}$ are the electron density at geocentric distance $r$ and at the lower boundary, $T_e$ is the electron temperature (assumed to be constant along the flux tube), $e$ and $k$ are the electronic charge and the Boltzmann constant, and $\phi$ is the ambipolar electrostatic potential. The electron density, $n_e(r)$, is obtained by assuming quasi-neutrality, i.e. the total ion and electron densities are the same everywhere. The ion density is calculated directly by counting the number of simulation particles in each cell. The length of the flux tube is divided into 200 equal cells. Each simulation ion represents a large number of real ions to make the computer program efficient.

We simulate the outflow of $H^+$ and $O^+$ ions along a geomagnetic flux tube from a lower boundary at 1.7 $R_e$ to an altitude of 8 $R_e$. Within this altitude range, the earth's magnetic field is assumed to follow that of a dipole field, that is $|\vec{B}| \sim 1/r^3$ where $r$ is the geocentric distance along the field line. The ions from the topside ionosphere are assumed to enter the base of the flux tube with velocities distributed functionally as the upgoing half of a drifting bi-Maxwellian. The parameters for the drifting bi-Maxwellian used for $H^+$ and $O^+$ respectively are 100 and 50 $cm^{-3}$ for the density; 20 and 0 km/s for the drift speed, and 3000° K for the parallel and perpendicular temperatures. Each simulation ion is injected into the base of the flux tube with a parallel ($v_\parallel$) and perpendicular ($v_\perp$) velocity component drawn randomly from the assumed distribution. As it is moved along $\vec{B}$ it is acted upon by the gravitational, ambipolar electric and magnetic mirror forces, while the effects of collisions are ignored. At each time step the ion's position is advanced by $v_\parallel \Delta t$, and its parallel velocity is advanced by $\frac{F_{\parallel tot}}{m_s}\Delta t$ ($s$ indicates ion species, and $F_{\parallel tot}$ is the sum of the parallel components of the above mentioned forces). These two processes are staggered by half a time step to improve the accuracy of the time integration. The perpendicular velocity is advanced in such a way so as to conserve the first adiabatic invariant. The computer code follows the guiding center motion of a large number of simulation particles as they move along a field line. With new particle positions a new total density profile is found, and from it the electric potential as a function of altitude using equation (1). The ions are then moved forward in the next time step in the same way and the process is continued until the desired time is reached. The distribution function $f_s(v_\parallel, v_\perp, r)$, as well as various bulk parameters such as drift speed and temperature which can be obtained from it, can be found at any time step by counting simulation particles in $v_\parallel$, $v_\perp$ and $r$ bins.

## Effect Of An Electron Temperature Impulse

In this paper, we will demonstrate some of the results the present model is able to produce by investigating the effect on the polar wind of an increase and then decrease in the electron temperature. Starting with an initially empty flux tube and using the parameters for the injected distribution indicated above, we first ran the code until a steady-state solution was achieved. After this we raised the electron temperature ($T_e$) from 3000° K to 50000° K linearly over a period of 100 seconds, maintained it at 50000° K for 300 seconds, and then allowed $T_e$ to decline back to 3000° K in the next 100 seconds. By 500 seconds into the simulation the electron temperature had returned to a value of 3000° K. Such a variation of the electron temperature may be similar to that experienced by a flux tube convecting through the polar cusp region. Since the model assumes a constant electron temperature along the flux tube we assume that at some point the flux tube connects to a region of hot plasma and because of rapid thermal conduction the electron temperature all along the flux tube quickly rises to some value intermediate to that of the hot plasma and the ionosphere. Figure 1 shows the time history of the electron temperature.

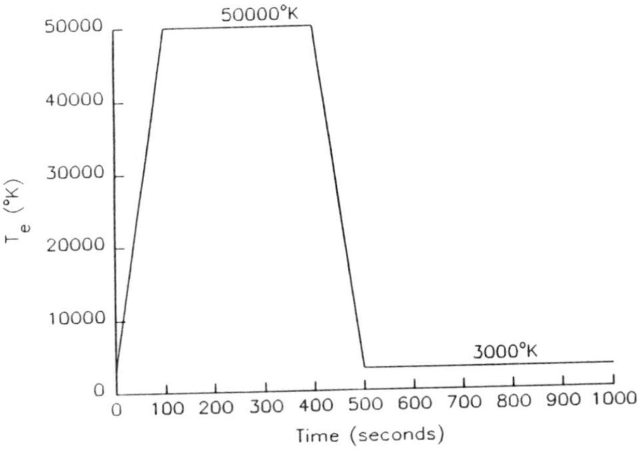


Fig. 1. The time history of the electron temperature ($T_e$) used in the simulation case reported here.

Figures 2 and 3 show a series of partial phase space plots for $H^+$ and $O^+$ ions at the indicated times. These are obtained by taking a given value of the function $f_s$ at time $t$ and integrating with respect to $v_\perp$. The result is plotted versus $r$ and $v_\parallel$. The time given in each panel is the time in seconds after the point when the electron temperature began its increase. The first panel in each figure is the initial distribution at $T_e$ = 3000° K when the polar wind is at its asymptotic steady state. By the time the electron temperature has returned to 3000° K (500 s), an indent structure is seen in the $H^+$ phase plots which indicates a decrease in the number of ions with lower velocities. By the same time a protrusion in the $O^+$ velocity distribution has developed, indicating an increase in $O^+$ ions with high velocities. Figure 2 shows how the $H^+$ parallel velocity distribution evolves with altitude and time. As the electron temperature is increased, an increase in the electric field accelerates the $H^+$ ions at the lower boundary, leaving a depletion of ions near the base. The electric field also accelerates the heavier and thus the slower $O^+$ ions (Figure 3). During the 300 seconds of high electron temperature, a density depletion of $H^+$ develops at the lower boundary, while a density enhancement for $O^+$ develops at the same place (see Figure 4). Since these ions enter with a spectrum of velocities, the density depletion and enhancement disperse as they travel up the flux tube, causing the indent and protruding structure in the $H^+$ and $O^+$ distribution functions to slant and become sharper and narrower (Figure 2 and 3). The dispersion effects can also be seen in Figure 4 where the density depletion and enhancement become smoother as the ions travel up the flux tube. It takes less than an hour for this irregular $H^+$ depletion structure to completely escape the flux tube and the velocity distribution to regain its original shape. The $O^+$ distribution took 15 hours to return to its original state for $T_e$ = 3000° K, due to the slower $O^+$ thermal speed.

During the dispersion of the depletion and enhancement structures there exist in certain altitude regions of the flux tube two populations of each species of ions (for example, $H^+$ at 800 s between 4.2 and 5 Re; and $O^+$ at 3500 s between 6 and 7 Re. See Figure 2 and 3). Such multi-populations produce a large 'pseudo-temperature' such as the apparent increase in $O^+$ parallel temperature as shown in Figure 5. This is due to the association of the concept of temperature with

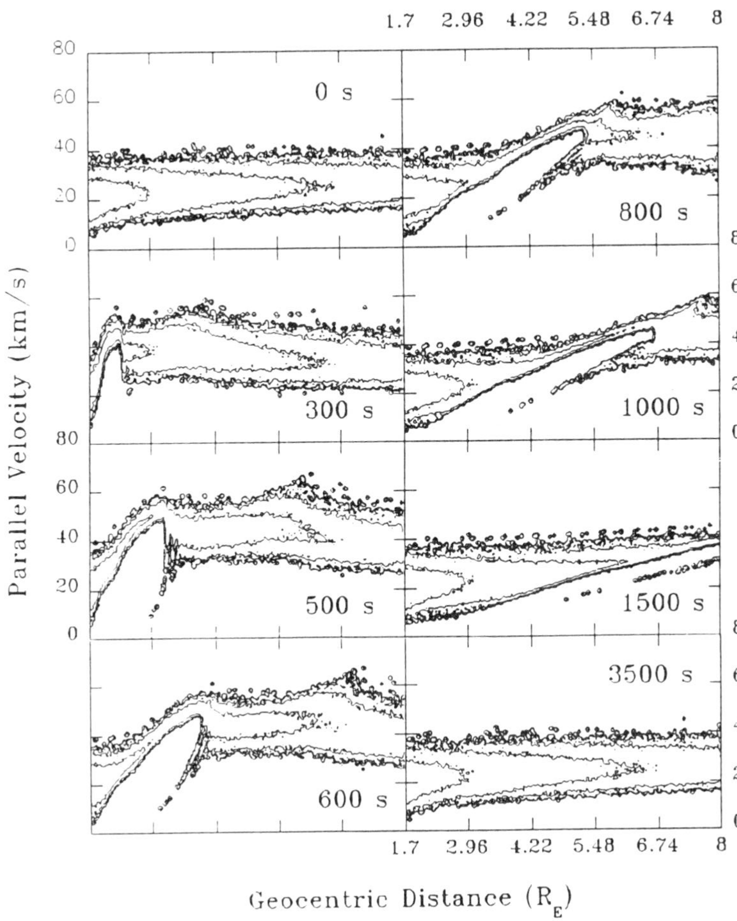


Fig. 2. Partial phase space contour plots for the $H^+$ ions at the indicated times. The first panel gives the asymptotic solution when $T_e$ is held at a value of 3000° K for a long time and represents the starting conditions. The contour levels are separated by one order of magnitude and the outermost, lowest level contour has a value of $10^{-4}$ s/cm$^4$.

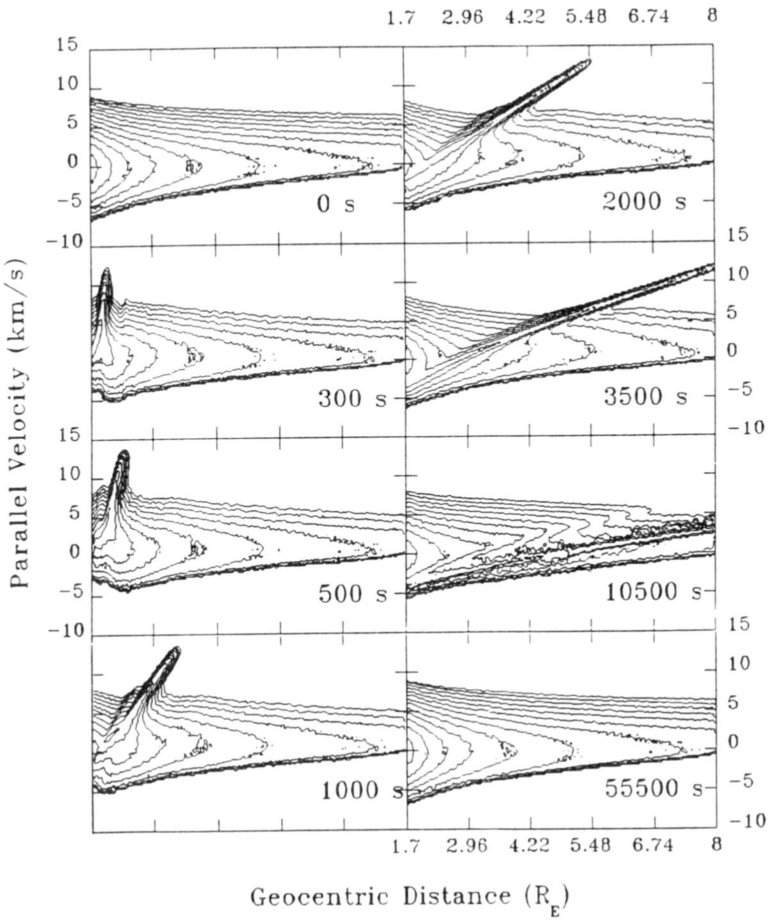


Fig. 3. Partial phase space contour plots for the $O^+$ with the same format as figure 2. The contour levels are separated by one order of magnitude and the outermost, lowest level contour has a value of $10^{-9}$ s/cm$^4$.

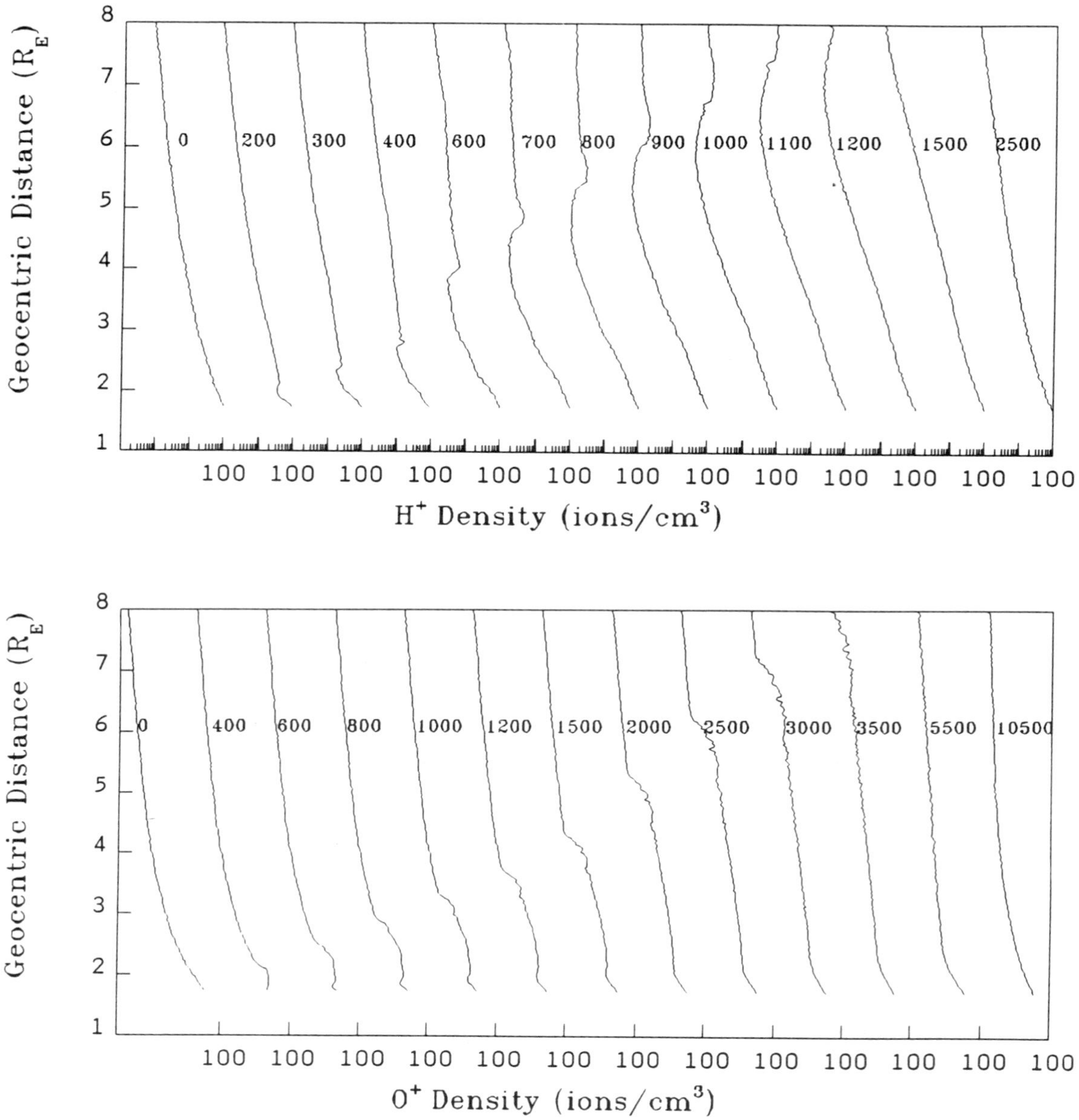


Fig. 4. A time series of $H^+$ and $O^+$ density profiles (times shown are in seconds).

the 'spread' of the velocity distribution. This is a disadvantage of all hydrodynamic models when the plasma is treated as a streaming fluid with no distinction between separate ion beams, and only bulk (or averaged) quantities are considered. The decrease in $H^+$ temperature in Figure 5 is a true decrease due to adiabatic expansion. The 'cool' $H^+$ travels up the flux tube much faster than the $O^+$ temperature enhancement.

Singh and Schunk [1985] have shown that artificial bumps or cavities in the ion density may produce shocks. In this paper, we have shown that the polar wind produces similar kinds of bumps and cavities in the density profile as it enters a region of short duration of high electron temperature. This kind of perturbation in the density profile is smoothed out by dispersion as it propagates along the flux tube.

## Conclusion

A semi-kinetic model shows that when a steady state polar wind flux tube encounters a sudden increase in electron temperature, it develops near its base a depletion in the $H^+$ density and an enhancement in the $O^+$ density. These structures subsequently travel up the flux tube, but because of dispersion the sharp bends in the density profiles in both ions are eventually smoothed out. Two streams of ions with different bulk drift speeds can be seen at certain altitudes. These give rise to an apparent increase in temperature.

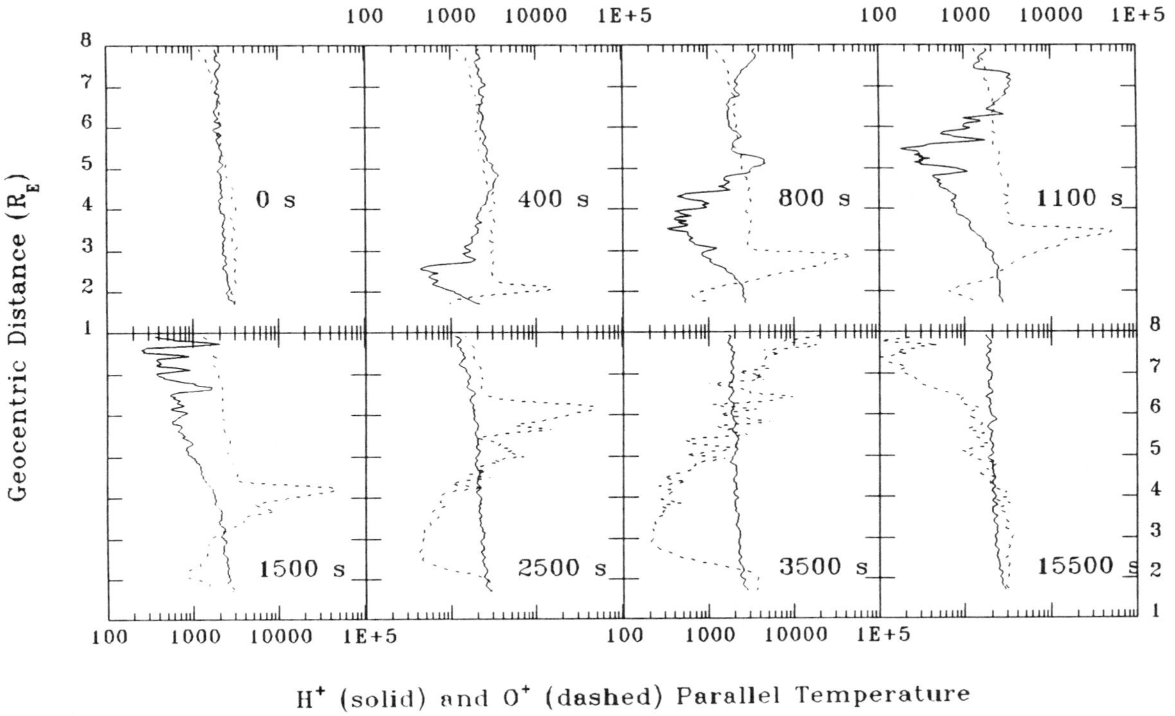


Fig. 5. A time series of $H^+$ (solid) and $O^+$ (dashed) parallel temperature profiles.

*Acknowledgements* This work was supported under NASA grant NAGW-1550 to the The University of Alabama in Huntsville.

References


Banks, P. M. and T. E. Holzer, The Polar Wind, *J. Geophys. Res.* *73*, 6846, 1968.

Barakat, A. R., and R. W. Schunk, $O^+$ ions in the polar wind, *J. Geophys. Res.* *88*, 7887, 1983.

Barakat, A. R., and R. W. Schunk, Effect of hot electrons on the polar wind, *J. Geophys. Res.* *89*, 9771, 1984.

Chappell, C. R., T. E. Moore, and J. H. Waite, Jr., The ionosphere as a fully adequate source of plasma for the Earth's magnetosphere, *J. Geophys. Res.*, *92*, 5896, 1987.

Dessler, A. J. and P. A. Cloutier, Discussion of the letter by Peter M. Banks and Thomas E. Holzer, 'The polar wind'. J. Geophys. Res., *74*, 3730, 1969.

Donahue, T. M., Polar ion flow : Wind or breeze? *Rev. Geophys. Space Phys.*, *9*, 1, 1971.

Ganguli S. B., and P. Palmadesso, Generalized fluid model of plasma outflow processes in the topside ionosphere, *Adv. Space Res.*, *8*, 69, 1988.

Gombosi T. I., and A. F. Nagy, Time-dependent polar wind Modeling, *Adv. Space Res.*, *8*, 59,1988.

Holzer, T. E., J. A. Fedder and P. M. Banks, A comparison of kinetic and hydrodynamic models of an expanding ion-exosphere, *J. Geophys. Res.*, *76*, 2453, 1971.

Horwitz, J. L., The ionosphere as a source for magnetospheric ions, *Revs. Geophys. Space Phys.*, *20*, 929, 1982.

Horwitz, J. L., Features of ion trajectories in the polar magnetosphere, *Geophys. Res. Lett.*, *11*, 1111, 1984.

Horwitz, J. L. and M. Lockwood, The cleft ion fountain: A two-dimensional kinetic model, J. Geophys. Res., *90*, 9749, 1985.

Horwitz, J. L., J. H. Waite, Jr., and T. E. Moore, Supersonic ion outflows in the polar magnetosphere via the geomagnetic spectrometer, *Geophys. Res. Lett.*, *12*, 757, 1985.

Horwitz, J. L., Velocity-filter mechanism for ion bowl distributions (bimodal conics), J. Geophys. Res., *91*, 4513, 1986a.

Horwitz, J. L., The tail lobe ion spectrometer J. Geophys. Res., *91*, 5689, 1986b.

Horwitz, J. L., M. Lockwood, J. H. Waite, Jr., T. E. Moore, C. R. Chappell, and M. O. Chandler, Transport of accelerated low-energy ions in the polar magnetosphere, in *Ion Acceleration in the Magnetosphere and Ionosphere*, Geophys. Monogr. Ser., *38*, edited by T. Chang, p. 56, AGU, Washington, D.C., 1986.

Horwitz, J. L., Parabolic heavy ion flow in the polar magnetosphere, J. Geophys. Res., *92*, 175, 1987.

Lemaire, J. and M. Scherer, Model of the polar ion-exosphere. *Planet. Space Sci.*, *18*, 103, 1970.

Lemaire, J. and M. Scherer, Simple model for an ion-exosphere in an open magnetic field. *Physics Fluids*, *14*, 1683, 1971.

Lemaire J., $O^+$, $H^+$ and $He^+$ ion distributions in a new polar wind model, *J. Atmos. Terr. Phys.*, *34*, 1647, 1972.

Lemaire, J., and M. Scherer, Kinetic models of the solar and polar winds, *Rev. Geophys. Space Phys.*, *11*, 427, 1973.

Li, P., G. R. Wilson, J.L. Horwitz, and T.E. Moore, Effect of mid-altitude ion heating on ion outflow at polar latitudes, *J. Geophys. Res.*, *93*, 9753, 1988.

Raitt, W. J., and R. W. Schunk, Composition and characteristics of the polar wind, in *Energetic Ion Composition in the Earth's Magnetosphere*, edited by R. G. Johnson, pp. 99-141, Terra Scientific Publishing Company, Tokyo, 1983.

Schunk R. W., Polar wind tutorial, in *SPI Conference Proceedings and Reprint Series, Vol. 8*, Scientific Publishers, Inc., Cambridge, MA 02139, pp. 81-134, 1988.

Singh, N., and R. W. Schunk, Numerical calculations relevant to the initial expansion of the polar wind, *J. Geophys. Res.*, *87*, 9154, 1982.

Singh, N., and R. W. Schunk, Numerical simulations of counter-streaming plasma and their relevance to interhemispheric flows, *J. Geophys. Res.*, *88*, 7867, 1983.

Singh, N., and R. W. Schunk, Temporal behavior of density perturbations in the polar wind, *J. Geophys. Res.*, *90*, 6487, 1985.

Wilson, G. R., C. W. Ho, J. L. Horwitz, N. Singh, and T. E. Moore, A new kinetic model for time-dependent polar plasma outflow: Initial results, *Geophys. Res. Lett.*, *17*, 263, 1990.

# Two-Spacecraft Charged Particle Observations Interpreted in Terms of Electrostatic Potential Drops Along Polar Cap Field Lines

C. J. POLLOCK,[1] C. R. CHAPPELL,[2] J. L. HORWITZ,[3] and J. D. WINNINGHAM[4]

[1] *Space Science Laboratory, NASA/MSFC, AL*
[2] *Associate Director for Science, NASA/MSFC, AL*
[3] *Department of Physics and Center for Space Plasma and Aeronomic Research The University of Alabama in Huntsville, Huntsville, AL*
[4] *Space Science & Instrumentation Division, Southwest Research Institute, P.O. Drawer 28510, San Antonio, TX*

We are studying the possible occurrence and magnitude of field-aligned electrostatic potential drops over the ionospheric polar caps. For this purpose, signatures in upgoing and downgoing photoelectrons, obtained in the topside ionosphere using the low altitude plasma instrument (LAPI) on the Dynamics Explorer 2 (DE 2), spacecraft are analyzed [*Winningham and Gurgiolo*, 1982]. These data are compared with positive ion data obtained at high altitude using the retarding ion mass spectrometer (RIMS) on the DE 1 spacecraft. Data were selected from intervals when DE 1 and DE 2 were approximately connected along polar cap magnetic field lines and when upflowing $O^+$ beams were observed in the RIMS data. We present here one case in which the comparison of data from the two DE spacecraft is quite favorable regarding its interpretation in terms of a field-aligned potential drop.

## INTRODUCTION

Ultraviolet solar radiation incident on the terrestrial atmosphere produces prolific ionization and a distribution of photoelectrons. Observations of such photoelectron distributions in the topside ionosphere by *Winningham and Heikkila* [1974] and *Winningham and Gurgiolo* [1982] have led these authors to conclude that large-scale electrostatic potential drops exist above the spacecraft (ISIS 1 and DE 2, respectively), along polar magnetic field lines connected to regions of photoelectron production. Potential drops of up to nearly 100 volts were inferred, based on the similarity in shape and magnitude of upgoing and downgoing photoelectron spectra. Since the electron observations were made in the topside ionosphere, above the photoelectron source region, no downgoing photoelectrons are expected to be observed on the open polar cap field lines in the absence of such an electrostatic reflection mechanism. The observation of such downgoing electrons, up to some cutoff energy per charge, is interpreted as evidence of a field-aligned potential drop, equal in magnitude to the energy per charge cutoff, along field lines above the spacecraft.

Modeling Magnetospheric Plasma Processes
Geophysical Monograph 62


If such electrostatic structures exist above the polar caps, the implications, regarding an ionospheric source of heavy plasma for the magnetosphere, are very important. Just as the electrons are reflected back toward Earth, any positive ions entering the potential drop region from below would be accelerated outward, away from the ionosphere. Such a mechanism would contribute strongly in determining both the nature and strength of the ionospheric plasma source. Other mechanisms for producing heavy ion outflow above the polar cap ionosphere include transverse ion heating in the cleft ionosphere, ejection by the magnetic mirror force and subsequent ion convection over the polar cap (e.g., *Horwitz and Lockwood,* 1985), and large-scale upward ambipolar electric fields associated with elevated electron temperature (e.g., *Barakat and Schunk,* 1984).

The RIMS on DE 1 often measures field-aligned $O^+$ streams flowing upward along polar geomagnetic field lines at high altitude. These streams are of ionospheric origin and have flow energies which are typically several eV, but which often exceed 10 eV. If the electrostatic potential drops described above exist, they could well be responsible for the acceleration of the $O^+$ ions observed by RIMS. In what follows, we present such high-altitude RIMS observations, along with near simultaneous low-altitude LAPI observations from day 81295, which are mutually consistent with the existence of field-aligned electrostatic potential drops of 10 volts and larger on polar cap field lines.

## OBSERVATIONS

We have examined data from intervals during which both DE spacecraft were on northern polar cap field lines in late 1981. Intervals were chosen so that upflowing $O^+$ streams were observed in the DE 1 RIMS data. Since this is polar cap data in the early winter hemisphere, the photoelectron signatures are not as clear as they might be since the ionosphere was not well illuminated. However, there are intervals when DE 2 was poleward of the auroral zone when photoelectron production was taking place below the spacecraft. Meanwhile, DE 1 was crossing the polar cap at high altitude and observing field-aligned $O^+$ streams flowing out of the ionosphere.

Figure 1 shows the ground tracks of the DE 1 and DE 2 spacecraft on a polar IL-MLT grid, as well as the plasma convection signature (courtesy of R. Heelis, University of Texas at Dallas), as determined using the ion drift meter (IDM) along the DE 2 track for a portion of the pass on day 81295. In this case, the UT coincidence and magnetic conjugacy were quite good. Although the UT coincidence is limited to a small region of space just north of the cleft, the spatial conjugacy is quite good, accepting UT differences of on the order of 20 minutes. Both spacecraft were moving poleward from the morning side on these passes. The UT associated with various spacecraft footprint positions are indicated. Note that the high-altitude DE 1 spacecraft crosses field lines, between 70° and 80° IL, which appear, based on the IDM convection signature, to be in the downstream flow from the (~0955-0958) position of the low-altitude DE 2 spacecraft.

In Figure 2, we present low-altitude electron flux observations for this event in grey scale spectrogram format. LAPI energy-time electron flux spectra are presented in the top panel for downgoing pitch angles and in the bottom panel for upgoing pitch angles. The entire DE 2 polar cap pass is shown in this figure, beginning at 0954 UT, when the spacecraft was approaching the morningside magnetospheric cleft region from the south. The cleft region was crossed between 0955 and 0956 UT, as is evidenced by the intense, soft electron precipitation which is characteristic of this ionospheric footprint of the low-latitude boundary layer. Photoelectron production was occurring below the spacecraft both north and south of the cleft (as evidenced by the upgoing photoelectrons below ~100 eV) until about 1006 UT when DE 2 moved across the terminator into the nightside. North of the cleft region, between 0956 and 1000 UT, downgoing electrons, in an energy range up to several tens of eV and less, as well as more energetic precipitating electrons are seen in the top panel, Figure 2. The more energetic precipitating electrons are in the several hundred eV range, extend almost all the way across the polar cap, and are identified as polar rain.

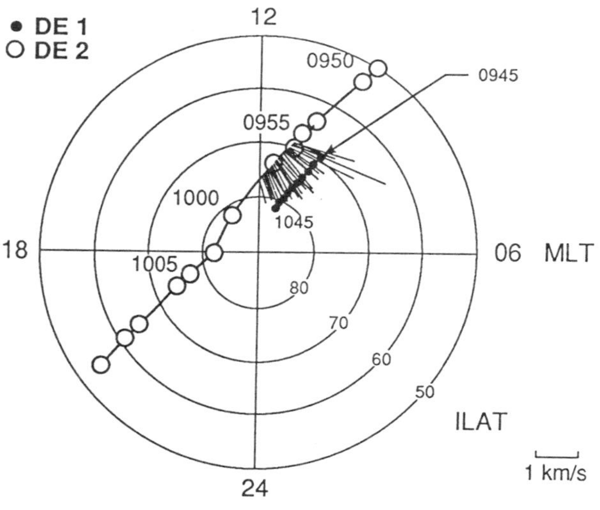


Fig. 1. Northern Hemisphere ground tracks of the DE 1 (filled circles) and DE 2 (open circles) spacecraft are shown in polar IL-MLT coordinates for the pass under study near 1000 UT, on day 81295. The spacecraft positions at various times are indicated. Ion drift signatures from a portion of the DE 2 pass are overlaid.

We take the point of view here that the lower-energy downgoing electrons are photoelectrons created in the ionosphere below which, after having moved upward past the spacecraft, have been reflected by field-aligned electrostatic potential structures located above the spacecraft. If this is correct, a potential drop of up to several tens of volts is inferred. A detailed examination of the characteristics of the low-energy downgoing electrons reveals significant temporal or spatial variation in their intensity and spectral shape, particularly within the energy range between 10 eV and 100 eV, during the interval 0956-1000 UT. Between 1001 and 1006 UT, photoelectron production was still taking place below the DE 2 spacecraft, as evidenced by the upgoing electrons shown in the bottom panel of Figure 2, however, only those in the lowest-energy channels (below ~10 eV) appear in the downgoing channel, providing an upper limit of the order of 10 V for any potential drop above DE 2 within this region.

In Figure 3, we show plots of upgoing (plusses) and downgoing (circles) electron distribution functions versus electron energy, from 1 to 100 eV. We show eight (denoted a-h) consecutive 60-second average distribution functions of upgoing and downgoing electrons, beginning at 0956 UT. The downgoing and upgoing electron spectra in the first four panels resemble each other more closely than do the downgoing and upgoing spectra in the last four panels. The distinction is subtle, however. It is most pronounced in the energy range between 10 and ~40 eV. In this energy range, downgoing electrons observed after 1000 UT are characterized by distribution functions which fall more steeply with energy than either their upgoing counterparts or the downgoing electron distributions measured prior to 1000 UT. There is, however, a decrease in the intensity in the electrons moving downward as compared to those moving upward during the interval covered by the first four panels of Figure 3. This intensity depletion may be accounted for by a spatial or temporal dependence in the potential structure above the spacecraft, allowing some electrons to escape, while reflecting others.

In Figures 4 and 5, we present RIMS $O^+$ data obtained on DE 1, at high altitude over the polar cap, in good coincidence with the DE 2 electron data presented above. The presentation in Figure 4 shows the $O^+$ data in RIMS energy-time (top panel) and spin-time (bottom panel) format, respectively. The bottom panel in Figure 4 shows the $O^+$ data in RIMS radial head spin-time format. In this display, the instrument count rate is plotted in grey scale, versus UT (and orbital parameters) along the abcissa and the radial head viewing angle (with respect to ram) along the ordinate. The dashed (dotted) line running from left to

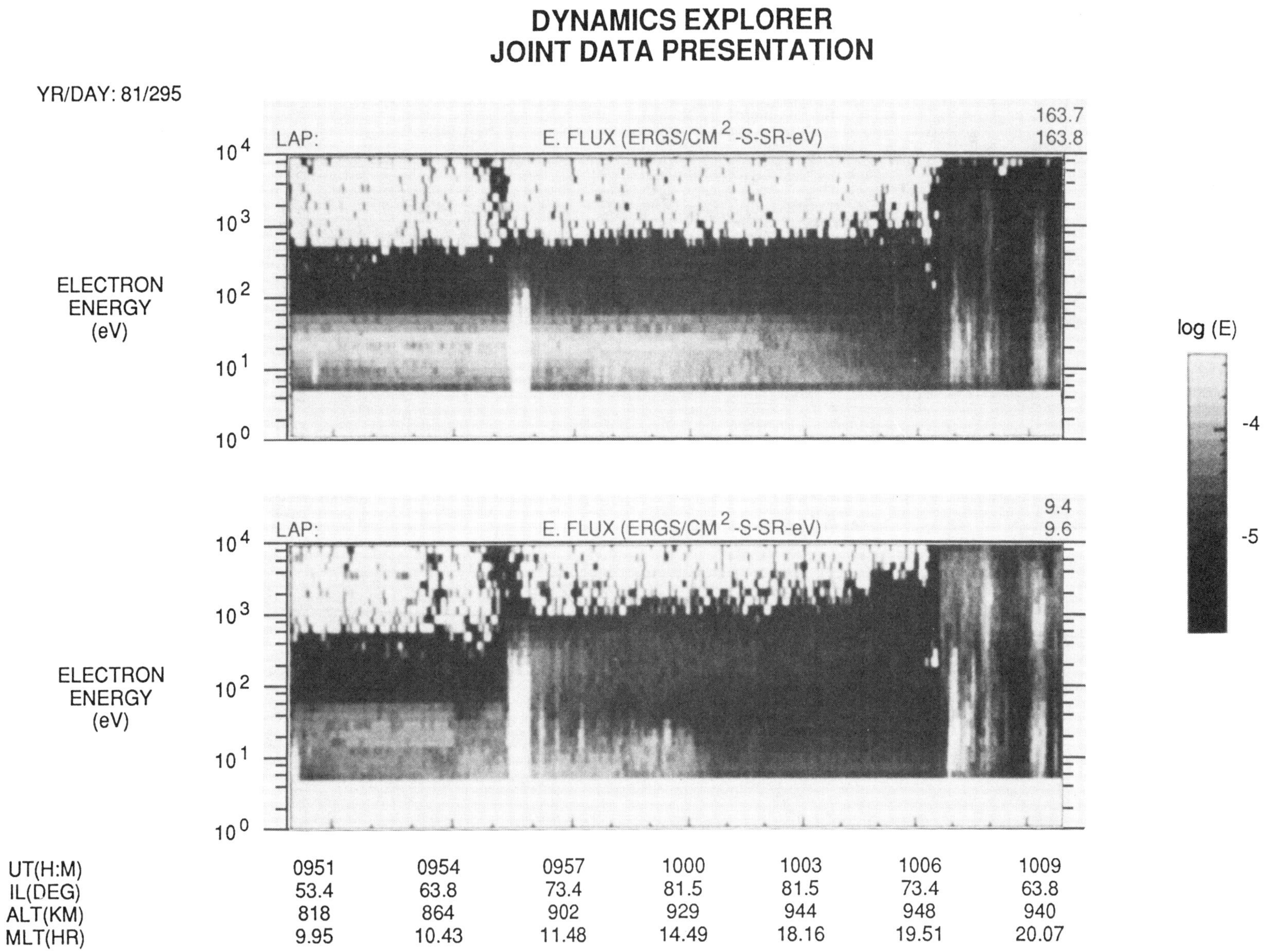


Fig. 2. Low-altitude electron observations, from the LAPI on the DE 2 spacecraft are shown in grey scale spectrogram format. Panels a and b display upgoing (pitch angle = 164°) and downgoing (pitch angle = 9°) electrons, respectively. In both panels, UT (along with various orbital parameters) is plotted along the abcissa, electron energy is plotted along the ordinate, and the value of the electron phase space density is plotted in grey scale.

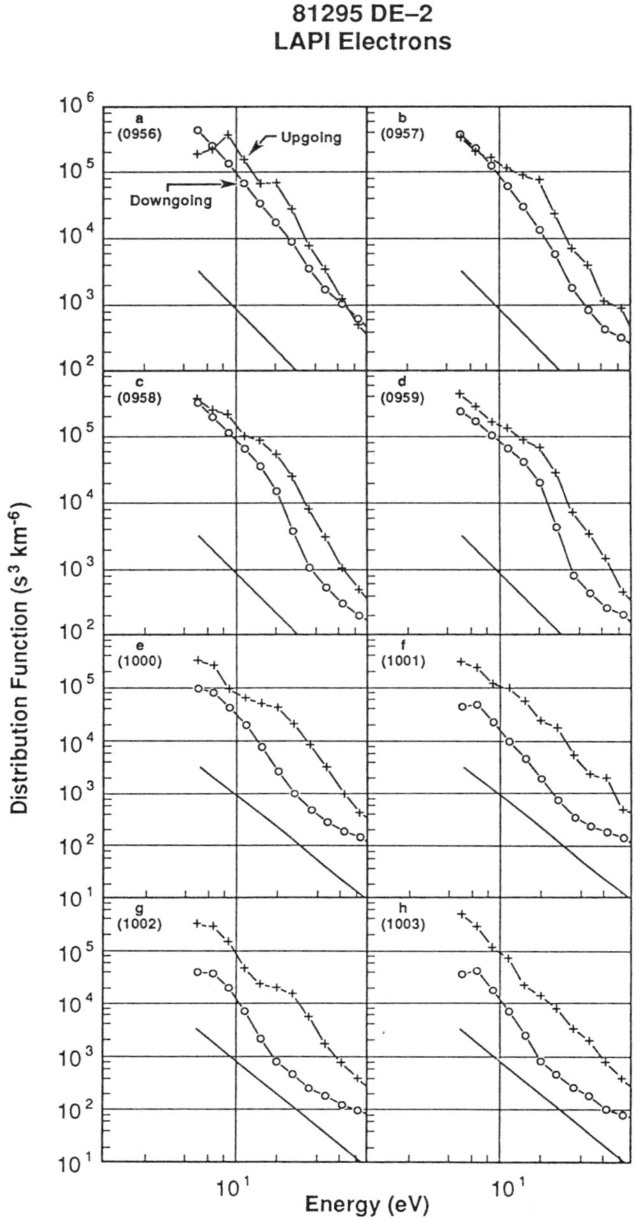


Fig. 3. Eight (denoted a-h) consecutive pairs of 60-sec average electron energy spectra, are shown, beginning at 0956 UT. Plotted is the phase space density along the ordinate and electron energy along the abcissa. Upgoing (plusses) and downgoing (circles) electron distributions are displayed in each panel.

right indicates spin phases at which RIMS viewed along the positive (negative) magnetic field direction, viewing ions moving upward (downward) at pitch angles of near 180° (0°). Upward streaming $O^+$ ions are seen, beginning near 1000 UT and extending until approximately 1050 UT, when the data end. The top panel shows the radial head $O^+$ data in RIMS energy-time format. In this case, the voltage applied to the retarding potential analyzer (RPA) is listed along the ordinate, rather than the viewing angle. Nominally, all ions with energy per charge greater than the applied RPA voltage are passed into the instrument. The magnetic mass spectrometer has a finite differential energy bandpass, however, which for $O^+$ is about 16 eV wide. This gives rise to a differential signature for the more energetic $O^+$ ions. This differential signature is present in the data in Figure 4, most evidently near 1015 UT. The energetics of the streaming $O^+$ ions may be gleaned from the RIMS energy-time signature shown in Figure 4 by noting the RPA voltage at which the RIMS count rates are shut off. This is typically near 10 volts, but extends to several tens of volts near 1015 UT.

Figure 5, contains line plots of the RIMS radial head $O^+$ count rate versus retarding potential. Data shown in panels a through e are 2-minute averages at selected times as indicated in the figure. The RIMS differential response to $O^+$, which was described above, is evident in the data at 1004 and 1015 UT. Vertical dashed lines indicate estimates of the ion flow energy. It can be seen that this energy extends from about 30 eV near 1015 UT to 7 eV near 1026 and 1048 UT. If these ions have been accelerated out of the ionosphere by an electrostatic potential drop, the data imply a voltage drop of between less than 10 volts to several tens of volts, in excellent agreement with the conclusions drawn from the LAPI electron data presented above. In this high-altitude region of low electron density, spacecraft charging often complicates interpretation of low-energy charged particle spectra. Positive spacecraft charging of several volts is common under such circumstances. The presence of such charging on DE 1 at the time the above measurements were obtained would add to the size of the potential drops inferred above, based on the RIMS ion measurements.

## DISCUSSION

We have presented correlated charged particle measurements from two spacecraft which are suggestive of the existence of magnetic field-aligned electric potential drops existing at intermediate altitudes on geomagnetic polar cap field lines. We believe this is the first time such polar cap measurements, showing simultaneously the effect of field-aligned electrostatic structure on charged particle distrubutions from above and below, have been presented. If these potential drops exist and are widespread over the magnetospheric polar caps, they represent an important mechanism for supplying the magnetosphere with ionospheric plasma. A scenario which involves feeding such a potential drop region with ions upwelling from the dayside auroral zone and subsequent acceleration of these ions to energies above the gravitational escape energy provides a plausible large-scale ionospheric plasma escape mechanism. Such upward potential drops could eject the otherwise gravitationally bound heavy ion flows characteristic of the cleft ion fountain [*Lockwood et al.*, 1985; *Horwitz and Lockwood*, 1985; *Horwitz*, 1987].

An alternative mechanism for producing the RIMS $O^+$ signature shown in Figure 3 involves transverse heating of the ions at some lower altitude, presumably within the dayside cleft, and subsequent conversion of perpendicular to parallel energy under the influence of the diverging geomagnetic field. Such a

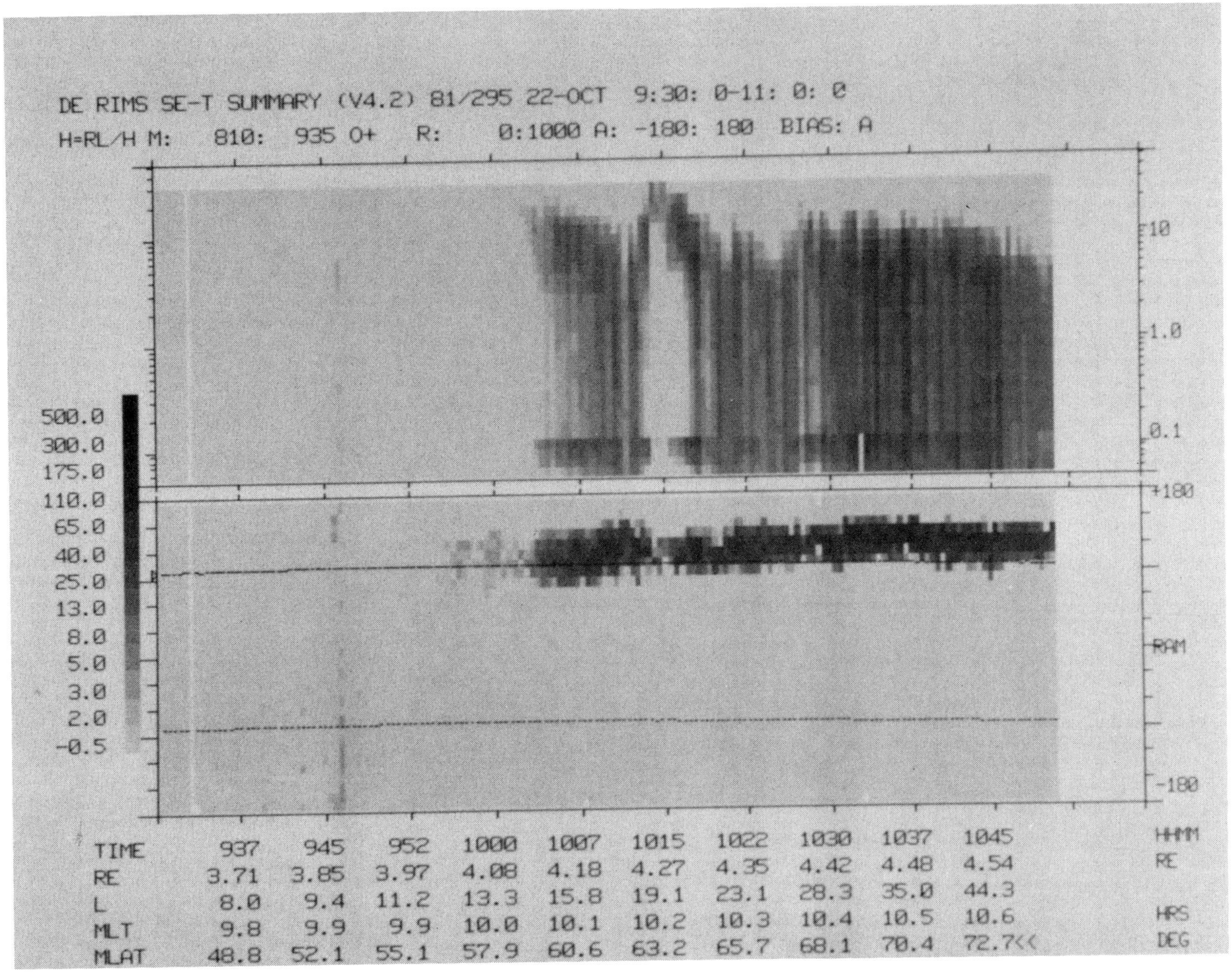


Fig. 4. RIMS $O^+$ data are presented in spectrogram format. In the top and bottom panels are shown RIMS energy-time and spin-time spectrograms, respectively. For both panels, UT and various orbital parameters are plotted along the abcissa. The applied RPA voltage is shown plotted along the ordinate for the energy-time spectrogram (top panel), while for the spin-time spectrogram (bottom panel) the ordinate displays the RIMS look direction spin angle. The grey scale denotes RIMS $O^+$ count rate plotted in grey scale. Viewing directions along the positive and negative geomagnetic field are indicated by dashed and dotted lines running from left to right in the bottom panel.

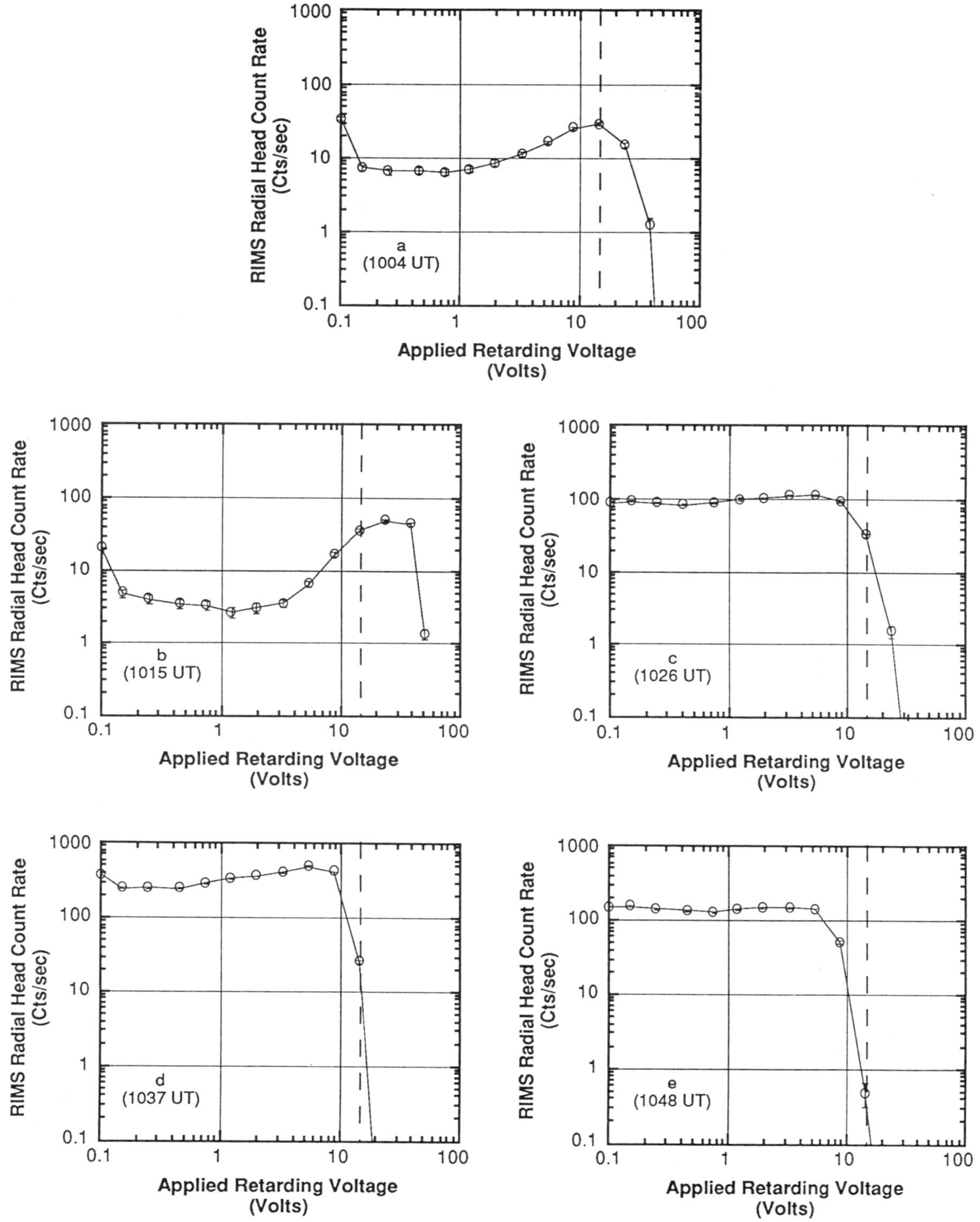


Fig. 5. Line plots of the average RIMS $O^+$ count rate versus applied RPA voltage for selected 2-minute intervals centered at the times shown. Vertical dashed lines represent estimates of the $O^+$ drift energy.

mechanism is believed to contribute strongly to the total ionospheric outflow, although it would be difficult, based on the data at hand, to distinguish which has produced the observed RIMS signature in this event. It is true that a potential drop which is large scale and homogeneous over the polar cap would produce a more steady energetic ion outflow than that seen in Figure 3 near 1015 UT. However, the electron data (Figure 2) give indication that the inferred potential drop above DE 2 is structured in space or time. Such structure could easily give rise to transient signatures at high altitude, as observed in this case with RIMS. Also, while it cannot be ruled out that the energetic $O^+$ ions seen in the RIMS data at 1015 UT may have originated within the morningside cleft and convected to the position of DE 1, without passing close to field lines connected to DE 2, the convection signature at DE 2 indicates drift directly toward the footprint of the high-altitude spacecraft. However, electric field data measured on DE 1 (courtesy of D. Gurnett, University of Iowa) show a burst of enhanced electric field at the time (~1015 UT) that the most energetic ions are observed on RIMS. The electric field is not only enhanced at this time, but is rotated with respect to the electric field direction both before and after the burst. A convection jet from the direction of the morningside cleft region may be inferred from this data. Thus this plasma may not have passed near the DE 2 track.

The core of the arguments made above lies in the observation (Figure 2) of electrons, in the 10-100 eV range, in the downgoing LAPI channel while on nominally open field lines. The most reasonable source for these is the photoelectron population moving upward, past the spacecraft, from below. Interpretation, in this context, of the spectra shown in Figure 3 is subtle, however, because neither the spectral similarities or the differences are striking. We believe that this is at least partially due to the variable nature of the downgoing electron flux. The 60-second averages we have used wash out distinct spectral features which evolve on a faster time scale in the DE 2 frame of reference. In Figure 6, we present four more LAPI electron spectra. Upgoing (plusses) and downgoing (circles) electron spectra are shown in higher time resolution during the intervals 0957:13-0957:18 UT (panel a) and 0957:32-0957:37 UT (panel b). The similarity of the upgoing and downgoing spectra shown in Figure *6a* is indeed striking, indicating a common source for these particles. The upgoing and downgoing electron distributions shown in Figure *6b* diverge above ~10 eV. These two pair of spectra were obtained only 20 seconds apart, yet both near 0957 UT, in the heart of the region where we have inferred that a large-scale electrostatic potential drop exists. If these potential drops exist above the DE 2 spacecraft, they appear, from the photoelectron data, to be structured, rather than homogeneous. This would account for the fact that the upflowing ions observed with RIMS at high altitude are not uniformly energetic, but sporadically so.

As to the establishment and maintenance of such potential drops, this is an open question. *Winningham and Gurgiolo* [1982] have suggested that the situation is analogous to the sheath formed around a probe which is immersed in a plasma and illuminated with a flux of extreme ultraviolet photons. Such a probe will charge positively, with respect to the surrounding plasma in order to prevent continual loss of negative charge in the form of photoelectrons. This sheath, however is expected to have spatial dimension which scales with the Debye length which, in the extreme case of $T_e = 10^6$ K and $n_e = 1$ cm$^{-3}$, is 200 m. Clearly, in cases like that considered here, the

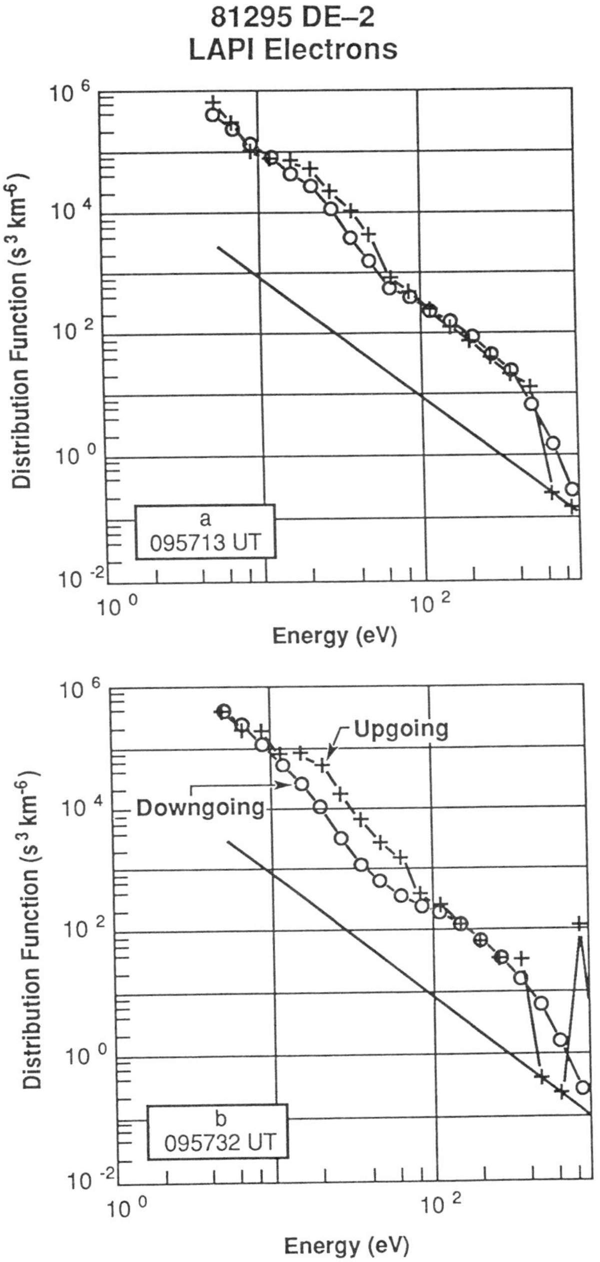


Fig. 6. Two (denoted a and b) pairs of 5-sec average electron energy spectra are shown centered near 0957:15 and 0957:35 UT. Plotted is the phase space density along the ordinate and electron energy along the abcissa. Upgoing (plusses) and downgoing (circles) electron distributions are displayed in each panel.

photoelectron production is taking place some hundreds of km below the DE 2 spacecraft, while the electron reflection is taking place above the spacecraft. It is noteworthy in this context that the apparent reflection of photoelectrons occurs in the presence of more energetic precipitating electrons, which we have identified as polar rain. The subject of the interaction of an ionospheric electron population, such as photoelectrons, with an incident flux of polar rain is one which merits further study.

## CONCLUSIONS

The LAPI electron and RIMS ion data shown above are not conclusive regarding the existence or effect of polar cap field-aligned electrostatic potential drops. However, at least in the case of day 81295, the data are strongly suggestive that such potential drops occur. They may further be responsible for accelerating the observed $O^+$ ions, to create outflowing field-aligned streams. In this case, any potential drop which may be inferred seems to be associated with the region just north of the dayside auroral zone. However, the proximity of the terminator and the presence of precipitating particles associated with the dayside auroral zone render such an association to be speculative.

Clearly, an excellent orbital configuration for the comparisons, such as these, of RIMS and LAPI data would be a noon-midnight orbit, with DE 1 at high altitude over the summer polar cap. Future work will involve searching for and analyzing such cases in the same manner as done in this work.

*Acknowledgments.* This research was supported by the Dynamics Explorer Program of the National Aeronautics and Space Administration, the United States National Research Council and, at the University of Alabama in Huntsville, by NASA grants NAG8-134, NAG8-822, and NAGW-1554. The authors are indebted to the engineering and science staff of the University of Texas at Dallas and the RIMS team at Marshall Space Flight Center. The LAPI data was accessed at MSFC using the Space Physics Analysis Network of the National Space Science Data Center and the Southwest Data Display/Analysis System, developed at Southwest Research Institute using the support of NASA Grant NAS5-33031. The authors are grateful to R. Heelis at the University of Texas at Dallas and to D. Gurnett at the University of Iowa for providing data used in preparing this work.

## REFERENCES

Barakat, A. R., and R. W. Schunk, J. Geophys. Res., 89, 9771, 1984.

Horwitz, J. L., J. Geophys. Res., 92, 175, 1987.

Horwitz, J. L., and M. Lockwood, J. Geophys. Res., 90, 9749, 1985.

Lockwood, M., M. O. Chandler, J. L. Horwitz, J. H. Waite, Jr., T. E. Moore, and C. R. Chappell, J. Geophys. Res., 90, 9736, 1985.

Winningham, J. D., and C. Gurgiolo, Geophys. Res. Lett., 9, 977, 1982.

Winningham, J. D., and W. J. Heikkila, J. Geophys. Res., 79, 949, 1974.

# ELECTRON DENSITY AND TEMPERATURE IN THE CUSP AND POLAR CAP REGIONS : CONTRIBUTIONS FROM THE WAVE AND PARTICLE EXPERIMENTS ON VIKING

**P.M.E. Décréau[1], S.Perraut[2], H.de Féraudy[2], L. Matson[3], O. Randriamboarison[1]**

Abstract. Two active experiments, a relaxation sounder and a mutual impedance probe on board the Viking satellite, are used to determine the plasma density in the cusp and polar regions up to 13,500km, the apogee of the spacecraft. The results are compared with density estimations derived from the upper frequency cut-off of auroral hiss. The latter estimations, valid most of the time in the polar cap, are generally not a right indicator of the total density value in the cusp region. Density data from a set of several passes indicate median values of about 40cm-3 in the cusp proper, 25cm-3 in the boundary cusp and 4cm-3 in the polar cap, for altitudes close to the apogee. The global electron temperature, estimated for few events by the combined wave and particle instrumentation, is much higher in the cusp (of the order of several 10eV) than in the polar cap (less than 1eV). In both regions, the active wave experiments are useful tools, which complement the informations provided by the particle analyzers and the passive wave measurements.

## 1. Introduction

Due to the difficulty of making in situ measurements in a tenuous medium where spacecraft charging occurs, little is known about the magnetospheric plasma outside the plasmasphere at the very low end of the energy scale [Hultqvist, 1985]. Wave methods can reveal the presence of the colder particles, but only in measuring the bulk characteristics of the plasma, mainly the total density. In the regions of high latitudes, where there is a lot of activity, the plasma is not only tenuous but also mostly outside of equilibrium. There, the emphasis has often been put on the study of structures (shape of the particle velocity distribution, gradients, dynamics), with the limitations in energy inherent to the instrumentation. It is still appealing to search for the more elusive low energy population, be it only through bulk characteristics, for at least two reasons : (1) the knowledge of the total density is a necessary step to interpret the wave activity, (one example being the generation of AKR [Louarn et al., 1990]) and, (2) the importance of the ionospheric source in the distribution of the core plasma inside the magnetosphere has seen a renewed attention lately [Horwitz, 1987; Chappell et al., 1987]; the survey of density is a way to check the importance of that source.

In the polar cap, measurements at high altitudes (above about 3500km) were quasi absent or at least unpublished before Persoon et al. [1983] presented statistical studies, showing interesting behavior of electron densities obtained via auroral hiss signatures detected on board DE 1. The Viking satellite is well equipped to explore the global plasma features inside polar regions. The active wave instruments carried by this spacecraft are an important complement to the measurement of passive emissions, which may at times be absent and at other times misleading [Perraut at al., 1990]. Under favorable conditions, the Mutual Impedance probe yields an estimation of the electron temperature. Moreover, the Viking orbit has interesting characteristics. It explores the polar cap at altitudes between about 7000km and 13500km (the apogee). In addition, an important number (>150) of passes cross the polar cleft. This is an interesting feature, as density data in the polar cleft are particularly scarce. The purpose of this paper is to present an initial view of the global plasma characteristics measured by Viking at the high latitudes. In the next section, the characteristics of the Viking measurements relevant to our analysis are summarized. In section 3 and 4, we present respectively the cleft and polar cap features. The last section gives a summary and prospectives.

[1] Laboratoire de Physique et Chimie de l'Environnement et Université d'Orléans, Orléans, France
[2] Centre de Recherches en Physique de l'Environnement, Issy les Moulineaux, France
[3] Swedish Institute of Space Physics, University of Umea, Umea, Sweden

Modeling Magnetospheric Plasma Processes
Geophysical Monograph 62

## 2. Characteristics of the measurements.

The data presented hereafter are derived mainly from the high frequency wave experiment package carried by Viking [Bahnsen at al.,1988]. The density is derived from the direct or indirect measurement of the plasma frequency in the natural emissions present in the medium, or actively triggered by the relaxation sounder (RS) or the mutual impedance experiment (MI) [Décréau et al.,1987]. The relaxation sounder stimulates the plasma at its natural resonance frequencies and provides a line spectrum. The measurement of the plasma parameters is derived from the location of these lines. The plasma frequency $F_{pe}$ is identified directly, or derived from the active $F_{uh}$ signature combined with the value of the electron gyrofrequency ($F_{ce}$) measured via the onboard magnetometer. The mutual impedance probe measures the impedance of a quadrupole immersed in the plasma. Its output is a continuous spectrum with broad resonances and antiresonances. The diagnosis of the plasma relies on a fit between experimental curves: the measured variations of the impedance versus frequency, and theoretical ones. The latter are calculated under the hypothesis of a Maxwellian velocity distribution, as a function of two parameters to be adjusted, the plasma frequency and the Larmor radius, $r_L$ and two known ones (measured by the magnetometer), the amplitude and the direction of the static magnetic field. In practice, the frequency position of the resonances are governed mainly by the $F_{pe}$ value, and the amplitude of the resonances by the $r_L$ value. The estimation of those two parameters provides respectively the electron density and temperature of the plasma.

A comparison of natural and stimulated emissions seen on Viking is shown in Figure 1. The active instruments are operated roughly every two minutes, in a cycle lasting about 16s. The resonance sounder sweep shows up as a vertical line in the spectrogram. It is followed by MI measurements in the time periods indicated by an enhanced signal level above 300kHz. Both instruments exhibit a significant intensification of the signal at the upper hybrid frequency $F_{uh}$. The salient feature in the passive spectrograms is the presence of intense natural emissions in the lower frequency range (below $F_{ce}$), displaying a clear cutoff outside the cusp. This noise is interpreted to be auroral hiss propagating in the whistler mode, its upper frequency cutoff being located at the plasma frequency for most cases taken at high latitudes [Persoon et al., 1983], here at the end of the pass.

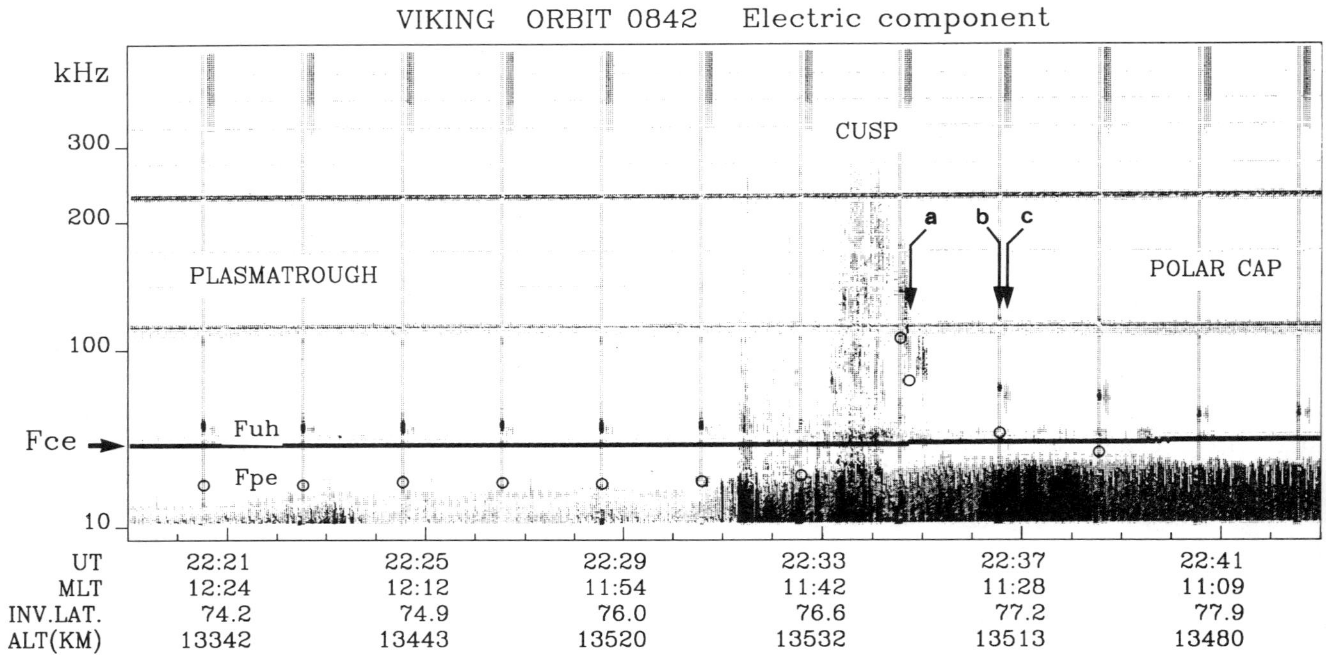


Fig. 1. Frequency time spectrogram of the electric field observed in a crossing of the cusp. The boundary cusp is met in the time periods 22:31:00-22:33:15 and 22:34:45-22:43:00 UT, and the cusp "proper" in the period 22:33:15-22:34:45 UT. The $F_{uh}$ resonance is triggered mostly by the active wave experiments. The derived $F_{pe}$ frequency position is indicated by open circles for each operatiof of the sounder, and also for one operation of the MI probe (at time a, which is, like times b and c, detailed in Figure 2).

It is not the case inside the cusp. There, one also observes bursts of broad band noise up to high frequencies, [Pottelette et al., 1990]. Several signal enhancements appear to follow the continuity of the actively triggered $F_{uh}$ time variations. The cutoff observed below the gyrofrequency could be related to the plasma frequency of the cold population, $F_{pc}$, as suggested for example by Tokar and Gary [1984]. In the plasmatrough, one sees also a discrepancy between the active and passive plasma signatures.

The detailed spectrograms plotted in Figure 2 are recorded by the high frequency wave experiments at three time periods (a,b,c) pointed in Figure 1. The natural wave spectra, dominated by the hiss emissions, remain flat in the vicinity of the $F_{uh}$ frequency. On the contrary, the operation of the two active instruments within a same short time slot (Figure 2b-c) provides a whole set of independent plasma signatures: the frequency position of the $F_{uh}$ resonance triggered by the sounder, the frequency position of the Bernstein mode emissions $F_{q2}$ and $F_{q3}$, also in the RS spectrum, and a broader resonance, of characteristic shape, displayed in the MI output. Furthermore, the comparison of the MI spectra to theoretical calculations illustrates how the MI resonances intensifies with decreasing $r_L$ values (here in case c, with respect to case a). The principles of the active wave techniques (Perraut et al., 1990, and references herein) are confirmed by the redundant information present in most Viking measurements, and illustrated in Figure 2. In such cases, it is reasonable to interpret the $F_{pe}$ signature as being related to the total electron density. As a consequence, the $F_{pc}$ hiss signature in the cusp region is clearly not $F_{pe}$ dependent, and must be interpreted differently than in the polar cap region.

It is interesting to compare plasma parameters derived from the wave observations to the density and energy estimations provided by the particle instrumentation. In the absence of ideal tools for a complete diagnosis of the ion and electron populations, we have used the electron spectrometer, which measures particles down to low energies (about 10eV) preferably to the ion instruments, limited generally to measurements above 40eV, and involving a more difficult plasma analysis at low energy. Above the threshold of 10eV, we can calculate moments of the electron distribution function, assuming stable plasma conditions for half a spin period. Because of the uncertainty in the efficiency of the particle counters, it is difficult to compare the absolute value of their density estimation to the total density $N_e$ derived from $F_{pe}$. However, in following and comparing the time variations of both data sets, one can get a good indication of how the particles are distributed in energy, namely above and under the 10eV level. Another useful information, in this area, is given by the apparent temperature measured by the MI probe. This parameter, i.e. the temperature value deduced from the amplitude of the characteristic

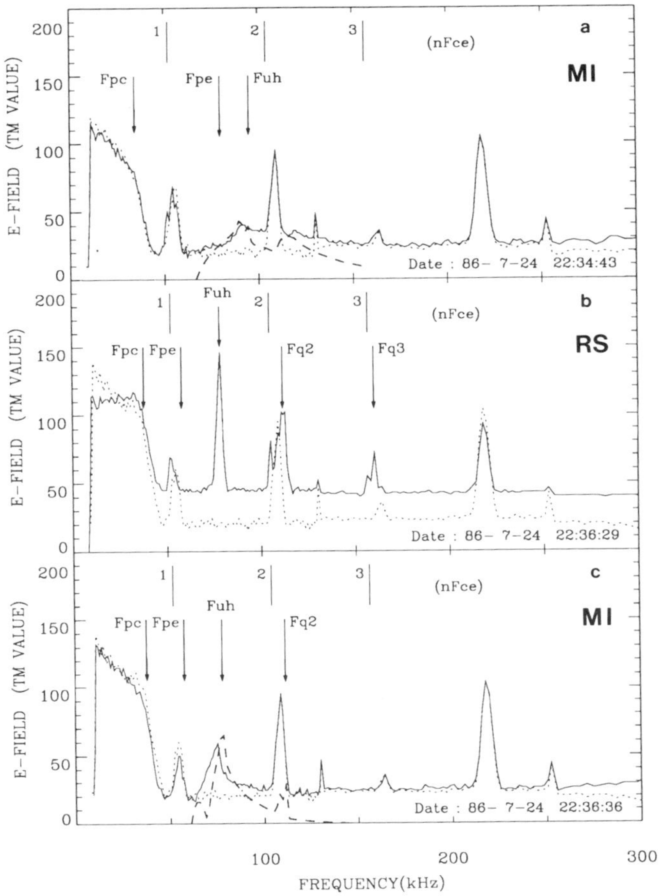


Fig.2. Frequency spectra of the electric field signal measured during operations of the active wave instruments (solid lines), as compared to the frequency spectra of natural emissions (dotted lines) recorded 3s apart. Spurious lines are seen at 54kHz and harmonics. The identification of $F_{pe}$, derived from the frequency position of the $F_{uh}$ resonance, yield to density values of $N_e = 76cm^{-3}$ (case a) and $N_e = 42cm^{-3}$ (case b-c), to be compared to a cold density $N_c$ of about $20cm^{-3}$, in an interpretation of $F_{pc}$ as a cold plasma signature. The theoretical calculations (dashed lines) compared to the MI observed impedance variations yield to an estimation of the Larmor radius of 8m in case a, 5m in case c, corresponding to apparent temperatures $kT_e$ of respectively 39eV and 12.5eV. Note that the theoretical calculations are available for input parameters varying in steps, which explains apparent unperfect fittings. The derived estimations use interpolations.

resonances observed in the MI spectrum, is effectively the Maxwellian temperature in case of a single well behaved population. In the event of a mixing of a cold and a hot component, it is an unknown function of the characteristics of the two populations. Qualitatively, an apparent MI temperature below the measured mean energy of the electron spectrometer reveals the presence of a cold (below 10eV) population. At the very least, this has been shown theoretically in the case of unmagnetized plasmas.

In conclusion, the active wave techniques complement the particle instruments at low energies. They also yield a more reliable estimation of the total density than a passive wave analysis. Another advantage of these techniques, shown in an example in Décréau et al. [1987], is that the auroral hiss emissions are seldom visible at very high latitudes. In such regions, active operations are able to reveal features which are not triggered naturally. However, the RS and MI instruments on Viking have their own limitations [Perraut et al., 1990]. A first limitation is that they were operated discontinuously with a time recurrence of 2 minutes. A second limitation is that the active signals can be blurred below intense passive emissions, like AKR radiations, or below the few intense spurious lines present in the electronic noise. In the former case, the hiss signature is generally still available for the density diagnosis. The sensitivity of the MI instrument is also often insufficient in a tenuous environment. Lastly, the instrument cannot detect plasma frequencies below a minimum value, fixed by the lower end of the frequency range for hiss signatures, by the filter bandwidth and magnetometer resolution for active wave measurements, since we need to discriminate $F_{uh}$ from $F_{ce}$. One gets a minimum measurable density between 1 and 2 particles $cm^{-3}$. The precision of the measurement increases with the measured densities. It is poor at the lower end of the measurable range.

TABLE 1. Time intervals of the data presented, and corresponding magnetic indices.

| Orbit Number | Date Y M D | Cusp Start-stop times(hh:mn) | Polar Cap Start-stop times(hh:mn) | Km | ΣKm Preceding 12 hours |
|---|---|---|---|---|---|
| 176 | 86 03 25 | | 20:45 - 20:51 | 5+ | 15 |
| 213 | 86 04 01 | | 14:24 - 14:40 | 1+ | 8- |
| 214 | 86 04 01 | | 18:34 - 19:02 | 3 | 9- |
| 225 | 86 04 03 | | 18:31 - 19:11 | 3 | 9 |
| 237 | 86 04 05 | | 22:46 - 23:13 | 3+ | 9+ |
| 272 | 86 04 12 | 08:16 - 08:32 | | 3- | 10+ |
| 340 | 86 04 24 | | 16:20 - 17:10 | 2+ | 10+ |
| 351 | 86 04 26 | | 16:20 - 16:58 | 0+ | 6 |
| 363 | 86 04 28 | | 20:30 - 21:12 | 1 | 8 |
| 381 | 86 05 02 | 03:26 - 03:52 | | 3 | 9- |
| 384 | 86 05 02 | | 16:17 - 17:11 | 5 | 17- |
| 416 | 86 05 08 | | 12:04 - 13:02 | 3 | 7+ |
| 417 | 86 05 08 | | 16:18 - 17:08 | 2- | 8 |
| 458 | 86 05 16 | 03:23 - 03:4 | | 1 | 4+ |
| 502 | 86 05 24 | 03:12 - 03:42 | | 2- | 7 |
| 792 | 86 07 15 | 20:22 - 20:27 | | 1- | 4- |
| 825 | 86 07 21 | 20:20 - 20:25 | 20:51 - 21:05 | 12- | 7 |
| 842 | 86 07 24 | 22:31 - 22:42 | 23:04 - 23:10 | 6 | 11 |
| 893 | 86 08 03 | 05:21 - 05:26 | 05:57 - 06:12 | 2+ | 5 |
| 944 | 86 08 12 | 11:48 - 11:57 | 12:06 - 12:29 | 2+ | 9+ |
| 1103 | 86 09 10 | 09:40 - 09:45 | 09:54 - 10:12 | 2 | 6+ |

## 3. The Polar cusp

The polar cusps are the two narrow throats of dayside geomagnetic flux tubes directly opened to the solar plasma streams. In the observations, they are revealed by the presence of soft electrons poured from the magnetosheath down to ionospheric altitudes (see Johnstone [1985] and references herein). On the other hand, one consequence of the coupling of the magnetosheath to the ionospheric cusp is to enhance the escape of ionospheric particles in this area [Lockwood et al., 1985], which can consequently contribute to the total density observed in the middle altitude range.

In order to present total electron density values measured in the cusp, we have used an independent signature of the crossing of this zone, on a case to case basis. On Viking, Kremser and Lundin [1988], have analyzed particle spectra to characterize the orbit according to the different high latitude regions explored. They distinguish the cusp "proper", with high fluxes of isotropic low-energy electrons and very low ion acceleration below Viking, the active boundary cusp with lower magnetosheath plasma densities and moderate acceleration of electrons and ions, and the mantle cusp, connected to the plasma mantle. The mapping of those different cusps in the longitude, invariant latitude plane [Lundin, 1988], results in a pattern similar to what is expected for the global cusp region [Formisano, 1980]. Of the several (ten) orbits we have analyzed, we selected the time periods attached to, respectively, the cusp "proper" and the boundary cusp, recognized by the particle signatures. The corresponding dates and geomagnetic indices are listed in Table 1, and the tracks of the cusp "proper" are shown on Figure 3. They are all located inside the statistical cusp "proper" area, or very close to it.

As apparent in Figure 1, the total density inside the cusp "proper" is higher than in the boundary cusp. The density versus time profiles plotted in Figure 4 show two other examples of the transition between those regions. This transition is clear for the "meridian" orbit (orbit 944, at almost constant local time). For the "parallel" orbit (orbit 458, at almost constant invariant latitude), the transition is smoother. In that particular case, there is actually no clear boundary, in terms of $N_e$ values, to the cusp "proper" for the afternoon side of the pass. Such a behavior is consistent with the elongated shape of the overall cusp section. The two other "parallel" orbits shown in Figure 3 (numbered 381 and 502) exhibit indeed a large cusp "proper" size in terms of MLT coverage. The Viking orbits were on the other hand such that they did not cover a large altitude range in the cusp region. Figure 5 displays

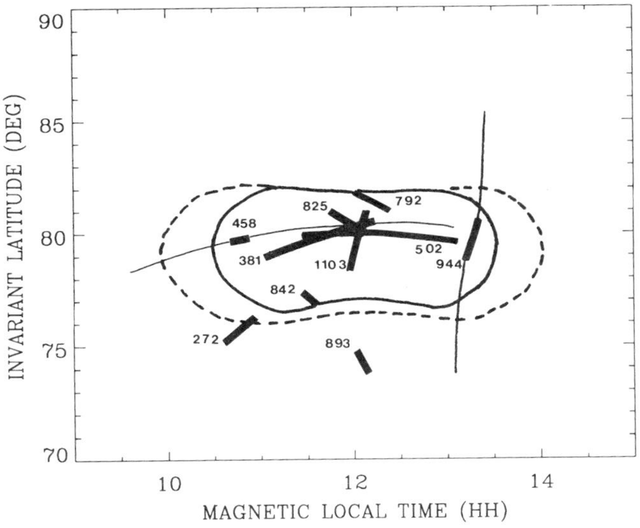


Fig.3. Schematics showing Viking trajectories in the cusp "proper" projected onto an invariant latitude, local time plot displaying the cusp"proper" (solid line) and the boundary cusp positions [Lundin, 1988].

suprathermal density given by the electron spectrometer follows the variations of the 'hot' electron density measured by the waves (in sustracting from the total density a cold density assumed to be indicated by the hiss signature), rather than the total density variations. Those results support indeed the view of a significant cold population of ionospheric origin, more visible when the spacecraft moves away from the cusp "proper", and here also more abundant at its polar edge. In the frame of that interpretation, it is tempting to relate the increase of cold electrons versus invariant latitude to the presence of a cleft ion fountain (see ion observations described in André et al., [1988]), known to spring up from the cusp toward the polar cap [Lockwood et al., 1985]. Such ions could carry with them cold electrons issued from the ionospheric reservoir. Moreover, the cusp densities measured by DE 1 in the suprathermal range by Burch [1985], are statistically lower (less than 10cm$^{-3}$) than the total density values recorded by Viking. This may simply be due to a geometrical effect, as the cusp narrows at Viking altitudes, with respect to its dimension at DE 1 altitudes (16,000 to 26,000km for the data set considered). Nevertheless, one has to consider the ionospheric contribution in analyzing a scattered plot of densities values versus geocentric distances for the ten orbits analyzed. It is clear that more data have to be collected to show an eventual trend of density variations versus radial distance. Nevertheless, those plots confirm a higher density in the cusp "proper" (with a median value of 41.5cm-3), than in the boundary cusp (with a median value of 25.3cm-3).

The main part of the total density measured in the cusp "proper" at Viking altitudes seems to come from the magnetosheath. Erlandson et al. [1988], have shown in one cusp event (displayed in Figure 1) that the amplitude of the diamagnetic depression observed was consistent with the hypothesis of the total density due to a massive injection of magnetosheath plasma. If one uses the hiss signature to estimate the density of the cold ionospheric population, one finds indeed a cold to total density ratio of about 25% at 22:34:43UT (Figure 2-a), and still lower, 10% or below in the center of this event [Perraut et al.,1990]. In the boundary cusp, however, that ratio will increase if the cold population remains stable for the whole event. For that particular orbit, the cold to total density value reaches 50% at 22:36:30UT (Figure 2-b), in the center of the boundary cusp. The estimation of the apparent electron temperature by the MI instrument at the two times quoted above, decreases from the cusp "proper" value (about 40eV) to the boundary cusp value (about 15eV), whereas the mean energy of the soft electrons (above 10eV) is about the same, decreasing from 130 eV to 100eV for the same time period. Moreover, inside the large density gradient observed from 22:34 to 22:37UT, the

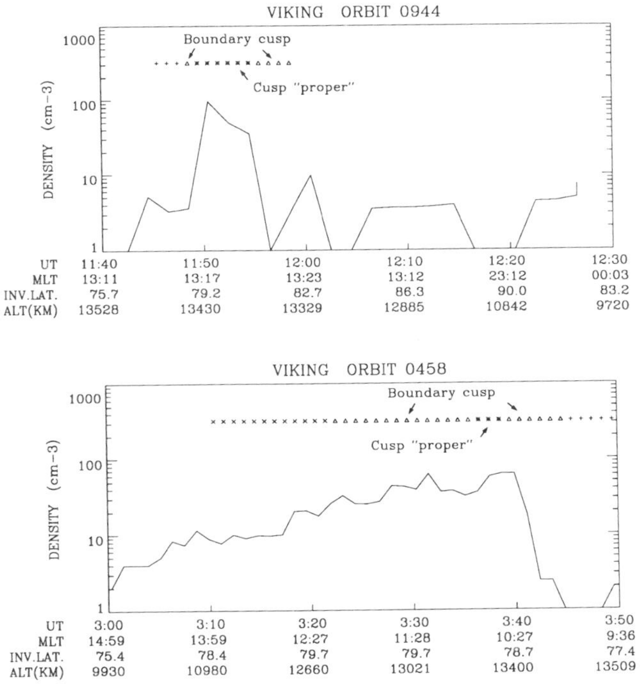


Fig.4. Density versus time profiles showing the transition from the cusp "proper" to the neighboring regions. The different regions crossed are symbolized respectively by asterisks for the cusp "proper", triangles for the boundary cusp, + signs for the cleft (connected to the low latitude boundary layer), and x signs for the mantle cusp.

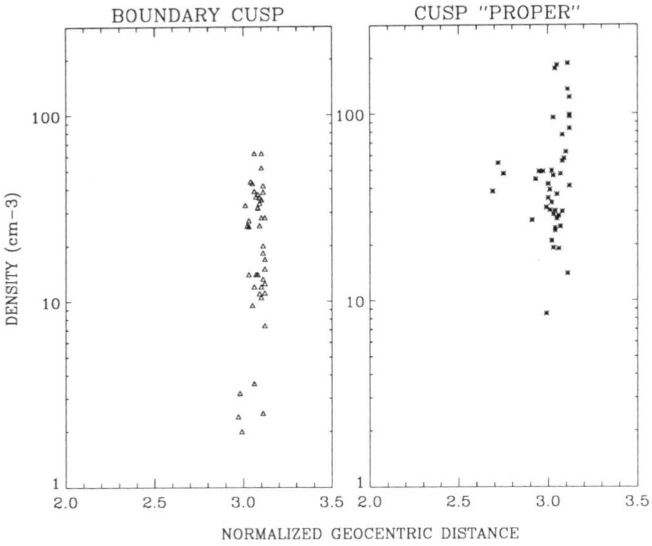


Fig.5. Scatter plot of the density versus radial distance obtained in the boundary cusp and in the cusp "proper".

Viking densities, whereas the magnetosheath electrons may be more strictly dominant at DE 1 altitudes, as argued in the Burch paper.

## 4. The polar cap

The polar caps refer to the low or mid-altitude regions of open field lines connected to the tail lobes of the magnetosphere. They are fed by the plasma mantle of solar wind origin, at least in the merging/reconnection model [Lundin, 1988], and by the polar wind and the cleft ion fountain, of ionospheric origin [Chappell et al.,1987], presumably dominant sources at Viking altitudes.

In their study of polar cap densities, Persoon et al, [1983], have analyzed a considerable number of DE 1 passes. We will present here a sampling of only a few (sixteen) Viking passes, taken hopefully sufficiently at random to provide a first idea of the order of magnitude of the polar cap densities measured. The corresponding dates and activity indices are given in Table 1. When present, the hiss signature was in agreement with the plasma frequency identified by the active instruments, however natural emissions were often non visible. Because we wanted to exclude the densities measured in the cusp, crossed by a number of the analyzed orbits, we chose to define the polar cap part of the pass a little differently of what was done on DE 1. We fixed the lower invariant latitude at 80°(instead of 76°), and furthermore excluded the dayside high latitude region covered by the statistical cusp and cleft shown in Lundin [1988].

The scattered plot of measured densities versus geocentric distances is shown in Figure 6-a. About 30% of data points are located below the indicated threshold of measurable densities. The median and quartile density values calculated in 4 bins of geocentric distances are displayed in Figure 6-b, and compared to the empirical law given by Persoon et al. [1983]. The Viking median density variation versus radial distance is very similar to the Persoon law, moreover there is an overlap between the upper to lower quartile ranges of the two data sets. However, the Viking median densities are systematically lower (by a factor 2 to 3) than the DE 1 ones. The reason of that discrepancy, which anyhow would have to be confirmed by a more substantial Viking data set, is not obvious. The characteristics of the DE 1 data set, as compared to the Viking one are (1) the absence of density values in periods (40% of the passes) without hiss emissions, (2) a broader range in invariant latitude, (3) a position in the solar cycle close to the maximum, whereas it is close to the minimum for Viking, and (4) relatively higher magnetic conditions (Kp ≈ 3, compared to Kp ≈ 2+). Concerning the first point, Viking measurements show that high density values, as well as very low ones, can be obtained in the absence of hiss; indeed, the bias which is introduced in our limited data set when ignoring the data without hiss emission is not significant. Second, it seems that higher latitudes are associated with lower densities; we need more data to confirm that point. Last, we know that the polar cap source is stronger both for solar minimum conditions, with respect to solar maximum

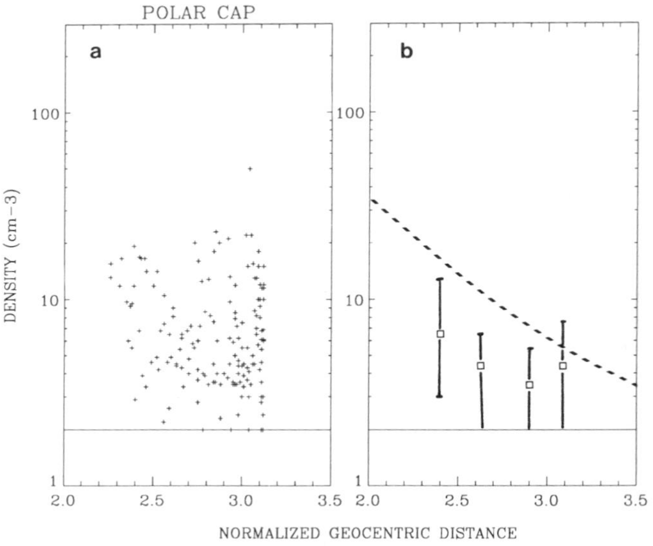


Fig.6. Density versus radial distance in the polar cap. The threshold in measurable densities is indicated.
a) Scatter plot
b) Medians and quartiles compared to the empirical law derived by Persoon et al., [1983] from DE 1 measurements : $n_e = \alpha R^{-3.85 \pm 0.32}$

ones, and for active conditions with respect to quiet ones [Chappell et al., 1987]. Consequently, the differences (3) and (4) would act in opposite ways. The analysis of the complete Viking data set should help to sktech more precisely the polar cap density under various conditions.

Density profiles taken along individual orbits in the polar cap show a high time variability, clear also in DE 1 studies. High densities of similar amplitude than those observed by DE 1 [Gallagher et al., 1987] are present on occasions [Décréau et al.,1987]. They show up as irregular clouds of limited size, and could be related to the presence of the cleft ion fountain, as envisaged in the above section, controlled by variable convection conditions. Due to the limitations of the MI instrument, one can estimate the electron temperature value only in the peak of a density irregularity. The apparent temperature estimated in such a case (orbit 237, at about 10:30MLT, 10,000km altitude, 84° Inv. Lat.), amounts to a value between 1000°K and 4000°K for a density of 16cm-3, measured by the two active instruments. In this time period, the very low count rate of the electron spectrometer corresponds to a density much less than 16cm-3, which confirms that the main part of the plasma is cold, in agreement with the temperature measured by MI, thus presumably of ionospheric origin.

## 5. Conclusion

The Viking mission (February to December 1986) includes a considerable number of passes over the cusp and polar cap regions at mi-altitudes. Its active wave instrumentation provides a reliable access to the total density number, difficult to measure in those regions. A first sampling of the results obtained show a predominance of high densities (a few 10cm-3) in the cusp region, mainly carried by magnetosheath-like plasma. Lower density values, similar to those measured on DE 1, are seen in the polar cap, where the ionospheric source is presumably dominant. Although those results are not at variance with what is already known, they reveal features that were hidden to other techniques. In particular (1) the irregular cold density clouds present over the polar cap, which could not be seen via the auroral hiss emissions at high invariant latitudes, and, (2) the ratio of cold to hot plasma in the cusp region, especially in the boundary cusp where the combination of the particle and the wave instrumentation is most useful. Further studies, including in particular the role of the local time and invariant latitude, of solar wind parameters and of geomagnetic conditions, are to be continued in order to better assess the dynamics and transport of plasma in those high latitude regions.

Acknowledgments. We wish to thank B.Aparicio, M.André, G.Kremser and R.Lundin for providing particle data, M.Hamelin, A.Morane and J.Paris for their assistance in analyzing wave data. The Viking project is managed by the Swedish Space Corporation under contract from the Swedish Board for Space Activities.

## References

André, M., H. Koskinen, L. Matson, and R. Erlandson, Local transverse ion energization in and near the polar cusp, *Geophys. Res. Lett., 15*, 107, 1988.

Bahnsen, A., M. Jespersen; E. Ungstrup, R. Pottelette, M. Malingre, P.M.E. Décréau, M. Hamelin, H. de Féraudy, S. Perraut, and B.M. Pedersen, First Viking results: high frequency waves, *Physica Scripta, 37*, 469, 1988.

Burch, J.L., Quasi-neutrality in the polar cusp, *Geophys. Res. Lett., 12,* 469, 1985.

Chappell, C.R., T.E. Moore, and J.H. Waite, Jr., The ionosphere as a fully adequate source of plasma from the earth's magnetosphere, *J. Geophys. Res., 92*, 5896, 1987.

Décréau, P.M.E., M. Hamelin, R. Massif, H. de Féraudy, E. Pawela, S. Perraut, R. Pottelette, and A. Bahnsen, Plasma probing by active wave experiments on the Viking satellite, *Ann. Geophysicae, 5A(4)*, 181, 1987.

Formisano, V., HEOS 2 observations of the boundary layer, Planet. *Space Sci., 28*, 245, 1980.

Gallagher, D.L., J.D. Menietti, J.L. Burch, A.M. Persoon, J.H.Waite, Jr. and C.R. Chappell, Evidence of high densities and ion outflows in the polar cap during the recovery phase, *J. Geophys. Res., 91*, 3321, 1986.

Horwitz, J.L., Core plasma in the magnetosphere, *Rev. Geophys. Phys., 25*, 579, 1987.

Hultqvist, B., Observations of low energy magnetospheric plasma outside the plasmasphere, *Space Sci. Rev., 42*, 275, 1985.

Johnstone, A.D., Electron injection in the polar cusp, in J.A. Holtet and A. Egeland, Ed., *The polar cusp, 47-65*, D.Reidel Publ. Co., Dordrecht, Holland, 1985.

Kremser, G. and R. Lundin, Viking observations of the ion composition in the cleft region at different local times, Cospar Conference in Helsinki, July 18-29, 1988.

Lockwood, M., M. O. Chandler, J.L. Horwitz, J.H. Waite, Jr., T.E. Moore, and C.R. Chappell, The cleft ion fountain, *J. Geophys. Res., 10*, 9736, 1985.

Louarn, P., A. Roux, H. de Féraudy, D. le Quéau, M. André, and L. Matson, Trapped electrons as a free energy sources for the AKR, *J. Geophys. Res.*, 95, 5983, 1990.

Lundin, R., On the magnetospheric boundary layer and solar wind energy transfer into the magnetosphere, *Space Sci. Rev., 48*, 263, 1988.

Perraut, S., H. de Féraudy, A. Roux, P.M.E. Décréau, J. Paris, and L. Matson, Density measurements in key regions of the earth's magnetopshere: cusp and auroral region, *J. Geophys. Res.*, 95, 5997, 1990.

Persoon A.M., D.A. Gurnett and S.D. Shawhan, Polar cap

electron densities from DE1 plasma and wave observations, *J.Geophys. Res., 88*, 10, 10,123, 1983.

Pottelette, R.,M. Malingre, N. Dubouloz, B. Aparicio, G. Holmgren, and G. Marklund, High frequency waves in the cusp/cleft regions, *J. Geophys. Res.*, 95, 5957, 1990.

Tokar, R.L. and Gary, S.P., Electrostatic hiss and the beam driven electron acoustic instability in the dayside polar cusp, *Geophys. Res. Lett., 11*, 1180, 1984.

# The Inner Magnetosphere/Plasmasphere

# MODELING OF THE STRUCTURE OF LONG-PERIOD ULF WAVES USING ENERGETIC PARTICLE OBSERVATIONS

Kazue Takahashi

The Johns Hopkins University Applied Physics Laboratory
Laurel, MD 20723-6099

*Abstract.* Long-period ULF waves (Pc 5 pulsations) in the magnetosphere are often modeled by a function $g(z)e^{i(m\phi - \omega t)}$, where g(z) represents the field-aligned eigenmode and m is the azimuthal wave number. The model is applicable to a wide variety of waves with different excitation mechanisms. Because each excitation mechanism has a preferred form of g(z) and a preferred value of m, observational determination of g(z) and m is crucial for distinguishing between different generation mechanisms for ULF waves observed at different locations, under different geomagnetic conditions, and with different magnetic field perturbations. In this paper, we discuss how energetic particle measurements from a single spacecraft can be used to determine the spatial structure of ULF waves. Phenomena related to finite Larmor radius effects and drift-bounce resonances provide the means with which the wave structure can be remote-sensed. Examples are taken from observations with the Active Magnetospheric Particle Tracer Explorers Charge Composition Explorer (AMPTE CCE) spacecraft in the near-geosynchronous region.

## Introduction

In the studies of long-period ULF pulsations (period $\gtrsim$ 100 s), determination of the spatial structure of the perturbed field is crucial for understanding how the waves are excited. Ideally, the structure may be determined from multisatellite observations, but this method obviously cannot be used in general, because only a small number of spacecraft are available at any given time and it is very unlikely that the spacecraft are appropriately located in the region of pulsation activity. Ground-based measurements have been successfully used for determining the longitudinal and latitudinal structures of a certain class of Pc 5 pulsations, but these observations provide little information on the field-aligned structure and the plasma condition related to pulsations.

In this paper we describe how single-spacecraft measurements can be used to determine or infer the spatial structure of Pc 5 pulsations (period > 150 s). The primary data for this technique are magnetic field and energetic ions (E $\geq$ 4 keV). The dynamics of the particles can be described by their zeroth order motion in the Earth's magnetic field plus their first-order perturbations arising from the electromagnetic field of ULF pulsations. Scale sizes related to both the gyromotion and guiding center drift-bounce motion of the particles are considered. Magnetic field data provides information of the field-aligned structure and polarization of the pulsations. After briefly describing a model for standing waves we examine representative examples from observations with AMPTE CCE.

Modeling Magnetospheric Plasma Processes
Geophysical Monograph 62


## Model Structure for ULF Waves

The magnetosphere is an inhomogeneous MHD system, and ULF pulsations, having a large scale size that is comparable to the dimension of the magnetosphere, cannot be described in terms of a single wave mode. The Alfvén mode and the magnetosonic mode are coupled and the coupling is described in terms of the radial variation of wave properties. However, in the following we shall ignore radial variations and assume, regardless of wave modes, that magnetic pulsations have a perturbation of the form

$$g(z)e^{i(m\phi - \omega t)} \tag{1}$$

where g(z) represents the field-aligned eigenmode and m is the (angular) azimuthal wave number, z is distance from the equator along the field

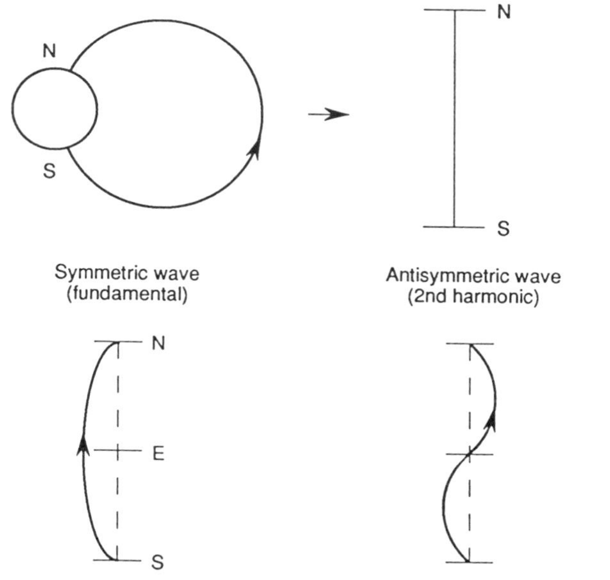


Fig 1. Field-aligned structure of standing waves on the geomagnetic field lines. For simplicity, the field lines are modeled by a stretched string, as shown at the top. If the field lines are fixed at the ionospheric foot points, field line distortions can be decomposed into symmetric waves such as the fundamental wave and antisymmetric waves such as the second harmonic wave. Here the symmetry refers to field line displacement.

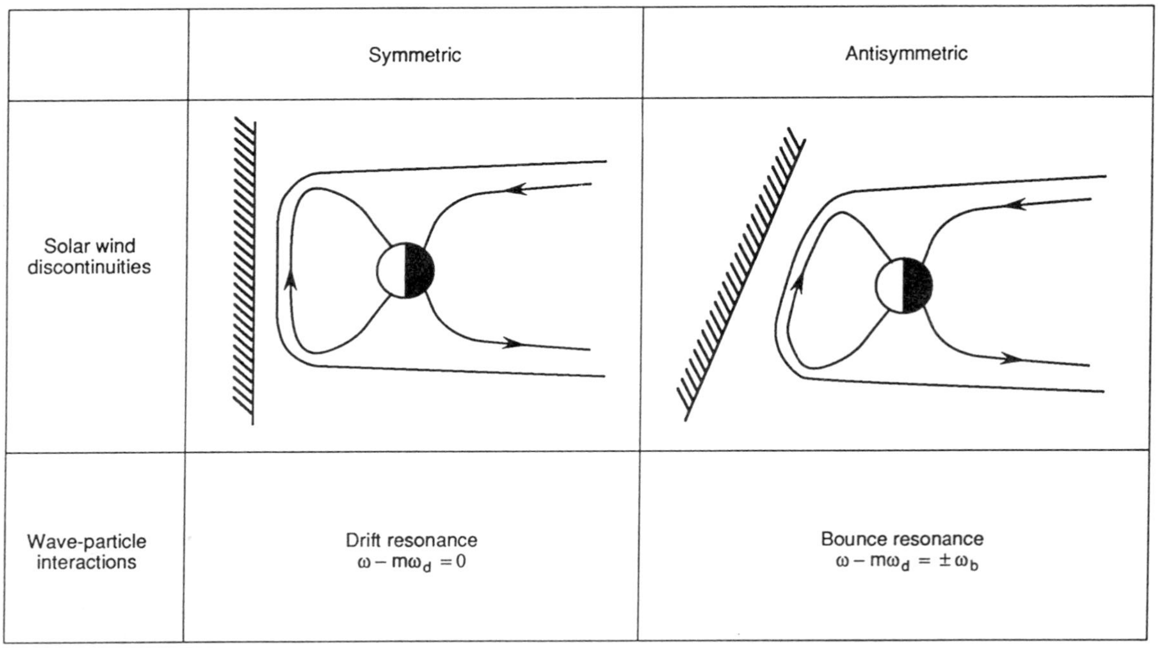


Fig. 2. Field-aligned symmetry of magnetic field disturbances related to the orientation of external (upper frames) or internal (lower frames) processes.

line, $\phi$ is azimuth, and $\omega$ is the frequency of the wave. Although the expression (1) implies an azimuthal symmetry of pulsation disturbances, the magnetosphere is not azimuthally symmetric. Therefore, the azimuthal wave number is usually used as a symbolic notation of the azimuthal filed variation in a limited longitudinal sector.

The function g(z) characterizes both the symmetry and the parallel wavelength of pulsations. Due to reflection at the ionosphere, long-period MHD disturbances often establish a standing wave structure on the geomagnetic field and in the case of perfect reflection the field line displacement should have either a symmetric structure (odd-mode harmonics) or an antisymmetric structure (even-mode harmonics) as illustrated in Figure 1.

The relationship between the symmetry of the standing waves and possible energy sources for exciting them is illustrated in Figure 2. As an external energy source for pulsations consider a pressure discontinuity in the solar wind. If the plane of the discontinuity is perpendicular to the magnetic equator, then the associated disturbances in the magnetosphere will consist of symmetric standing waves. If, on the other hand, the discontinuity is oblique to the equatorial plane, then antisymmetric waves will be excited as well.

Next consider wave-particle interactions. Because the zeroth-order guiding center motion of energetic particles consists of north-south bounce motion and azimuthal drift, transfer of energy between standing waves and the particles can occur if the resonance condition

$$(\omega - m\bar{\omega}_d)^2 = (N\omega_b)^2 \qquad (2)$$

is satisfied, where $\bar{\omega}_d$ is the bounce-averaged drift frequency, N is an integer (0, 1, 2, ...) and $\omega_b$ is the bounce frequency. From geometrical consideration one can see that symmetric waves (antisymmetric waves) interact with particles through the N = 0, 2, 4, ... (N = 1, 3, 5, ...) resonances [Southwood and Kivelson, 1982]. Usually the lowest order interactions, N = 0 and N = 1, are considered to be significant.

We can also give a qualitative discussion on the relationship between m and source disturbances. A brief summary is given in Table 1. Since m is inversely proportional to the azimuthal wavelength, m is small for large-scale external disturbances, whereas it is large for instabilities occurring in the ring current. Note that for large m the radial amplitude profile of MHD waves becomes evanescent. For this case we expect that the location of the energy source is close to where the waves are observed.

The sign of m is related to the wave propagation direction, eastward or westward, hence it often reflects the motion or propagation of the energy

Table 1. Relationship Between the Azimuthal Wave Number (m) and Various Physical Processes

| | Magnitude of m | |
|---|---|---|
| | 10 | 100 |
| $\lambda_\perp$ | 4 $R_E$ (25000 km) | 0.4 $R_E$ (2500 km) |
| Source location | External (solar wind) | Internal (ring current) |
| | Sign of m | |
| | + | − |
| Propagation direction | Eastward | Westward |
| K-H wave | Afternoon | Morning |
| Magnetic drift | Eastward (electrons) | Westward (ions) |
| Diamagnetic drift | $\nabla p > 0$ | $\nabla p < 0$ |

source. For example, disturbances in the solar wind, as well as Kelvin-Helmholtz waves on the magnetopause, lead to $m < 0$ in the morning and $m > 0$ in the afternoon, that is, tailward propagation [Olson and Rostoker, 1987]. On the other hand, pulsations excited within the magnetosphere often have their propagation directions tied to the drift motion of energetic particles. For example, waves generated by ion drift resonance $\omega - m\bar{\omega}_d = 0$ will propagate westward. Similarly, waves related to the diamagnetic drift (e.g., the drift mirror wave of Hasegawa [1969]) will propagate westward (eastward) if the pressure gradient is inward (outward), as is the case at the outer (inner) edge of the ring current.

In the following sections we shall focus on waves with large m (20–100). These pulsations are of great interest in terms of instabilities in the ring current. Spacecraft observations are essential for the waves because large-m pulsations are invisible to ground-based observers due to ionospheric masking effects [Hughes and Southwood, 1976]. We shall describe two types of magnetic pulsations as a demonstration of the importance of particle data for understanding the physical properties of ULF pulsations. They are compressional Pc 5 waves and radially polarized Pc 5 pulsations.

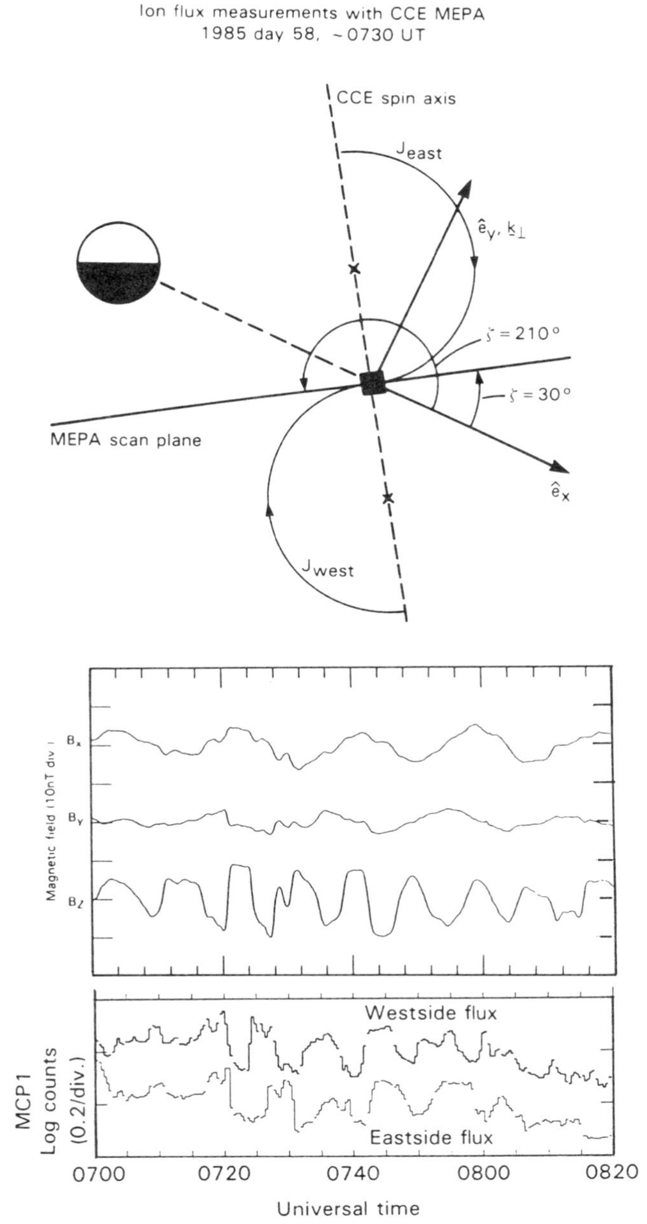


Fig. 3. An example of simultaneous measurements of a magnetic pulsation and an ion flux pulsation [after Takahashi et al., 1990a]. The location of the spacecraft and the geometry of ion measurement are shown at the top.

## Compressional Pc 5 Waves

Compressional Pc 5 waves are generated in association with enhanced fluxes of energetic ions and are a classical example of internally-excited long-period pulsations [Sonnerup et al., 1969]. Figure 3 illustrates a compressional Pc 5 wave observed by the AMPTE CCE spacecraft near 0300 magnetic local time [Takahashi et al., 1990a]. The event was selected on the basis of the $B_z$ component that oscillated at half the period of the transverse oscillation (period ~ 1000 s). Previous studies have shown that this "frequency doubling" occurs because the wave consists of an antisymmetric low-frequency mode and a symmetric high-frequency mode [Takahashi et al., 1987a, b]. The two traces at the bottom show the modulation of ion fluxes ($E > 4$ keV) entering the particle detector from opposite directions near 90° pitch angle, as illustrated at the top of Figure 3. Important points of the flux variations are (1) that they follow the oscillation in $B_z$ and (2) that the oscillation in the west side fluxes leads that in the east side flux.

An approximate expression for the perturbed phase space density, $\delta f$, in the presence of a strong magnetic field compression is given by Takahashi et al. [1990a] as

$$\delta f = -\frac{q}{MB_0}\frac{\partial f}{\partial \mu}\frac{v_\perp \delta B_z}{ck_\perp} J_1\left(\frac{k_\perp v_\perp}{\Omega}\right)\exp\left[\frac{ik_\perp v_\perp}{\Omega}\sin(\zeta - \theta)\right] \qquad (3)$$

where q is electric charge, M is particle mass, $B_0$ is the magnitude of the ambient field, $\mu$ is the magnetic moment, $v_\perp$ is velocity perpendicular to the ambient field, $\delta B_z$ is the compressional field perturbation, $J_1$ is the Bessell function of order one, $k_\perp$ is the perpendicular wave number, $\Omega$ is the cyclotron frequency, $\zeta$ is the angle between $\hat{e}_x$ (radial component) and $\underline{v}_\perp$ and $\theta$ is the angle between $\hat{e}_x$ and $\underline{k}_\perp$. The phase factor $\exp\left[\frac{ik_\perp v_\perp}{\Omega}\sin(\zeta - \theta)\right]$, which can be rewritten as $\exp[i\rho k_\perp \sin(\zeta - \theta)]$, where $\rho$ is the Larmor radius, represents a finite Larmor-radius effect. Having measurements of $\delta f$ at a number of energy steps and at different $\zeta$ angles is equivalent to having a number of subsatellites distributed around the spacecraft. For such a case, unambiguous determination of $k_\perp$ is possible [Lin et al., 1988]. If only integral fluxes are measured, as was the case in Takahashi et al. [1990a], the phase lag can be related to $k_\perp$ by integrating $\delta f$ over energy. The latter technique was used for the event shown in Figure 3, and the result of calculation is shown in Figure 4. The observed phase lag corresponds to an azimuthal wave length of 5200 km (m ~ 60). The direction of propagation, as seen from the satellite frame, is found to be eastward.

## Radially Polarized Pc 4-5 Waves

Another type of long-period pulsations is radially polarized Pc 5. From early magnetic filed measurements near the magnetic equator, these waves were considered to have an antisymmetric (second-harmonic) standing Alfvén wave structure [Cummings et al., 1969]. Theoretical work by Southwood [1976] suggested that these waves are excited by drift-bounce

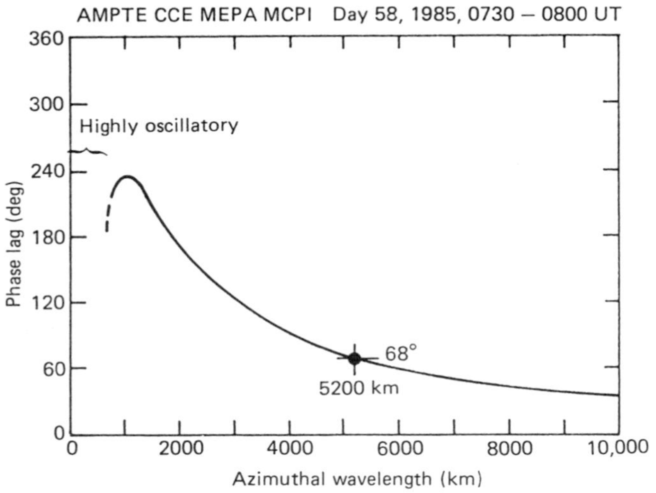


Fig. 4. The relationship between the east/west phase lag of ion flux oscillations and the azimuthal wave length [after Takahashi et al., 1990a]. For the example shown in Figure 3, the phase lag is 68° and the inferred wavelength is 5200 km.

resonance of ring current ions. Because the theory predicts a large m value (m ~ 100) and a propagation direction that depends on the relative distance from the center of the ring currents, attempts have been made to determine m from satellite observations. For example, Hughes et al. [1979] used multisatellite magnetic field measurements to estimate m. They were unable to do so because of poor coherence between satellites.

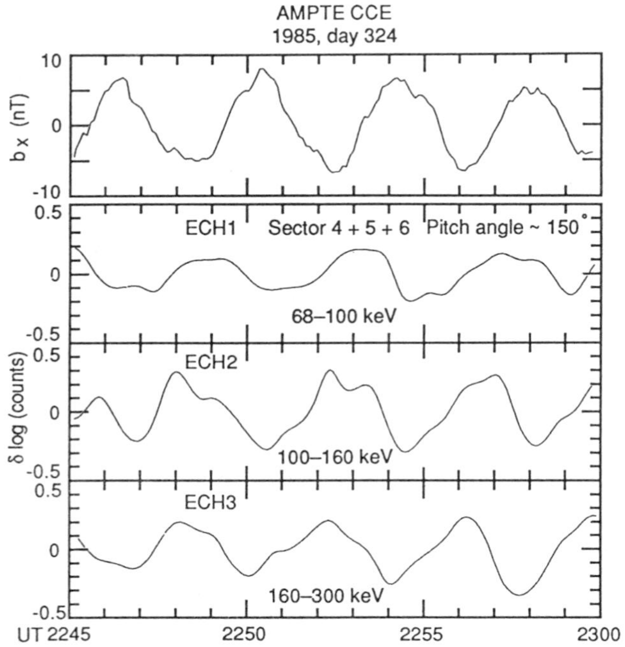


Fig. 5. Relationship between the radial magnetic field pulsation and ion flux oscillations for a Pc 5 event [after Takahashi et al., 1990b]. The phase of the ion flux oscillation advances with increase in energy.

On the other hand, energy-dependent response of ions to magnetic pulsations was used to estimate m by Kivelson and Southwood [1985] and by Kokubun et al. [1989]. Although these studies obtained m ~ 100, the limited energy and pitch angle coverage of their detectors did not allow a comprehensive determination of m. We have done an in-detail analysis of an ion flux pulsation event using CCE observations. We interpret the data after briefly reviewing the adiabatic theory of wave-particle interactions.

In the drift-kinetic approach the particle flux response to a second-harmonic standing Alfvén wave with a radial field perturbation can be formulated as follows. First, assume that individual ions experience energy modulation

$$\delta W(t) = \int_{orbit} q\underline{E} \cdot \underline{V}_d dt' \tag{4}$$

where $\underline{E}$ is the electric field of an Alfvén wave and $\underline{V}_d$ is the guiding center drift velocity. For a semi-qualitative argument it suffices to assume an electric field of the form

$$E_y = -i \frac{\omega}{ck_z} b_{eq} \sin(k_z z) e^{i(m\phi - \omega t)} \tag{5}$$

where $b_{eq}$ is the equatorial amplitude of the magnetic pulsation. The perturbed distribution function near the equator is given as

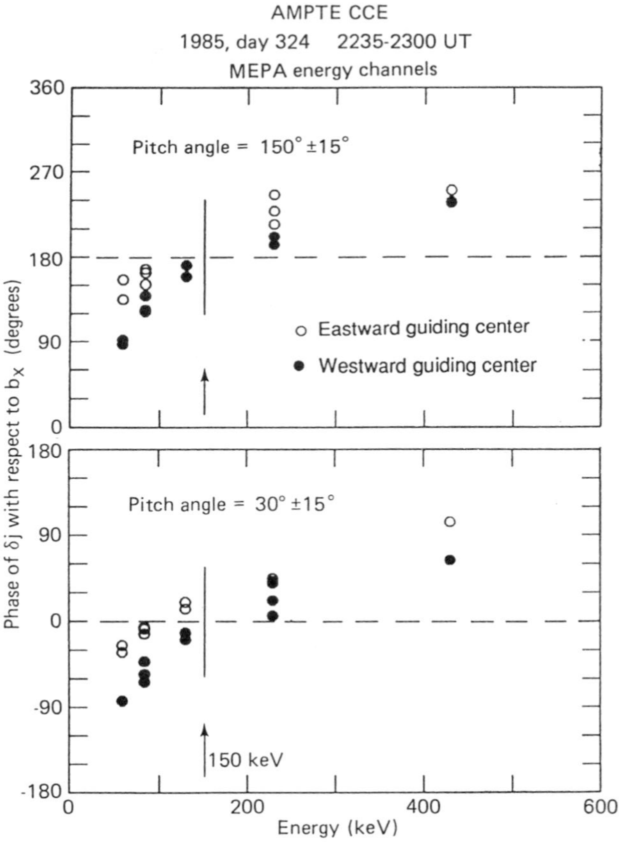


Fig. 6. Summary of the energy dependence of the phase of the ion flux oscillations shown in Figure 5 [after Takahashi et al., 1990b]. The phase becomes either 0° or 180° at 150 keV, and this energy is interpreted as corresponding to the high-energy bounce-drift resonance.

$$\delta f = -\delta W\left(1 - \frac{m\omega^*}{\omega}\right)\frac{\partial f}{\partial W}$$

$$= -i\left(\frac{qLR_E}{c}\right)\left[\frac{\partial f}{\partial W}\right](\omega - m\omega^*)$$

$$\times\left[\frac{\bar{\omega}_d\ \omega_b \cos\Phi}{\omega_b^2 - (m\bar{\omega}_d - \omega)^2}\right] z_m b_{eq} e^{i(m\phi - \omega t)} \tag{6}$$

where $\omega^* \equiv -\left(\frac{c}{qB_{eq}LR_E^2}\right)\left(\frac{\partial f}{\partial L}\right)\left(\frac{\partial f}{\partial W}\right)^{-1}$, $\Phi$ is bounce phase, and $z_m$ is the distance of the mirror point from the equator [Kivelson and Southwood 1985, Takahashi et al., 1990b].

A few predictions can be made from (6):

1. A singular behavior (resonance) occurs when the denominator $\omega_b^2 - (m\bar{\omega}_d - \omega)^2$ approaches zero. Across the resonance the phase of $\delta f$ changes by 180°. Away from the resonance $\delta f$ is in quadrature with the local radial field perturbation.
2. The energy at which resonance occurs depends on particle equatorial pitch angle. This is because both $\omega_b$ and $\bar{\omega}_d$ are a function of the pitch angle [Hamlin et al., 1961]. For given m and $\omega$, the resonance energy is lower (at the high-energy resonance $\omega_b^2 \sim (m\bar{\omega}_d)^2$) for equatorial pitch angles closer to 90°.

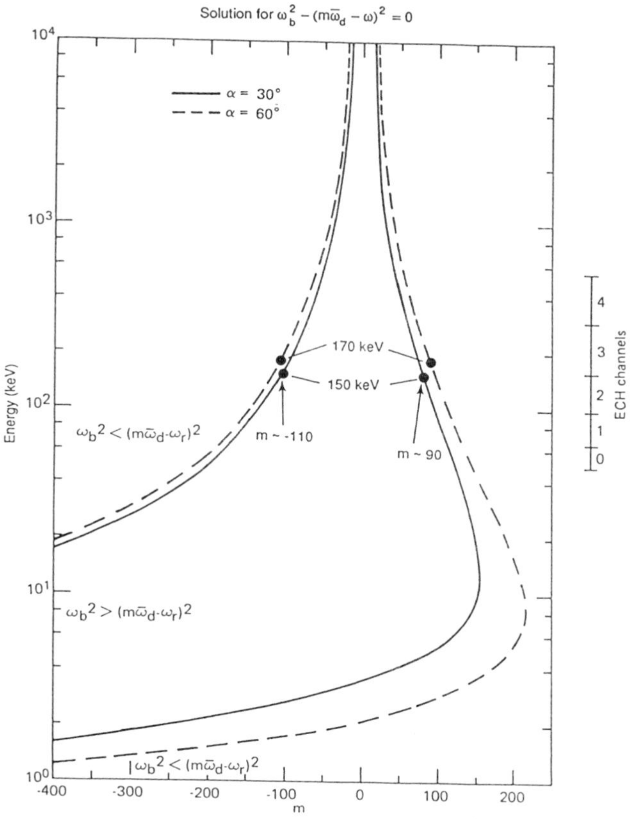


Fig. 7. A diagram showing the solution to the lowest-order bounce-drift resonance equation as a function of energy for a given set of frequency $\omega$ and azimuthal wave number m [after Takahashi et al., 1990b].

3. Particle fluxes from the north and south oscillate in antiphase. This is because the cos $\Phi$ factor in (6) gives opposite signs for these fluxes. The ultimate reason is that the electric field of an antisymmetric wave has opposite signs at north and south.
4. The amplitude of $\delta f$ is proportional to $z_m$ for pitch angles not too away from 90°. This is because the electric field increases with magnetic latitude.

We now apply the theory to an actual observation. Figure 5 shows oscillations seen by the AMPTE CCE spacecraft in the radial magnetic field ($b_x$), and in the fluxes of ions at three energy bands [Takahashi et al., 1990b]. The oscillations have a period of ~ 200 s (a Pc 5 wave), and characteristic time lag that depends on energy. A summary of the energy dependence of the lag is given in Figure 6 for two pitch angle groups. The upper (lower) panel shows particles moving southward (northward). In both groups the phase advances with energy, crossing 180° or 0° phase at ~ 150 keV. From the argument presented above, the energy is taken to represent the resonance between the particles and the wave.

The implication of this observation can be seen in Figure 7, where the solution for the resonance energy is illustrated as a function of m for two equatorial pitch angles. The energy bands of the observed particles are shown at the right. For a given m, the resonance condition $\omega_b^2 - (m\omega_d - \omega)^2 = 0$ yields two solutions of ion energy, provided m is smaller than a certain value. They are the high-energy resonance $\omega_b^2 \sim (m\bar{\omega}_d)^2$ and the low-energy resonance $\omega_b^2 \sim \omega_r^2$, where $\omega_r$ is the real part of the wave frequency. The solid circles in the diagram indicate where a resonance behavior was observed in the particle data. Although the magnitude of m (~ 100) obtained from this diagram is reasonable in comparison to previous estimates, we are left with the ambiguity of the sign of m. A careful look at each term of equation (6) predicts, however, that the observed phase shift with energy is consistent with m < 0 (i.e., westward propagation).

Because the present data set has a full pitch angle coverage, we can test the prediction of equation (6) regarding the north/south phase lag. Figure 8 illustrates flux oscillations of two groups of particles whose zeroth order guiding center orbits are mirror images of each other about the magnetic equator. Clearly, these fluxes oscillate in antiphase as is predicted.

Lastly, we comment on the pitch angle dependence of the amplitude of the flux oscillations. Figure 9 shows the amplitude as a function of the equatorial pitch angle for five energy bands. In each band, the amplitude is minimum at 90° and is maximum at ~30° and ~150°. This indicates that the electric field amplitude has a maximum at some distance from the equator and that the flux of particles spending a long time near the field maximum is modulated most. If a particle mirrors far away from the electric field maximum, i.e. at the equator or at a very high latitude, then equation (4) implies that the energy modulation for the particle is smaller

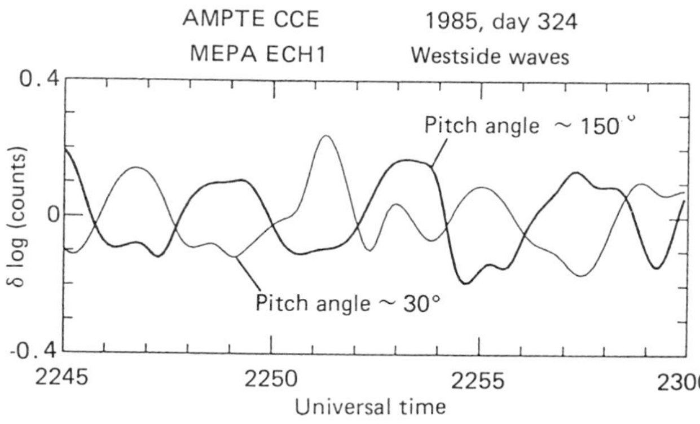


Fig. 8. Ion flux pulsations indicating an antiphase relationship between fluxes directed northward and southward [after Takahashi et al., 1990b].

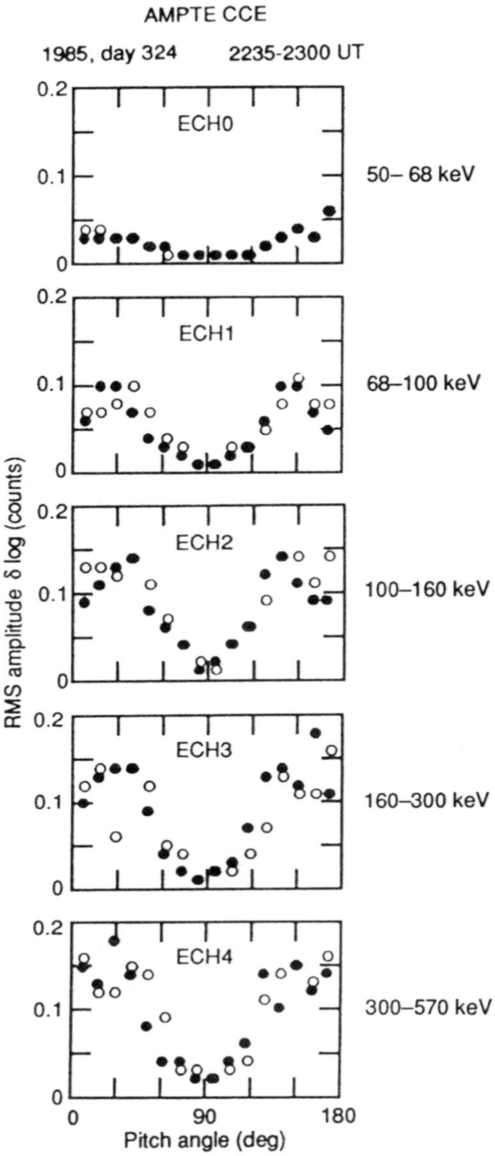


Fig. 9. Equatorial pitch angle dependence of the amplitude of ion flux oscillations associated with an antisymmetric standing Alfvén wave [after Takahashi et al., 1990b].

than that for particles mirroring near the field maximum, which is perhaps located ~±30° in dipole latitude. Thus, the particle data are also useful for determining the field-aligned amplitude structure of standing waves.

## Conclusions

In conclusion, we have outlined a simple model for ULF waves in the magnetosphere and have described how this model can be used for organizing the behavior of energetic particles during magnetic pulsation events. Two exemplary ULF wave events were used to demonstrate that particle data are quite useful for determining the parameters in the model. In many cases we know the zeroth order orbits of the particles, and when flux oscillations are interpreted as due to ULF perturbations the energy and pitch angle of individual particles, a great deal of information on the structure of the waves can be obtained from the observed relationship between particle fluxes and the magnetic field.

*Acknowledgments.* Much of the work presented in this paper was taken from published results on AMPTE/CCE observations. The author gratefully acknowledges the contributions of the CCE investigators, T. A. Potemra, L. J. Zanetti, and M. H. Acuña for the magnetometer data, and R. W. McEntire, E. P. Keath, D. E. Fort, A. T. Y. Lui, and S. M. Krimigis for the particle data. He also thanks C. Z. Cheng for theoretical support. This work was supported by NASA under Task I of Navy Contract N00039-89-5301.

## References

Cummings, W. D., R. J. O'Sullivan, and P. J. Coleman, Jr., Standing Alfvén waves in the magnetosphere, *J. Geophys. Res.*, *74*, 778, 1969.

Hamlin, D. A., R. Karplus, R. C. Vik, and K. M. Watson, Mirror and azimuthal drift frequencies for geomagnetically trapped particles, *J. Geophys. Res.*, *66*, 1, 1961.

Hasegawa, A., Drift mirror instability in the magnetosphere, *Phys. Fluids*, *12*, 2642, 1969.

Hughes, W. J., and D. J. Southwood, The screening of micropulsation signals by the atmosphere and ionosphere, *J. Geophys. Res.*, *81*, 3234, 1976.

Hughes, W. J., R. L. McPherron, J. N. Barfield, and B. H. Mauk, A compressional Pc 4 pulsation observed by three satellites in geostationary orbit near local midnight, *Planet. Space Sci.*, *27*, 821, 1979.

Kivelson, M. G., and D. J. Southwood, Charged particles behavior in low-frequency geomagnetic pulsations, 4, Compressional waves, *J. Geophys. Res.*, *90*, 1486, 1985.

Kokubun, S., K. N. Erickson, T. A. Fritz, and R. L. McPherron, Local time asymmetry of Pc 4-5 pulsations and associated particle modulations at synchronous orbit, *J. Geophys. Res.*, *94*, 6607, 1989.

Lin, N., R. L. McPherron, M. G. Kivelson, and D. J. Williams, An unambiguous determination of the propagation of a compressional Pc 5 wave, *J. Geophys. Res.*, *93*, 5601, 1988.

Olson, J. V., and G. Rostoker, Longitudinal phase variations of Pc 4-5 pulsations, *J. Geophys. Res.*, *83*, 2481, 1978.

Sonnerup, B. U. Ö., L. J. Cahill, Jr., and L. R. Davis, Resonant vibration of the magnetosphere observed from Explorer 26, *J. Geophys. Res.*, *74*, 2276, 1969.

Southwood, D. J., A general approach to low-frequency instability in the ring current plasma, *J. Geophys. Res.*, *81*, 3340, 1976.

Southwood, D. J., and M. G. Kivelson, Charged particle behavior in low-frequency geomagnetic pulsations 2. Graphical approach, *J. Geophys. Res.*, *87*, 1707, 1982.

Takahashi, K., J. F. Fennell, E. Amata, and P. R. Higbie, Field-aligned structure of the storm time Pc 5 wave of November 14-15, 1979, 1979, *J. Geophys. Res.*, *92*, 5857, 1987a.

Takahashi, K., L. J. Zanetti, T. A. Potemra, and M. H. Acuña, A model for the harmonic of compressional Pc 5 waves, *Geophys. Res. Lett.*, *14*, 363, 1987b.

Takahashi, K., C. Z. Cheng, R. W. McEntire, and L. M. Kistler, Observation and theory of Pc 5 waves with harmonically related transverse and compressional components, *J. Geophys. Res.*, *95*, 977, 1990a.

Takahashi, K., R. W. McEntire, A. T. Y. Lui, and T. A. Potemra, Ion flux oscillations associated with a radially polarized transverse Pc 5 magnetic pulsation, *J. Geophys. Res.*, *95*, 3717, 1990b.

# THE ELECTROSTATIC DRIFT WAVE IN THE INNER MAGNETOSPHERE

T. S. Huang

Laboratory for Extraterrestrial Physics, NASA, Goddard Space Flight Center
Code 692, Greenbelt, MD 20771

Abstract. We investigate electrostatic drift waves with **k·B** = 0 in the Birkeland current flow region of the inner magnetosphere. In the evening sector of Region I (and perhaps the morning sector of Region II) associating with parallel currents away from the ionosphere there exists a field-aligned potential drop. It modifies the magnetosphere-ionosphere coupling relation in those regions, and effectively reduce the field line-tying effect of the ionosphere. Using the theory of individual particle motion and the magnetosphere-ionosphere coupling relation, we formulate the linear theory for an azimuthally propagating electrostatic drift wave in the above regions. The results show that an instability may arise from resonance of the wave with the drift of hot plasma particles (ring current plasma and sheet plasma). This instability is related to the difference between the distributions of hot electrons and ions in the plasma. The growth rate of the wave is linear in the azimuthal wave number.

## 1. Introduction

Guiding center drifts (other than the **E**×**B** drift) commonly produce currents and excite drift waves in planetary magnetospheres. Since the **k·B** = 0 drift wave is directly related to the flux interchange instability, it is important in the study of plasma transport process. A growing drift wave of this type may be excited in a magnetospheric region (the gradient/curvature drift or centrifugal drift dominates) which has a negative radial gradient of plasma content, for example the Earth's plasmapause [Richmond, 1973; Huang et al., 1990, and references therein] or the Io plasma torus [Huang and Hill, 1990]. In the Earth's inner magnetosphere the hot plasma has a positive radial gradient of plasma content [e.g. Smith and Hoffman, 1973; Strangeway and Kaye, 1986; and Spence et al., 1989]. By the simple MHD energy principle consideration, this configuration in isolation is stable against the **k·B** = 0 electrostatic drift wave. However, in Birkeland current flow regions the existence of a field-aligned potential drop might make the configuration unstable. In this paper, we attempt to study the effect of the field-aligned potential drop on the drift waves in Birkeland current flow regions.

In the evening sector of Region I, the Birkeland current flows away from the ionosphere and causes a field-aligned potential drop. Also, the potential drop is thought with some controversies to occur in the morning sector of Region II. The potential drop modifies the relationship between the electric potential and the drift current within the magnetosphere. In the magnetosphere the electric potential and the drift current are related though magnetosphere-ionosphere coupling. Without a field-aligned potential drop, the electric potential on each field line can simply be mapped from the ionosphere to the magnetosphere, and the variation in the electric potential of the magnetosphere is then related to the ionospheric conductivity. In the presence of the field-aligned potential drop, in addition to the ionospheric conductivity, an electric resistivity along field lines must also be included in the magnetosphere-ionosphere coupling system. In fact, the variation in the potential drop dominates the variation in the electric potential of the magnetosphere, and the line-tying effect of the ionosphere is effectively reduced. This reduction of the line-tying effect provides a condition under which the resonance of the wave with particle drift excites unstable drift waves.

We will study the drift waves using the theory of individual particle motion and including the line-tying effect of the ionosphere and the effect of the field-aligned potential drop. In section 2 the drift wave equation is derived. The dispersion relations are obtained in section 3. In section 4 we give the numerical results.

## 2. Wave equation

For simplicity we approximate the Earth's magnetic field in the region of interest by a dipole aligned with the spin axis. In terms of the coupling between ionosphere and magnetosphere [Vasyliunas, 1970], the electric potential in the ionosphere is related to the divergence of the magnetospheric current. In the magnetic curvilinear coordinate $\alpha$ and $\beta$ system we have [Huang et al., 1990]

Modeling Magnetospheric Plasma Processes
Geophysical Monograph 62

$$\Sigma_p R^2 \frac{\partial}{\partial\alpha}\left(\frac{(B^i)^2}{B_r^i} sin^2\theta \frac{\partial}{\partial\alpha}\Phi^{(i)}\right) + \frac{\Sigma_p}{R^2 B_r^i sin^2\theta}\frac{\partial^2}{\partial\beta^2}\Phi^{(i)} = \frac{j_{\|}}{B^i} \tag{1}$$

where $R$ is the Earth's radius, $\theta$ is the colatitude of the field line on the Earth's surface, $B^i$ and $B_r^i$ are the total strength and radial component of the magnetic induction at the Earth's surface, $\Sigma_p$ is the height-integrated Pedersen conductivity of the Earth's ionosphere, $j_{\|}$ is the Birkeland current density, and $\Phi^{(i)}$ represents the electric potential in the ionosphere. Unlike the convention used by Huang et al [1990], the sign of $j_{\|}$ is positive for currents flowing away from the ionosphere. For a dipolar magnetic field

$$B^i = B_E(4 - \frac{3}{L})^{1/2} \tag{2}$$

$$B_r^i = 2B_E(1 - \frac{1}{L})^{1/2} \tag{3}$$

$$sin^2\theta = \frac{1}{L} \tag{4}$$

where $B_E$ is the equatorial magnetic field at the surface, and $L$ is McIlwain's parameter. If we take longitude $\phi$ as coordinate $\beta$, then the other coordinate (by definition $\mathbf{B} = \nabla\alpha\times\nabla\beta$)

$$\alpha = -\frac{R^2 B_E}{L} \tag{5}$$

Because the waves we study are relatively slow in comparison with the transmission of Alfvén wave along field lines from the ionosphere to the magnetosphere, the potential on each field line is approximately equal except for the potential drops in the evening sector of Region I and possibly in the morning sector of Region II. Including the field line-aligned potential drop, the electric potential in the magnetosphere can then be written in the form

$$\Phi^{(m)} = \Phi^{(i)} - \Phi_{\|} \tag{6}$$

where $\Phi_{\|}$ represents the parallel electric potential drop.

It has been confirmed that in the evening sector of Region I a parallel current flows away from the ionosphere, and there exists a potential drop along the field lines. In the morning sector of Region II also a parallel current flows away from the ionosphere, but the existence of a potential drop is still questionable. In the following study, we assume the relationship between the potential drop and the parallel current in the morning sector of Region II is the same as that in the evening sector of Region I. Namely, in both regions the potential drop is roughly proportional to the current density [Knight, 1973; Lyons, 1980]. The feet of the field lines in those sectors in the magnetosphere are approximately distributed in two half annular sheets, as shown in Figure 1. The actual Birkeland current and field-aligned potential drop on the boundaries of those region decrease gradually [e.g., Kan and Cao, 1988]. Without introducing large error, we ignore the transitions of potential drop on the boundaries. Thus the electric potential drop jumps at the boundaries of the sheets. It is obvious that for an azimuthally propagating wave the effect of the potential jump at the short radial boundaries is less important than that at the arc boundaries. We simply replace the half annular sheets by a whole annular sheet (see Figure 1). Each region has an electric potential drop in its interior, but no drop exists inside and outside it. We can express the potential drop as

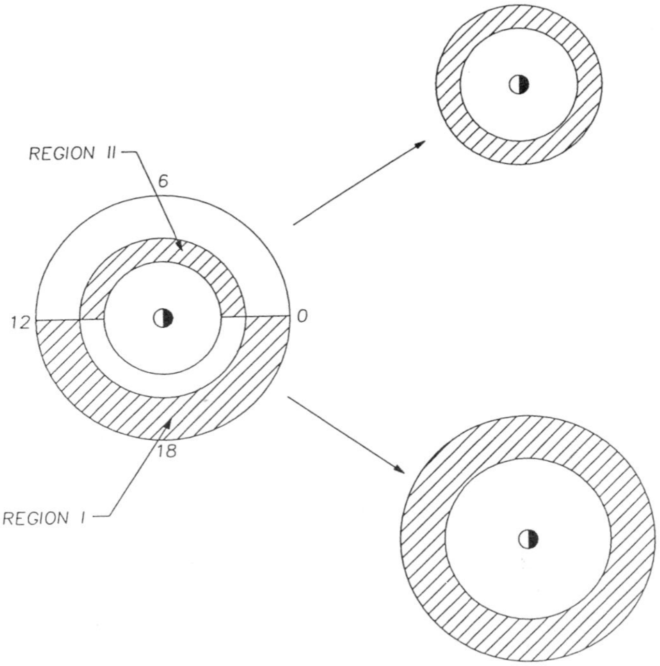


Fig. 1. Illustrations of the morning sector of Region II and the evening sector of Region I in the equatorial plane (on the left side), and their model for the calculation of the drift waves (on the right hand side).

$$\Phi_{\|} = \begin{cases} 0 & (L < L_1; L > L_2) \\ \chi\,(j_{\|} - j_c) & (L_1 < L < L_2) \end{cases} \tag{7}$$

where $j_c$ and $\chi$ are constants, and $L_1$ and $L_2$ represent the inside and outside boundaries of annular sheets. The second line of (7) is valid only for $j_{\|} > j_c$. According to Lyons' calculation [1980], $j_c$ and $\chi$ are ~ $10^{-7}\ A/m^2$ and ~ $3\times10^9 V/(A/m^2)$, respectively. During the geomagnetically inactive period, the average current densities in the evening sector of Region I and in the morning sector of Region II are ~ $1.1\times10^{-6} A/m^2$ and ~ $7\times10^{-7} A/m^2$, respectively [Potemra et al., 1979]. Both are greater than $j_c$. Thus in these regions any variation in $j_{\|}$ gives rise to a variation in $\Phi_{\|}$. Including the perturbation parallel current $\Delta j_{\|}$ caused by the waves, we express $j_{\|}$ in (7) as

$$j_{\|} = j_{\|o} + \Delta j_{\|} \tag{7A}$$

where $j_{\|o}$ represents the unperturbed current density. The value of $j_{\|o}$ is a constant greater than the critical value $j_c$ for $L_1 < L < L_2$, and zero for $L < L_1$ and $L > L_2$.

We linearize the electric potential $\Phi^{(m)}$ as

$$\Phi^{(m)}(L,\phi,t) = \Phi_m(L)\exp(im\phi - i\omega t) \tag{8}$$

where integer $m$ is the azimuthal wave number. According to the theory of adiabatic particle motion the parallel current $\Delta j_{\|}$ can be expressed [Huang et al, 1990] as

$$\Delta j_{\|} = i\frac{m^2L^2\left(1-\frac{3}{4L}\right)^{1/2}}{R^2}\sum_j q^j \frac{d\eta_o^j}{dL} G^j(m,\omega)\,\Phi_m \tag{9}$$

(here the polarization drift is less important, and is ignored) where $j$ represents the species of the particles, $q^j$ is the particle's charge, $\eta_o^j$ is the plasma content (particle's number per magnetic flux tube) in the equilibrium state, and $G^j(m,\omega)$ represents

$$G^j(m,\omega) = \int \frac{\omega_d^j f^j}{-\omega + m\omega_d^j}\, d\mu dJ \tag{9A}$$

In (9A) $\mu$ and $J$ are the first and second invariants, $f^j$ is the particle distribution function in the space ($\alpha$, $\beta$, $\mu$, $J$), and $\omega_d^j$ is the bounce average of the particle's angular drift velocity. Here we have ignored the effect of plasma convection.

In this paper we consider waves with $|m| >> 1$. In comparison with the wavelength, the scale lengths of azimuthal variations in the actual plasma (equilibrium) distribution and in the ionospheric Pedersen conductance $\Sigma_p$ are so large that the effects of these variations on the wave are unimportant. Thus, we can assume the plasma distribution to be axisymmetric and treat $\Sigma_p$ as a constant in the wave equation. For regions $L < L_1$ and $L > L_2$, we replace $\Phi^{(i)}$ by $\Phi^{(m)}$ and $j_{\|}$ by $\Delta j_{\|}$, and insert (2) – (5), (8) and (9) into (1). Also, we neglect the minor terms with $1/L$ in (2), (3) and (9). Then we obtain the wave equation

$$\frac{d^2\Phi_m}{dL^2} + \frac{1}{L}\frac{d\Phi_m}{dL} - \frac{m^2}{4L^2}[1+D]\,\Phi_m = 0 \qquad (L < L_1;\ L > L_2) \tag{10}$$

where

$$D = i\frac{L}{\Sigma_P}\sum_j q^j \frac{d\eta_o^j}{dL} G^j(m,\omega) \tag{10A}$$

For the region $L_1 < L < L_2$, performing the same procedure but replacing $\Phi^{(i)}$ by $\Phi^{(m)} + \Phi_{\|}$ and using (7) and (7A), we have

$$\frac{d^2(P\Phi_m)}{dL^2} + \frac{1}{L}\frac{d(P\Phi_m)}{dL} - \frac{m^2}{4L^2}\left(1+\frac{D}{P}\right)P\Phi_m = 0 \qquad (L_1 < L < L_2) \tag{11}$$

where

$$P = 1 + i\frac{m^2\chi L^2}{R^2}\sum_j q^j \frac{d\eta_o^j}{dL} G^j(m,\omega) \tag{11A}$$

We will see that including the electric resistivity along field lines in these region leads to a reduction in the line-tying effect of the ionosphere, which is represented by term $D$.

Outside the plasmapause, the ionosphere-origin plasma has a number density comparable to that of the ring current plasma and the sheet plasma. But its mean particle energy of ~ 1eV is much smaller than that of the latter plasmas, which is ~ 1 to 100 keV. So its effect on the drift wave is negligible, and we need to consider only the ring current plasma and the sheet plasma. For the ring current plasma and sheet plasma, Equation (9A) can be approximately expressed as

$$G^j(m,\omega) \approx \int_0^\infty \frac{\omega_d^j(W) f^j(W)}{-\omega + m\omega_d^j(W)}\, dW + Q \tag{12}$$

where $W$ is the kinetic energy of particle, and $f_j(W)$ is the distribution function with variable of $W$, and $Q$ is the resonance term [Huba and Wu, 1976]. Here $f_j(W)$ satisfies normalization condition

$$\int_0^\infty f^j(W)dW = 1 \tag{12A}$$

$Q$ can be written as

$$Q \approx \begin{cases} 0 & (mq^j\omega_R > 0;\ mq^j\omega_R < 0 \text{ and } mq^j\omega_I < 0) \\ -i\dfrac{\pi\omega}{m^2\overline{\omega}_d^j}\exp\left(-\dfrac{\omega}{m\overline{\omega}_d^j}\right) & (mq^j\omega_R < 0 \text{ and } mq^j\omega_I > 0) \end{cases} \tag{12B}$$

where $\overline{\omega}_d^j$ is the average angular drift velocity within a flux tube, and $\omega_R$ and $\omega_I$ represent the real and imaginary parts of $\omega$.

## 3. Dispersion relation

We note that term $D$ in the square brackets of (10) is $L$-dependent. However, because $\Phi_m$ decreases fast as $L$ goes away from the boundaries of the annular sheet, the wave is important only in the region near the boundaries. Thus, to first approximation, we can solve (10) in regions $L < L_1$ and $L > L_2$ by replacing variable $L$ in term $D$ by $L_1$ and $L_2$, respectively. The solution is

$$\Phi_m(L) = \begin{cases} C_1 L^{|m|\lambda_1/2} & L < L_1 \\ C_2 L^{-|m|\lambda_2/2} & L < L_2 \end{cases} \tag{13}$$

where

$$\lambda_1 = [1 + D(L_1)]^{1/2} \tag{13A}$$

$$\lambda_2 = [1 + D(L_2)]^{1/2} \tag{13B}$$

In (11) not only term $D$ in the brackets is $L$-dependent, but $P$ is also. However, since

$$|P| >> |D| \tag{14}$$

term $D$ in (11) can be negligible, and the equation becomes a simple one for function $P\Phi_m$. The solution to the wave equation in region $L_1 < L < L_2$ is

$$\Phi_m(L) = \frac{C_3 L^{|m|/2} + C_4 L^{-|m|/2}}{P} \qquad L_1 < L < L_2 \tag{15}$$

In (13) and (15) $C_1$, $C_2$, $C_3$ and $C_4$ are constants.

The continuity requirements of $\Phi_m$ and $d\Phi_m/dL$ at $L_1$ and $L_2$ lead to dispersion relation

$$\frac{\left(1 + \lambda_1 + \frac{2L_1}{|m|P(L_1)}\frac{dP(L_1)}{dL}\right)\left(1 + \lambda_2 - \frac{2L_2}{|m|P(L_2)}\frac{dP(L_2)}{dL}\right)}{\left(1 - \lambda_1 - \frac{2L_1}{|m|P(L_1)}\frac{dP(L_1)}{dL}\right)\left(1 - \lambda_2 + \frac{2L_2}{|m|P(L_2)}\frac{dP(L_2)}{dL}\right)} = \left(\frac{L_1}{L_2}\right)^{|m|} \tag{16}$$

Because $L_1/L_2 < 1$ and $|m| >> 1$, the right side of (16) is much smaller than 1. It leads to

$$\frac{|m|(1 + \lambda_1)}{2} P(L_1) + L_1 \frac{dP(L_1)}{dL} = 0 \tag{17A}$$

and

$$\frac{|m|(1 + \lambda_2)}{2} P(L_2) - L_2 \frac{dP(L_2)}{dL} = 0 \tag{17B}$$

For our case, terms $D$ in expressions (13A) and (13B) are much smaller than 1. Replacing $\lambda_1$ and $\lambda_2$ by 1, (17A) and (17B) can be written as

$$P(L_1) + \frac{L_1}{|m|}\frac{dP(L_1)}{dL} = 0 \tag{18A}$$

and

$$P(L_2) - \frac{L_2}{|m|}\frac{dP(L_2)}{dL} = 0 \tag{18B}$$

Both (18A) and (18B) have many roots. In general, they need to be solved numerically. However, for two extreme cases: $|\omega| << |m\overline{\omega}_d^j|$ and $|\omega| >> |m\overline{\omega}_d^j|$, we can obtain analytic solutions. In fact, the solutions to the two cases correspond to two branches of the solutions.

For $|\omega| << |m\overline{\omega}_d^j|$, the factor of the exponential function in the resonance terms (the second line of (12B)) is near 1, and the effect of the resonance term is significant. For simplicity, we consider plasma composed of one species of hot ions and hot electrons. Thus, to the second order approximation, we have

$$\sum_j q^j \frac{d\eta_o^j}{dL} \int_0^\infty \frac{\omega_d^j(W) f^j(W)}{-\omega + m\omega_d^j(W)} dW = \frac{e}{m}\left(\frac{d\eta_o^i}{dL} - \frac{d\eta_o^e}{dL}\right) - \frac{e}{m^2}\left(\frac{d\eta_o^i}{dL}\frac{1}{|\overline{\omega}_d^i|} + \frac{d\eta_o^e}{dL}\frac{1}{|\overline{\omega}_d^e|}\right)\omega \tag{19}$$

and

$$P = 1 + i\frac{e\chi L^2}{R^2}\left[m\left(\frac{d\eta_o^i}{dL} - \frac{d\eta_o^e}{dL}\right) - \omega\left(\frac{d\eta_o^i}{dL}\frac{1}{|\overline{\omega}_d^i|} + \frac{d\eta_o^e}{dL}\frac{1}{|\overline{\omega}_d^e|}\right)\right] \quad \text{(no resonance)} \tag{20A}$$

$$P = 1 + i\frac{e\chi L^2}{R^2}\left[m\left(\frac{d\eta_o^i}{dL} - \frac{d\eta_o^e}{dL}\right) - \omega\left(\frac{d\eta_o^i}{dL}\frac{1}{|\overline{\omega}_d^i|} + \frac{d\eta_o^e}{dL}\frac{1}{|\overline{\omega}_d^e|} + i\pi\frac{d\eta_o^i}{dL}\frac{1}{|\overline{\omega}_d^i|}\right)\right] \quad \text{(ion resonance)} \tag{20B}$$

$$P = 1 + i\frac{\chi L^2}{R^2}\left[m\left(\frac{d\eta_o^i}{dL} - \frac{d\eta_o^e}{dL}\right) - \omega\left(\frac{d\eta_o^i}{dL}\frac{1}{|\overline{\omega}_d^i|} + \frac{d\eta_o^e}{dL}\frac{1}{|\overline{\omega}_d^e|} + i\pi\frac{d\eta_o^e}{dL}\frac{1}{|\overline{\omega}_d^e|}\right)\right] \quad \text{(electron resonance)} \tag{20C}$$

Inserting (20A) – (20C) into (18A) and (18B) we have

$$m g_1(L_1) - g_2(L_1)\,\omega + \frac{L_1}{|m|}\left[m\frac{dg_1(L_1)}{dL} - \frac{dg_2(L_1)}{dL}\omega\right] = i\frac{R^2}{e\chi} \tag{21A}$$

$$m g_1(L_2) - g_2(L_2)\,\omega$$

$$-\frac{L_2}{|m|}\left[m\frac{dg_1(L_2)}{dL} - \frac{dg_2(L_2)}{dL}\omega\right] = i\frac{R^2}{e\chi}$$

for the no resonance case,

$$mg_1(L_1) - (g_2(L_1) + i\pi g_3(L_1))\omega$$
$$+\frac{L_1}{|m|}\left[m\frac{dg_1(L_1)}{dL} - \left(\frac{dg_2(L_1)}{dL} + i\pi\frac{dg_3(L_1)}{dL}\right)\omega\right] = i\frac{R^2}{e\chi} \quad (21B)$$

$$mg_1(L_2) - (g_2(L_2) + i\pi g_3(L_2))\omega$$
$$-\frac{L_2}{|m|}\left[m\frac{dg_1(L_2)}{dL} - \left(\frac{dg_2(L_2)}{dL} + i\pi\frac{dg_3(L_2)}{dL}\right)\omega\right] = i\frac{R^2}{e\chi}$$

for the ion resonance case, and

$$mg_1(L_1) - (g_2(L_1) - i\pi g_4(L_1))\omega$$
$$+\frac{L_1}{|m|}\left[m\frac{dg_1(L_1)}{dL} - \left(\frac{dg_2(L_1)}{dL} + i\pi\frac{dg_4(L_1)}{dL}\right)\omega\right] = i\frac{R^2}{e\chi} \quad (21C)$$

$$mg_1(L_2) - (g_2(L_2) - i\pi g_4(L_2))\omega$$
$$-\frac{L_2}{|m|}\left[m\frac{dg_1(L_2)}{dL} - \left(\frac{dg_2(L_2)}{dL} + i\pi\frac{dg_4(L_2)}{dL}\right)\omega\right] = i\frac{R^2}{e\chi}$$

for the electron resonance case. In (21A), (21B) and (21C)

$$g_1(L) = L^2\left(\frac{d\eta_o^i}{dL} - \frac{d\eta_o^e}{dL}\right) \quad (21D)$$

$$g_2(L) = L^2\left(\frac{d\eta_o^i}{dL}\frac{1}{|\overline{\omega}_d^i|} + \frac{d\eta_o^e}{dL}\frac{1}{|\overline{\omega}_d^e|}\right) \quad (21E)$$

$$g_3(L) = L^2\frac{d\eta_o^i}{dL}\frac{1}{|\overline{\omega}_d^i|} \quad (21F)$$

$$g_4(L) = L^2\frac{d\eta_o^e}{dL}\frac{1}{|\overline{\omega}_d^e|} \quad (21G)$$

Solving (21A), (21B) and (21C) we obtain solutions

$$\omega_{l1} = \frac{mg_1(L_1)\left(1 + \frac{G_1(L_1)}{|m|}\right)}{g_2(L_1)\left(1 + \frac{G_2(L_1)}{|m|}\right)} - i\frac{R^2}{e\chi g_2(L_1)\left(1 + \frac{G_2(L_1)}{|m|}\right)}$$

$$\omega_{l2} = \frac{mg_1(L_2)\left(1 - \frac{G_1(L_2)}{|m|}\right)}{g_2(L_2)\left(1 - \frac{G_2(L_2)}{|m|}\right)} - i\frac{R^2}{e\chi g_2(L_2)\left(1 - \frac{G_2(L_2)}{|m|}\right)} \quad (22A)$$

(no resonance)

and

$$\omega_{l1} = \frac{mg_1(L_1)\left(1 + \frac{G_1(L_1)}{|m|}\right) - i\frac{R^2}{e\chi}}{g_2(L_1)\left(1 + \frac{G_2(L_1)}{|m|}\right) + i\pi g_3(L_1)\left(1 + \frac{G_2(L_1)}{|m|}\right)}$$

$$\omega_{l2} = \frac{mg_1(L_2)\left(1 - \frac{G_1(L_2)}{|m|}\right) - i\frac{R^2}{e\chi}}{g_2(L_2)\left(1 - \frac{G_2(L_2)}{|m|}\right) + i\pi g_3(L_2)\left(1 - \frac{G_2(L_2)}{|m|}\right)} \quad (22B)$$

(ion resonance)

$$\omega_{l1} = \frac{mg_1(L_1)\left(1 + \frac{G_1(L_1)}{|m|}\right) - i\frac{R^2}{e\chi}}{g_2(L_1)\left(1 + \frac{G_2(L_1)}{|m|}\right) + i\pi g_4(L_1)\left(1 + \frac{G_2(L_1)}{|m|}\right)}$$

$$\omega_{l2} = \frac{mg_1(L_2)\left(1 - \frac{G_1(L_2)}{|m|}\right) - i\frac{R^2}{e\chi}}{g_2(L_2)\left(1 - \frac{G_2(L_2)}{|m|}\right) + i\pi g_4(L_2)\left(1 - \frac{G_2(L_2)}{|m|}\right)} \quad (22C)$$

(electron resonance)

where

$$G_1 = \frac{L}{g_1(L)}\frac{dg_1(L)}{dL} \quad (22D)$$

$$G_2 = \frac{L}{g_2(L)}\frac{dg_2(L)}{dL} \quad (22E)$$

$$G_3 = \frac{L}{g_3(L)}\frac{dg_3(L)}{dL} \quad (22F)$$

$$G_4 = \frac{L}{g_4(L)}\frac{dg_4(L)}{dL} \quad (22G)$$

Waves $\omega_{l1}$ and $\omega_{l2}$ are related to the inside and outside boundaries, respectively. The non-resonant solutions have only negative growth rates, and the waves decay. The solutions for the ion resonance and electron resonance cases have positive growth rates when $g_1 < 0$ and $m > 0$, and $g_1 > 0$ and $m < 0$, respectively. In other words, if $g_1 < 0$, the ion resonance excites a growing wave; if $g_1 < 0$, the electron resonance excites a growing wave.

To determine simplified expressions for the frequencies of growing waves, we ignore the minor terms with $1/|m|$ in (22B) and (22C), and assume that the hot ions and electrons have the same temperature. Also, we take $d\eta_o^e/dL \approx d\eta_o^i/dL = d\eta_o^h/dL$ ($d\eta_o^h/dL$ represents the average of $d\eta_o^e/dL$ and $d\eta_o^i/dL$) in the calculations except when their difference is involved. For the latter case we take $d\eta_o^i/dL - d\eta_o^e/dL = y d\eta_o^h/dL$. Then we have wave frequencies

$$\omega_{l1} = \frac{\left|\overline{\omega}_d^i(L_1)\right|}{(4+\pi^2)}\left[-\left(2|my| + \frac{\pi R^2}{e\chi L_1^2}\frac{d\eta_o^h(L_1)}{dL}\right)\right.$$
$$\left. + i\left(\pi|my| - \frac{2R^2}{e\chi L_1^2}\frac{d\eta_o^h(L_1)}{dL}\right)\right]$$
$$\omega_{l2} = \frac{\left|\overline{\omega}_d^i(L_2)\right|}{(4+\pi^2)}\left[-\left(2|my| + \frac{\pi R^2}{e\chi L_2^2}\frac{d\eta_o^h(L_2)}{dL}\right)\right.$$
$$\left. + i\left(\pi|my| - \frac{2R^2}{e\chi L_2^2}\frac{d\eta_o^h(L_2)}{dL}\right)\right] \quad (23)$$

When $y < 0$, the unstable wave is excited by ion resonance; when $y > 0$, the unstable wave is excited by electron resonance. The growth rates are proportional to the particle drift velocities, and linear in the product of $m$ and $y$. The second terms of the imaginary parts in (23) impede the excitation of drift waves. They come from the effect of field-aligned resistivity, which has effectively reduced the line-tying effect of the ionosphere.

For the case $|\omega| >> |m\overline{\omega}_d^j|$, the contribution of the resonance terms to the dispersion relation is negligible. We still consider a plasma consisting of one hot ion species and a hot electron population. Thus, we have

$$P = 1 + i\frac{m^2\chi L^2}{R^2}\sum_j q^j \frac{d\eta_o^j}{dL}\int_0^\infty \frac{\omega_d^j(W)f^j(W)}{-\omega + m\omega_d^j(W)}dW$$
$$\approx 1 + i\frac{m^2 e\chi L^2}{\omega R^2}\left(\frac{d\eta_o^i}{dL}\left|\overline{\omega}_d^i\right| + \frac{d\eta_o^e}{dL}\left|\overline{\omega}_d^e\right|\right) \quad (24)$$

Substituting (24) into (17A) and (17B) then gives

$$\omega + i\frac{m^2 e\chi}{R^2}h(L_1) + i\frac{|m|e\chi L_1}{R^2}\frac{dh(L_1)}{dL} = 0 \quad (25A)$$

and

$$\omega + i\frac{m^2 e\chi}{R^2}h(L_2) - i\frac{|m|e\chi L_2}{R^2}\frac{dh(L_2)}{dL} = 0 \quad (25B)$$

where

$$h(L) = L^2\left(\frac{d\eta_o^i}{dL}\left|\overline{\omega}_d^i\right| + \frac{d\eta_o^e}{dL}\left|\overline{\omega}_d^e\right|\right) \quad (25C)$$

Solving (25A) and (25B) gives solutions to (17A) and (17B)

$$\omega_{h1} = -i\frac{m^2 e\chi}{R^2}\left(1 + \frac{H(L_1)}{|m|}\right)h(L_1) \quad (26A)$$

$$\omega_{h2} = -i\frac{m^2 e\chi}{R^2}\left(1 - \frac{H(L_2)}{|m|}\right)h(L_2) \quad (26B)$$

where

$$H = \frac{L}{h(L)}\frac{dh(L)}{dL} \quad (26C)$$

Both waves $\omega_{h1}$ and $\omega_{h2}$ have negative growth rates. Therefore, the waves with $|\omega| >> |m\overline{\omega}_d^j|$ decay.

In summary, in the evening sector of Region I and the morning sector of Region II the effect of field-aligned resistivity reduces the line-tying effect of the ionosphere. The reduction of the line-tying effect makes possible the excitation of a growing drift wave in the Birkeland current flow region by the resonance of wave with particle drift. The frequencies of the unstable drift waves satisfy $|\omega| << |m\overline{\omega}_d^j|$. They are excited by ion resonance in the case that the radial gradient of hot ion distribution is smaller than the radial gradient of hot electron distribution, and excited by electron resonance in the reverse case.

## 4. Numerical results

The Birkeland current flow at the ionosphere is located at latitudes of 66° – 74° [Potemra et al, 1979]. The locus of field line footprints connected to the evening sector of Region I and the morning sector of Region II in the equatorial plane can be represented by two half annular sheets with inside boundaries of $L$ = 8.5, 6 and outside boundaries of $L$ = 13 and 8.5. Therefore, in our calculations, we have $L_1 = 8.5$ and $L_2 = 13$ for the evening sector of Region I, and $L_1 = 6$ and $L_2 = 8.5$ for the morning sector of Region II.

Observations determining the plasma configuration in the magnetosphere region of $L = 5 - 15$ have been collected for many years. The energy density (or plasma pressure) and particle number density vary widely, and are closely related to geomagnetic activity. According to the results given by Spence et al [1989], the plasma pressure in this region is roughly proportional to $L^{-4}$. We ignore particles' parallel motion, and assume that the magnetic moment of the particles remains constant during the plasma inward convection process. Thus the plasma density in this region is inversely proportional to $L$, and the plasma content is proportional to $L^3$. In the following calculations we take two groups of typical values for ion temperature and average number density of hot ion and electron at $L$ = 5: (A) 2 $keV$ and $1\times10^6\ m^{-3}$ [Lennartsson, 1989]; (B) 100 $keV$ and $2\times10^5\ m^{-3}$ [Smith and Hoffman, 1973]. Also, we assume the electron temperature is equal to the ion temperature. Based on above data and assumptions we have (in MKS units):

$$\eta_o^h = 1\times10^{18}L^3$$
$$\frac{d\eta_o^h}{dL} = 3\times10^{18}L^2$$
$$\overline{\omega}_d^e = \overline{\omega}_d^i = 6\times10^{-4}L^{-2}$$

for temperature and particle density values (A), and

$$\eta_o^h = 2\times10^{17}L^3$$

$$\frac{d\eta_o^h}{dL} = 6\times10^{17}L^2$$

$$\overline{\omega}_d^e = \overline{\omega}_d^i = 0.03L^{-2}$$

for temperature and particle density values (B).

In terms of values (A) we can represent expression (23) as

$$\omega_{I1} = 1.9\times[(-(2|my| + 51) + i(\pi|my| - 32)]\times10^{-6}sec^{-1}$$

$$\omega_{I2} = 7.9\times[(-(2|my| + 9.4) + i(\pi|my| - 6.0)]\times10^{-7}sec^{-1}$$

for the evening sector of Region I, and

$$\omega_{I1} = 3.8\times[(-(2|my| + 210) + i(\pi|my| - 132)]\times10^{-6}sec^{-1}$$

$$\omega_{I2} = 1.9\times[(-(2|my| + 51) + i(\pi|my| - 32)]\times10^{-6}sec^{-1}$$

for the morning sector of Region II. Similarly, values (B) give

$$\omega_{I1} = 9.4\times[(-(2|my| + 51) + i(\pi|my| - 32)]\times10^{-5}sec^{-1}$$

$$\omega_{I2} = 4.0\times[(-(2|my| + 9.4) + i(\pi|my| - 6.0)]\times10^{-5}sec^{-1}$$

for the evening sector of Region I, and

$$\omega_{I1} = 1.9\times[(-(2|my| + 210) + i(\pi|my| - 132)]\times10^{-4}sec^{-1}$$

$$\omega_{I2} = 9.4\times[(-(2|my| + 51) + i(\pi|my| - 32)]\times10^{-5}sec^{-1}$$

for the morning sector of Region II.

Observational hot electron and hot ion number densities in the inner magnetospheric plasma differ by a factor 1.5 ~ 2.0 [e.g., Garrett et al., 1981a, 1981b]. As an example, we assume a ratio $y = 0.3$. Then the critical wave numbers for the growing wave are 32 and 6 for the evening sector of Region I, and 132 and 32 for the morning sector of Region II. In the evening sector of Region I, a wave mode with $|m| = 300$ (corresponding to wavelengths 0.06 ~ 0.13 *R*) has a growth rate 0.8 ~ 1.7 hour$^{-1}$ for values (A) and 40 ~ 85 hour$^{-1}$ for values (B). It is obvious that the flux interchange instability caused by these waves is significant.

Acknowledgement. The author is grateful to R. A. Hoffman and C. S. Wu for helpful discussions. Also, comments by R. A. Wolf and B. H. Mauk were appreciated. This work was completed while the author was an NAS/NRC Research Associate at Laboratory for Extraterrestrial Physics, Goddard Space Flight Center.

## References

Garrett, H. B., D. C. Schwank, and S. E. DeForest, A statistic analysis of the low-energy geosynchronous plasma environment, I, Electrons, Planet Space Sci., 29, 1021, 1981a.

Garrett, H. B., D. C. Schwank, and S. E. DeForest, A statistic analysis of the low-energy geosynchronous plasma environment, II, Ions, Planet Space Sci., 29, 1045, 1981b.

Huang, T. S., R. A. Wolf, and T. W. Hill, Interchange instability of the Earth's plasmapause, J. Geophys. Res., 95, 17187, 1990.

Huang, T. S., and T. W. Hill, Drift wave instability in the Io plasma torus, submitted to *J. Geophys. Res.*, 1990.

Huba, J. D., and C. S. Wu, Effect of a magnetic field gradient on the lower hybrid drift instability, Phys. Fluids, 19, 988, 1976.

Kan, J. R., and F. Cao, Effect of field-aligned potential drop in a global magnetosphere-ionosphere coupling model, J. Geophys. Rev., 93, 7571, 1988.

Knight, L., Parallel electric field, Planet. Space Sci., 21, 741, 1973.

Lennartsson, W., Energetic (0.1- to 16- keV/e) magnetospheric ion composition at different levels of solar F10.7, J. Geophys. Res., 94, 3600, 1989.

Lyons, L. R., Generation of Large-scale regions of auroral currents, electric potentials, and precipitation by the divergence of the convection electric field, J. Geophys. Res., 85, *17*, 1980.

Potemra, T. A., T. Iijima, and N. A. Saflekos, Large-scale characteristics of Birkeland currents, in Dynamics of the Magnetosphere, edited by S.-I. Akasofu, D. Reidel, Hingham, Mass., pp. 165-199, 1979.

Richmond, A. D., Self-induce motions of thermal plasma in the magnetosphere and the stability of the plasmapause, Rad. Sci., 8, 1019, 1973.

Smith, P. H., and R. A. Hoffman, Ring current particle distributions during the magnetic storms of December 16-18, 1971, J. Geophys. Res., 78, 4731, 1973.

Spence, H. E., M. G. Kivelson, R. J. Walker, and D. J. McComas, Magnetospheric plasma pressures in the midnight meridian: observations from 2.5 to 35 $R_E$, J. Geophys. Res., 94, 5264, 1989.

Strangeway, R. J., and S. M. Kaye, Quiet time mass composition at near geosynchronous altitudes, J. Geophys. Res., 91, 7105, 1986.

Vasyliunas, V. M., Mathematic models of magnetospheric convection and its coupling to the ionosphere, in Particles and Fields in the Magnetosphere, edited by B. M. McComas, D. Reidel, Dordrecht-Holland, pp. 60-71, 1970.

RING CURRENT $0^+$ INTERACTION WITH PC 5 MICROPULSATIONS

S. Qian and M. K. Hudson

Department of Physics and Astronomy, Dartmouth College
Hanover, New Hampshire 03755

I. Roth

Space Sciences Laboratory, University of California at Berkeley
Berkeley, California, 94720

Abstract. Pc 5 ULF micropulsations in the few mHz frequency range have been measured by satellite observations during and after a geomagnetic storm. Rapid loss of ring current oxygen during the recovery phase of the storm was observed by ISEE-1. The loss of oxygen ions was attributed to an interaction of oxygen with micropulsations, which have the same period as the oxygen bounce time. A test particle model is used here to study this drift-bounce resonant interaction. In this model test particles move in the dipole magnetic field of the earth, with corotation and convection electric fields included, along with compressional Pc 5 wave fields. Studies of $H^+$ and $O^+$ orbits show that the wave interacts strongly with low energy ring current $O^+$ (tens of keV) and has little effect on $H^+$ below 100 keV. The convection electric field plays an important role in tuning the resonance condition $\omega_b=\pm(m\omega_d-\omega)$, since the bounce and drift frequencies are energy dependent. The drift-bounce resonant interaction and oscillations of the particle energy, $L$ value and first adiabatic invariant are studied for different wave modes, ion energies and pitch angles. It is suggested that this interaction may be responsible for loss from the ring current of low energy $O^+$ satisfying the resonance condition.

## Introduction

Compressional Pc 5 magnetic pulsations have been observed by multiple satellites during quiet to moderate geomagnetic activity. The events have been studied by *Higbie et al.* [1982], *Nagano and Araki* [1983], *Takahashi et al.* [1985; 1987] and others. Those pulsations excited during the recovery phase of a magnetic storm are observed over the entire dayside. The compressional Pc 5 associated with recovery phase has a frequency of about 1–5 mHz, and propagates westward with azimuthal wavelength in the range from 2000 to 7000 km in the reference frame of the plasma. Unlike stormtime Pc 5s which occur immediately after an injection of ring current particles, compressional Pc 5s are observed when geomagnetic activity becomes quiet with $K_p$ around one. Compressional Pc 5 pulsations have significant wave magnetic fields parallel to the earth's magnetic field, in contrast to transverse waves characteristic of stormtime Pc 5s.

The $H^+$ and $O^+$ phase space density measurements made by the ISEE-1 satellite show significant loss of 5–15 keV ring current oxygen ions during the recovery phase of a storm. The densities were measured while the satellite was inbound at 0900 local time and then while it was outbound around 0600 local time during the recovery phase of the storm of December 11, 1977 [*Cladis and Lennartsson*, 1986]. Using inbound phase space densities as the initial condition, Cladis and Lennartsson studied the adiabatic drift motion of hydrogen ions and found that the phase space densities in the two measurements along the drift paths were consistent with the observations. However, the measured phase space density of oxygen ions was lower on the outbound leg than calculated from the drift of ions measured on the inbound leg, which implies nonadiabatic behavior of oxygen ions. They presented simultaneously measured Pc 5 wave data from the electric field instrument on ISEE-1, and suggested that a bounce-resonant interaction of oxygen ions with standing Alfvén waves might be responsible for the enhanced loss of ring current oxygen observed.

The transverse Pc 5 waves which were studied by *Cummings et al.* [1969] were modeled as a standing Alfvén wave. The drift-mirror instability [*Hasegawa*, 1969] and drift-bounce resonant interaction of ring current protons with Alfvén waves [*Southwood et al.*, 1969; *Southwood*, 1973; 1976; *Chen and Hasegawa*, 1988] have been proposed as possible excitation mechanisms. In this paper we concentrate on the effect of Pc 5s on ring current ions, rather than on the generation mechanism of the waves. The particle motion is studied with a test particle code. Convection and corotation electric fields are included in the calculations where noted, and these play an important role in tuning the drift-bounce resonance with Pc 5 waves by changing the particle energy and therefore its drift and bounce frequencies. This study shows that oxygen ions in the 15 keV energy range undergo strong oscillations in equatorial radial position $L$, perpendicular and parallel energy, and first adiabatic invariant $\mu$, due to a drift-bounce resonant interaction with the wave. The test particle calculations show that all four quantities, $W_\perp$, $W_\parallel$, $L$ and $\mu$ undergo oscillations at the same frequency. For ions which are near but not exactly at resonance, the oscillation frequency obtained analytically is the mismatch from resonance $\Delta\omega=\omega_b\pm(m\,\omega_d-\omega)$, where $\omega_b$ is the bounce frequency, $\omega_d$ is the bounce averaged drift frequency, m is the azimuthal mode number and $\omega$ is the wave frequency. The results indicate that only antisymmetric waves can have a resonant interaction with ring current ions for measured wave parameters. The condition for resonant interaction is discussed, and it is shown that hydrogen in the 100 keV energy range can also satisfy the drift-bounce resonance condition. These higher energy hydrogen ions which constitute the bulk of the ring current may be responsible for exciting compressional Pc 5s via drift-bounce resonance [*Southwood*, 1976; *Chen and Hasegawa*, 1988], while the waves so generated affect the highly time variable low energy oxygen component.

Modeling Magnetospheric Plasma Processes
Geophysical Monograph 62

## Test Particle Model and Results

We use a test particle model to study ion behavior in a compressional Pc 5 pulsation. The electromagnetic field is assumed to consist of a time-independent part, which is the dipole magnetic field with superimposed corotation and convection electric fields, and time-dependent Pc 5 wave fields. We follow *Cladis and Francis* [1985], prescribing a corotation and convection electric field as follows:

$$\mathbf{E}_{cor} = -(\Omega_E \times \mathbf{r}) \times \mathbf{B}$$

$$\mathbf{E}_{con} = -\nabla \Psi ,$$

where $\mathbf{B}$ is the earth's dipole magnetic field and a $K_p$ dependent potential is assumed [*Maynard and Chen,* 1975]

$$\Psi = 0.045\,(1 - 0.159\,K_p + 0.0093\,K_p^2)^{-3} L^2 \sin(\phi) .$$

Here $\Psi$ is in kV, $\Omega_E$ is the angular velocity of the earth, $r$ is the distance from the center of the earth to the field point, $R_E$ is an earth radius and $\phi$ is azimuthal angle. $K_p$ is about $1-2$ during the recovery phase of a magnetic storm. The resulting convection field whose magnitude increases with $r$ is mapped along a magnetic field line assuming no parallel potential drop and is only valid out to $L = 10$ [*Volland,* 1973]. The compressional Pc 5 is assumed to be a standing wave traveling in the azimuthal direction. The wave fields are specified by the time varying components $b_\parallel$, $b_\alpha$ and $e_\phi$ which are the compressional and radial perturbed magnetic field components and the azimuthal perturbed electric field component, collectively satisfying Faraday's Law and $\nabla \cdot \mathbf{b} = 0$. The fields are determined from a vector potential which has only an azimuthal component $\mathbf{A}_\phi$ such that

$$\mathbf{b} = \nabla \times \mathbf{A}_\phi$$

$$\mathbf{e}_\phi = \frac{-1}{c}\frac{\partial}{\partial t}\mathbf{A}_\phi$$

where $\mathbf{A}_\phi = R_E\, B_0 \beta_\parallel \alpha^n \sin(k\lambda)\, e^{i(m\phi - \omega t)}$ for the antisymmetric mode, with $\cos(k\lambda)$ replacing $\sin(k\lambda)$ for the symmetric mode. Here $\lambda$ is magnetic latitude in radians, $k$ is a dimensionless parallel wavenumber, m is azimuthal mode number, and $B_0 = 0.31$ G. Dipole coordinates used in the calculation are

$$\alpha = \frac{R_E}{r}\cos^2\lambda , \quad \delta = \left[\frac{R_E}{r}\right]^2 \sin\lambda$$

and longitude $\phi$, where $\hat{\alpha}$, $\hat{\phi}$, $\hat{\delta}$ form a right hand coordinate system with $\hat{\alpha} = -\hat{r}$ and $\hat{\delta} = -\hat{\theta}$ relating dipole and spherical coordinates at the equator. Guided by satellite measurements, we choose azimuthal mode number m = 40 to 140 for a westward propagating wave, and the wave frequency varies from 2 to 5 mHz. In all cases except otherwise mentioned, the antisymmetric wave is used and particles are started at 0900 local time in the equatorial plane with $r = 6\,R_E$, in order to model the ISEE-1 observations. Assuming $n = 2$ results in $b_\parallel$, $b_\alpha \propto B \propto 1/L^3$, where $B$ is the equatorial dipole magnetic field strength, while $n = -1$ yields $b_\parallel$, $b_\alpha$ independent of $L$. The general form of the wave fields, including other $n$ values, is discussed in the Appendix. $\beta_\parallel$ is an amplitude factor in the range 0.0015 to 0.008 in the simulations, chosen to give wave amplitudes comparable to the measured amplitudes of $b_\parallel/B \sim 10\%$, which results in $b_\alpha \sim 20$ nT and $e_\phi \sim 5$ mV/m for $n = 0$. Typically we take $n = 0$, $k = 2$ and m = 40, which results in the latitudinal dependence of $b_\parallel$, $b_\alpha$, and $e_\phi$ plotted in Figure 1. The model takes into account the curvature of the dipole magnetic field geometry, resulting in perturbed fields which are in reasonable agreement with Chen and Hasegawa's eigenmode calculation. More recently a Gaussian envelope has been employed in the $n = 0$ case to confine the imposed electromagnetic fields in $L$ (Hudson, *et al.,* 1991).

The particle orbit is numerically integrated from the equation of motion

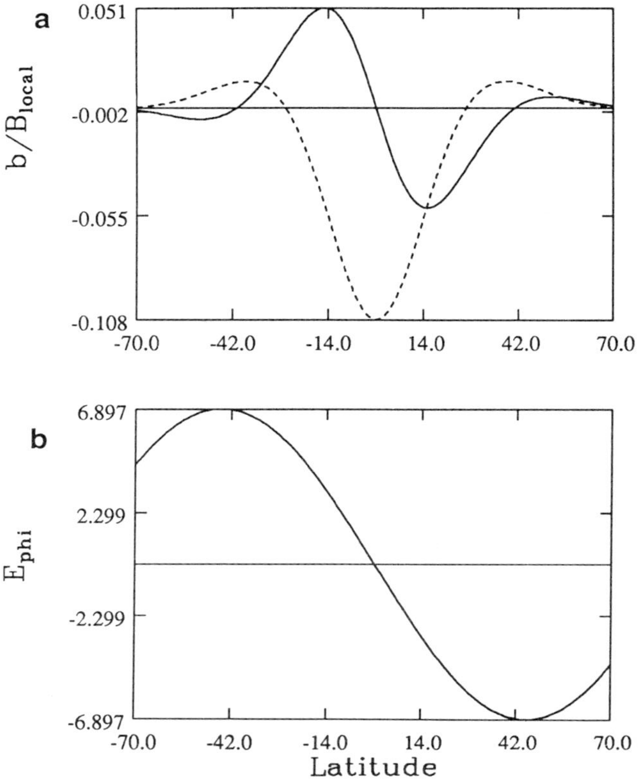


n=0 rkz=2. bd=0.0015 f=0.0037 l=6. l0=6. a=1.

Figure 1. *(a)* Antisymmetric wave $b_\parallel$ and $b_\alpha$ vs. $\lambda$ *(b)* $e_\phi$ vs. $\lambda$ in mV/m for $n = 0$, $\beta_\parallel = 0.0015$, $k = 2$, $L = 6$ and $f = 3.7$ mHz.

$$m\frac{d\mathbf{v}}{dt} = q\,(\mathbf{E} + \mathbf{v} \times \mathbf{B}) \quad \text{and} \quad \frac{d\mathbf{x}}{dt} = \mathbf{v} .$$

We normalize time to the gyrofrequency at the equator, length to the initial gyroradius at the equator and velocity to the initial thermal velocity of hydrogen. We use the Richardson extrapolation with modified midpoint method [*Press et al.,* 1988] to push the particles in phase space. At each step of the extrapolation procedure, we decompose the perturbed fields as well as the convection and corotation fields into Cartesian components. After evaluation of the new positions and velocities of the particle, we calculate the fields at the new location as functions of the dipole coordinates. Our diagnostics consist of plotting macroscopic quantities (e.g., energy, magnetic moment) as functions of time, and trajectories in dipole coordinates.

Figures 2*a* and 3*a* show orbits of 15 keV hydrogen and oxygen ions injected with 50 degree pitch angle at 0900, projected onto the equatorial plane. The hydrogen ion drifts westward along the L shell toward the duskside and its orbit is almost identical to that without a wave perturbation; there is a small change in $L$ due to the convection electric field. The projection of the particle trajectory onto the equatorial plane is such that the inner radius of the plot is the mirror point and the outer radius is the equatorial crossing point. The oxygen orbit in Figure 3 is strongly affected by the wave. In Figures 2*b* and 3*b*, parallel, transverse and total kinetic energy of the $H^+$ and $O^+$ ions are plotted when the ions cross the equatorial plane. When $H^+$ drifts westward from 0900 to 0600 in Figure 2, it drifts toward an electric potential barrier due to the con-

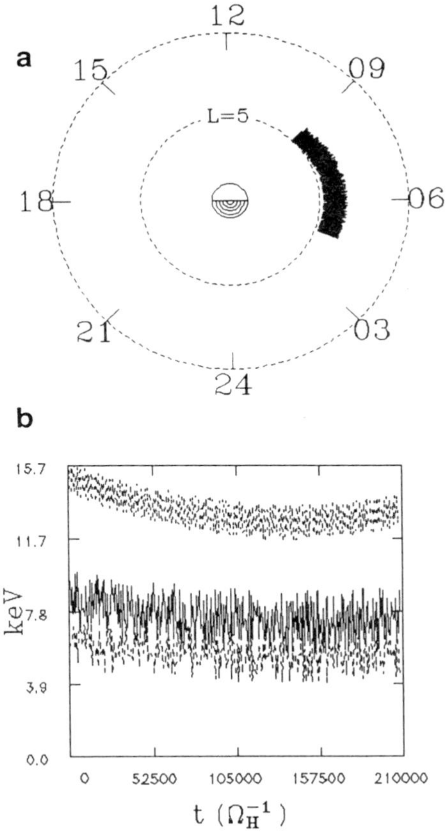


Figure 2. Trajectory time history for a single ion injected into prescribed dc and ac wave fields. Ion is injected at the equator at 0900 LT and gyro, bounce and drift motion is followed numerically. Inner and outer radius of the equatorial projection correspond to mirror and equatorial crossing points respectively. *(a)* Projection of a hydrogen ion orbit onto the equatorial plane for $K_p = 1.5$. Initial ion energy is 15 keV with 50 degree equatorial pitch angle at t=0. The Pc 5 wave has mode number m=40, $n = 0$, $k = 2$, $\beta_\parallel = .0015$, and frequency of 3.7 mHz; *(b)* Energy history of the hydrogen ion is also shown. The solid line is perpendicular kinetic energy, and the long-dashed line is parallel kinetic energy; the short-dashed line is total kinetic energy.

vection electric field. The potential barrier is 3–6 keV high for ions drifting at $L = 6-8$ for $K_p = 1.5$. If a 15 keV ion drifts westward adiabatically, its kinetic energy decreases until it reaches 0600 and then it increases. The motion of the hydrogen ion is adiabatic, while it is apparent that the motion of the oxygen ion is nonadiabatic. For the later case, we have turned off the convection and corotation electric fields, added in subsequent examples, in order to focus on drift-bounce resonance effects.

Figure 4 shows the effect of adding the convection and corotation electric fields to the $O^+$ case shown in Figure 3. Now the oxygen is first decelerated, then accelerated by the convection electric field, such that $\Delta\omega$ changes continuously as $\omega_b$ and $\omega_d$ vary with changing total energy $W$. Thus there is a slow oscillation in quantities which depend on total energy $W$, and a tuning of the oscillation at $\Delta\omega$. Very large excursions in $L$ result, and oxygen may be lost to the magnetopause under such circumstances.

## Analysis

From the preceding test particle calculations we find that the displacement of $O^+$ from an $L$ shell is due to a drift-bounce resonant interaction of the oxygen ion with an antisymmetric wave when $\omega_b + m\omega_d \sim \omega$. In the case shown in Figure 3, the oxygen bounce frequency $\omega_b$ plus $m\omega_d$ is approximately 3.7 mHz, the frequency of the compressional wave. Because of the different mass, the hydrogen bounce frequency is about four times larger and the resonance cannot occur for the parameters of Figure 2. Physically, the drift-bounce resonance occurs when a particle drifts in one bounce period by an integral number of azimuthal wavelengths Doppler shifted by the azimuthal phase velocity of the wave. The particle then sees approximately static fields. The resonance is also affected by the corotation electric field and drift of the plasma, which shifts the resonant frequency for a given particle energy, mass, and wave mode number m. Finally, the convection electric field continuously changes the particle energy, hence affecting the resonance condition through the dependence of $\omega_b \propto W^{1/2}$ and $\omega_d \propto W$. We see in Figure 4 that a particle near resonance at 0600 LT moves away from resonance as it gains energy from the convection electric field across the nightside.

In Figure 3 it is seen that a resonant particle has a drift path which oscillates between different $L$ shells at the same frequency as the oscillations in perpendicular, parallel and total particle kinetic energy. This frequency can be calculated from the bounce frequency $\omega_b$ and the drift frequency $\omega_d$ using formulas from *Hamlin et al.* [1961] applied to the initial particle energy (15 keV) and pitch angle (50°) and average $L \cong 7$, yielding $|\Delta\omega| = |\omega_b + m\omega_d - \omega| \cong .106\omega$ corresponding to a period $\tau_{\Delta\omega} = 2.55 \times 10^3$ s, or $3.5 \times 10^4\ \Omega_H^{-1}$, which agrees approximately with the slow time scale oscillation in $L$, $W_\perp$, $W_\parallel$ and $W$ in Figure 3.

The change in total kinetic energy is due to the static electric field and wave electric field. The contribution from the convection electric field is simply $\Psi(r_0) - \Psi(r)$. The corotation electric field changes particle kinetic energy when particles drift in the radial direction [*Southwood and Kivelson,* 1981; *Kivelson and Southwood*, 1985]

$$\delta W_r \cong q \int_0^t \delta \mathbf{v}_L \cdot \mathbf{E}_{\rm cor}(\mathbf{r}(t'))\, dt'. \tag{1a}$$

where

$$\delta \mathbf{v}_L = c\,(-\mu \nabla_\phi b_\parallel \times \mathbf{B} + q\, \mathbf{e}_\phi \times \mathbf{B})/qB^2 \tag{1b}$$

and $\mathbf{r}(t')$ is the particle trajectory. When particles drift at nearly constant $L$, $\delta W_r$ is small, but it may become significant near resonance when particles undergo radial oscillations in their drift paths. Similarly, the convection electric field modifies particle kinetic energy

$$\delta W_c = q \int_0^t (\mathbf{v}_d + \delta \mathbf{v}_L) \cdot \mathbf{E}_{\rm con}(\mathbf{r}(t'))\, dt'. \tag{1c}$$

The convection electric field also modifies the radial particle position as

$$R_E \frac{d}{dt} L = \frac{c\, \mathbf{E}_{\rm conv} \times \mathbf{B}(L)}{B^2} \tag{1d}$$

which yields

$$\frac{\Delta L}{L} = \frac{3}{4} \left[ \frac{c\, E_0}{B} \right] \frac{1}{\omega_d}$$

where $E_0$=0.045 kV for $K_p$=0, and is otherwise obtained from $\Psi$. The effect of the convection electric field is apparent in the slow oscillation of $W, W_\perp, W_\parallel, L$ and $\mu$ seen in Figure 4. This oscillation comes from the $\cos\phi \equiv \cos\omega_d t$ dependence of the convection electric field, which through (1*c*) affects the particle energies and through (1*d*) the $L$ value.

We will estimate $\delta W$ due to wave fields for the case where $E_{\rm cor} = E_{\rm conv} = 0$ for comparison with Figure 3. The average acceleration

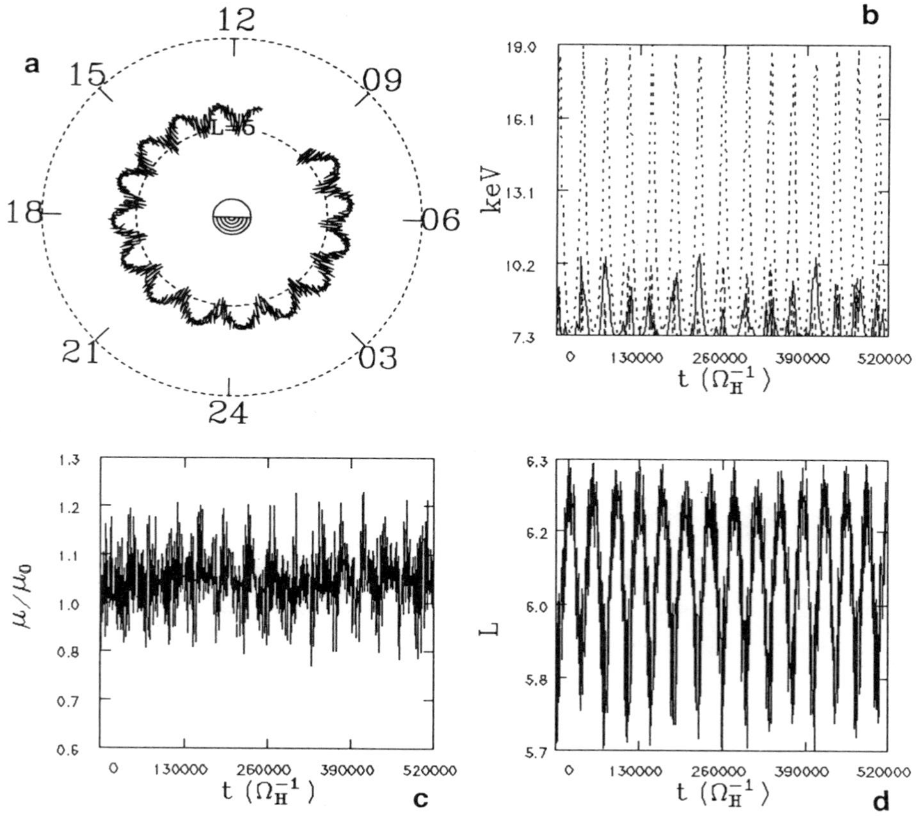


Figure 3. Same as Figure 2 for 15 keV oxygen with 50 degree equatorial pitch angle at $t=0$ and no convection or corotation electric field. The radial oscillation period in *(a)* and the energy oscillation period in *(b)* correspond to the mismatch frequency $|\Delta\omega| = |\omega_b + m\omega_d - \omega|$. *(c)* Magnetic moment time history and *(d)* L value time history.

by the transverse wave electric field over many gyro and bounce periods yields

$$\delta W = q\int_0^t v_d\, e_\phi(\alpha,\lambda,\phi;t')\,dt'. \tag{1e}$$

where $\delta W = \delta W_\perp + \delta W_\parallel$. Explicitly,

$$\dot W_\parallel = -\mu(\mathbf{v}\cdot\nabla)B \tag{1f}$$

$$\dot W_\perp = \mu(\mathbf{v}\cdot\nabla)B + q\,\mathbf{v}_d\cdot\mathbf{e}_\phi \tag{1g}$$

The mirror force exchanges energy back and forth between $W_\parallel$ and $W_\perp$, while the second term in (1g) integrated in (1e) produces an oscillation in total energy.

The time integration in (1*e*) is along the particle trajectory. The particle guiding center is approximated by [*Schulz and Lanzerotti*, 1974],

$$\alpha = L^{-1} \tag{2a}$$

$$\lambda(t) \cong \frac{p\sqrt{1-B/B_{max}}}{L\,R_E\,M\,\omega_b}\sin(\omega_b t + \lambda_0) \equiv \Lambda \sin(\omega_b t + \lambda_0) \tag{2b}$$

$$\phi(t) \cong \omega_d t + \phi_0 \tag{2c}$$

where $p$ is the particle momentum at the equator, $\Lambda$ is the mirror point magnetic latitude, $B$ is the dipole magnetic field strength at the equator and $B_{max}$ is the dipole magnetic field strength at the mirror point. Eqs. (2*a*)−(2*c*) assume a constant $L$ value and constant drift velocity for the particle trajectory. Eq. (2*b*) follows from the sinusoidal behavior of particle motion along a field line arc length $s$ [*Schulz and Lanzerotti*, eq. 1.26), which is related to parallel dipole coordinate $\delta$ by $ds = R_E\,(B_0/B)\,d\delta$. We have also used the near equatorial approximation $\delta \cong \lambda/L^2$. The energy change due to the antisymmetric transverse wave electric field can be calculated from the unperturbed particle trajectory, keeping only the leading terms in the Bessel function summations for small argument $k\Lambda$

$$\frac{\delta W}{W_\perp} = -i\,3\omega\,(1+2v_\parallel^2/v_\perp^2)\,\beta_\parallel J_1(k\Lambda)\left\{\frac{\exp[i(\omega_b + m\omega_d - \omega)t]-1}{\omega_b + m\omega_d - \omega} - \frac{\exp[i(-\omega_b + m\omega_d - \omega)t]-1}{-\omega_b + m\omega_d - \omega}\right\}\exp(i m\phi_0) \tag{3a}$$

This expression assumes $n=2$ in the wave field model, and should be multiplied by $1/L^{n-2}$ for other $n$ values. Here we have substituted the particle guiding center coordinates (2*a*)-(2*c*) with $\lambda_0 = 0$ into $e_\phi$ given in the Appendix and then into (1*e*), expanding $\sin k\lambda$ into a Bessel func-

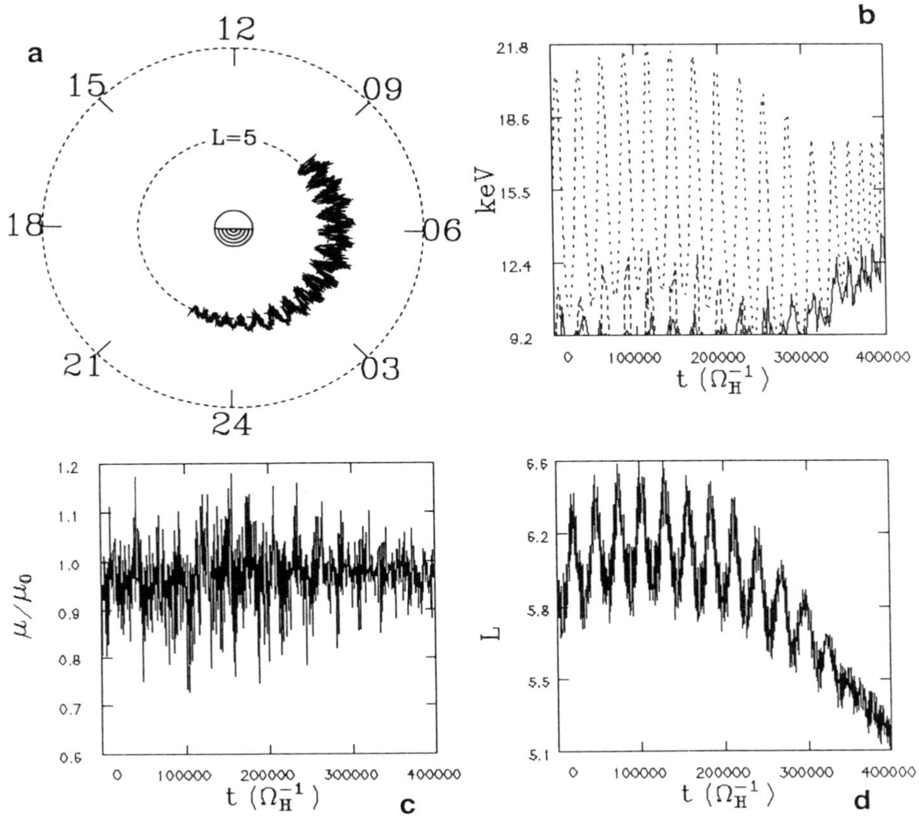


Figure 4. Same as Figure 3 with convection and corotation electric field included for $K_p = 1.5$, wave frequency of 3.7 mHz and other parameters the same.

tion summation and setting $\cos\lambda \cong 1$. For the symmetric wave we obtain for $n = 2$

$$\frac{\delta W}{W_\perp} = 3\omega\,(1+2v_\parallel^2/v_\perp^2)\,\beta_\parallel\left\{J_0(k\Lambda)\,\frac{\exp[i\,(m\,\omega_d-\omega)t\,]-1}{m\omega_d-\omega} + J_2(k\Lambda)\left[\frac{\exp[i\,(2\omega_b+m\omega_d-\omega)t\,]-1}{2\omega_b+m\omega_d-\omega} - \frac{\exp[i\,(-2\omega_b+m\omega_d-\omega)t\,]-1}{-2\omega_b+m\omega_d-\omega}\right]\right\}\exp(i\,m\phi_0) \qquad (3b)$$

The denominators in eq. ($3a$) show resonant behavior whenever $\omega_b = \pm(m\,\omega_d-\omega)$ is satisfied for the antisymmetric wave. The analogous expression obtained for the symmetric wave has resonant behavior for $m\omega_d = \omega$ or $2\omega_b = \pm(m\omega_d-\omega)$. These conditions relate particle energy and mass to wave parameters m and $\omega$ at resonance, as discussed below. We will not further analyze the symmetric case since the resonance condition for symmetric waves is not satisfied for ring current ions and measured wave parameters m and $\omega$.

The oscillation in $L$ apparent in Figures 3 and 4 can be calculated as follows. The radial displacement is obtained by integrating (1b) along the trajectories ($2a$)-($2c$)

$$\frac{\delta L}{L} = \frac{1}{L\,R_E}\int_0^t \delta v_L\,(\alpha,\delta,\lambda\,;\,t')\,dt' = i\,\beta_\parallel\left\{\omega J_1(k\Lambda) - \frac{\mu}{(L\,R_E)^2}\,\frac{mc}{q}\,[-J_1(k\Lambda)+k\,J_1(2\Lambda)]\right\} \left\{\frac{e^{i(\omega_b+m\omega_d-\omega)t}-1}{(\omega_b+m\omega_d-\omega)} + \frac{e^{i(-\omega_b+m\omega_d-\omega)t}-1}{(-\omega_b+m\omega_d-\omega)}\right\}\exp(im\,\phi_0) \qquad (4)$$

Comparing (4) with ($3a$), we see that $\delta L$ is out of phase with $\delta W$, consistent with Figure 3.

When an oxygen ion is resonant with the wave, the first adiabatic invariant or the magnetic moment $\mu$ also undergoes oscillation at the mismatch frequency $\Delta\omega$, as seen in Figures 3 and 4. Assuming vacuum fields and neglecting the contribution from the wave electric field, the change in $\mu$ per unit mass $M$ can be approximated by [*Cohen et al.*, 1978]

$$\frac{\delta\mu}{M} = \int d\left(\frac{v_\perp^2}{2B}\right) \equiv \delta\mu_s + \delta\mu_t \qquad (5a)$$

where the static field contribution to $\delta\mu$ is

$$\frac{\delta\mu_s}{M} = -\int dt\left[\left(2\mu/B\right)^{1/2}\left([W-\Psi-2\mu B\,]\rho_\perp\cos\psi - \frac{q}{m}E_\perp\cos\psi_E\right)\right] \qquad (5b)$$

and the time varying field contribution is

$$\frac{\delta\mu_t}{M} \cong -\int dt\left[\frac{v_\perp^2}{2B^2}\,\frac{\partial(B+b_\parallel)}{\partial t} + \frac{v_\parallel}{B}\,\mathbf{v}_\perp\cdot\frac{\partial\hat{\mathbf{b}}}{\partial t}\right] \qquad (5c)$$

The unit vector $\hat{\mathbf{b}} = (\mathbf{B}+\mathbf{b})/(B+b) \cong (\mathbf{B}+\mathbf{b})/B$, where $\mathbf{B}$ is the dipole field and $\mathbf{b}$ is the perturbation field, $\rho_\perp = |\nabla_\perp B|/B$, and $\cos\psi = \mathbf{v}_\perp \cdot \nabla B/(v_\perp |\nabla_\perp B|)$. Spatial gradients of the dipole field are assumed to dominate those of the perturbed fields, $\cos\psi_e = \mathbf{v}\cdot\mathbf{E}_\perp/(v_\perp E_\perp)$, and $W$ is the total kinetic energy. $\delta\mu_s$ can be calculated by the method of steepest descent and the change in one bounce period generally has the form [*Birmingham,* 1984]

$$\frac{\delta\mu_s}{\mu} \cong A\cos(\psi_0)\, e^{-\varepsilon F}, \tag{5d}$$

where $A$ is a function of $\varepsilon = \frac{1}{\rho}\left[\frac{\nabla B}{B}\right]^{-1}$, the ratio of the magnetic field gradient scale length to the gyroradius, $F$ is an integral of order unity and $\psi_0$ is the initial gyrophase angle. In the above expressions, $\mu = mv_\perp^2/2B$ is taken as the lowest order term in a series expansion for the magnetic moment [*Catto et al.*, 1981; *Hastie et al.*, 1969]. Figures 3c and 4c include the next order correction. Our numerical calculations show that the change in the magnetic moment due to the static field is small. The contribution due to the time dependence of the wave can be calculated easily since the second term in (5*c*) will be symmetric when the first is antisymmetric, so the first term will dominate when the antisymmetric resonance condition is satisfied.

Eq. (5*c*) can be approximately rewritten as

$$\delta\mu_t(t)/\mu = -\int_0^t \frac{1}{B}\frac{\partial B}{\partial t'}\,dt'$$

which gives for antisymmetric waves and $n=2$

$$\delta\mu_t/\mu = -i\,\omega\,\beta_\parallel\left[-J_1(k\Lambda) + k\,J_1(2\Lambda)\right] \tag{5e}$$

$$\left\{\frac{e^{i(m\omega_d-\omega+\omega_b)t}-1}{(m\omega_d-\omega+\omega_b)} - \frac{e^{i(m\omega_d-\omega-\omega_b)t}-1}{(m\omega_d-\omega-\omega_b)}\right\}\exp(im\phi_0)$$

We obtain the same resonance condition as expected from the simulations. The drift-bounce resonance causes oscillations in $\mu$. Figure 3c and 4c show the time history of $\mu = W_\perp/B(\mathbf{r}) + \mu_1$, where $W_\perp$ is perpendicular energy at the particle position $\mathbf{r}$, $B$ is evaluated at $\mathbf{r}$ and $\mu_1$ is the next order correction to the magnetic moment (Catto *et al.*, 1981). The $\Delta\omega$ oscillation of $\delta\mu$ has the same form as $\delta W$ and $\delta L$. $\delta\mu_t$ is in phase with $\delta L$ given by (4) since the $J_1$ bracket in (5e) is negative for $k$=2. This result is consistent with Figures 3c and 3d. The slow oscillation on the drift time scale in Figure 4 has the same period as that seen in $W$ and $L$, reflecting the change in particle energy associated with the convection electric field.

## Discussion

The drift-bounce resonant interaction explains many features of ion behavior in the Pc 5 wave simulations. We choose m=40 to 140 because this is the range most frequently observed. From the resonance condition, it is clear that the azimuthal mode number is an important factor. The resonance condition is satisfied when

$$K\,\omega_b = (m\,\omega_d - \omega) \tag{6}$$

where $K$ is an integer, $\omega_b \sim \sqrt{W/M}$ is the bounce frequency and $\omega_d \sim W$ is the bounce averaged drift frequency of an ion of energy $W$ and mass $M$. When the resonance condition is satisfied strong interaction is observed for initial pitch angles ranging from 35° to 85° for 15 keV oxygen ions over the range of frequency and mode number m observed. For 5 keV and 20 keV oxygen ions, strong interaction occurs in a narrower range of pitch angles. The resonant parameter space has been organized by *Takahashi et al.* [1990], who have plotted the drift-bounce resonance condition (6) for $K=\pm1$ in energy $W$ vs. azimuthal mode number m space. The fact that $\omega_b \sim \sqrt{W/M}$ and $\omega_d \sim W$ means that the resonance condition $\omega_b^2 = (m\omega_d - \omega)^2$, which appears in the common denominators of $\delta W$, $\delta L$ and $\delta\mu_t$ for the antisymmetric wave, has two roots $W$ for a fixed value of m, corresponding to $K=\pm1$ for $m>0$. In Figures 5a and 5b we have plotted the condition for $H^+$ and $O^+$ from L=5.5-8.5 at $f=2$ mHz. Figure 5c shows the corresponding

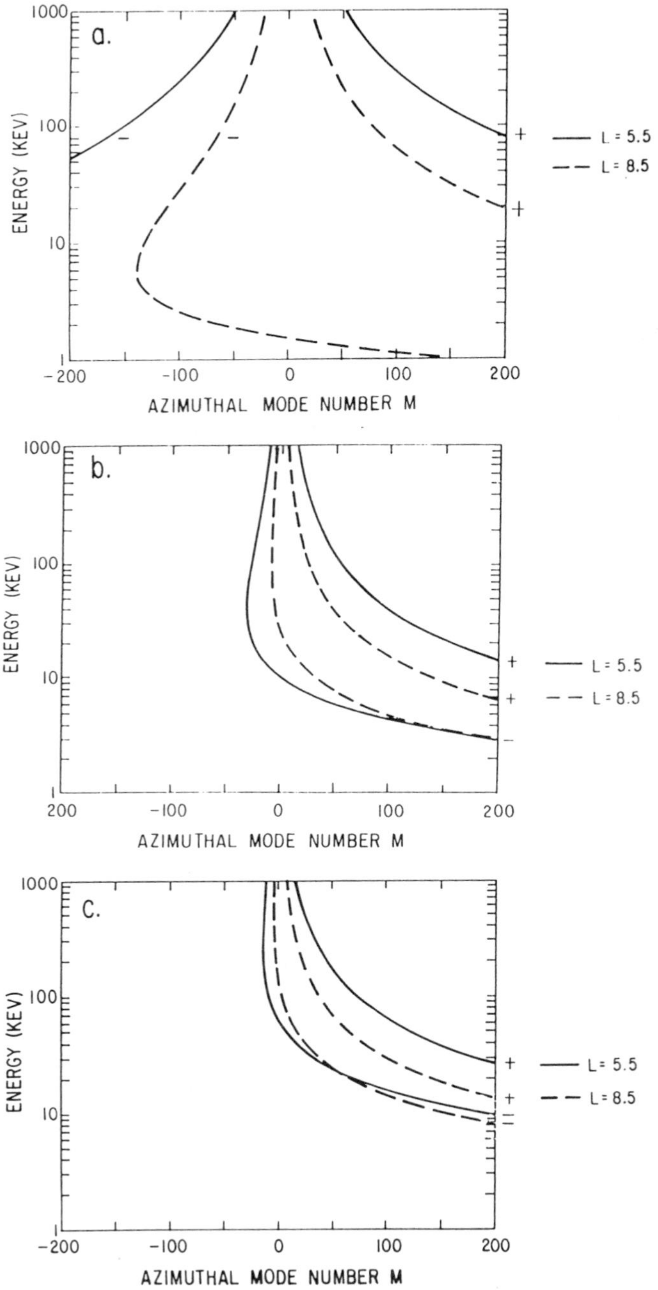


Figure 5. Energy vs. azimuthal mode number plots for resonant conditions of *(a)* hydrogen and *(b,c)* oxygen for L= 5.5 (solid line) and L= 8.5 (dashed line). The + indicates the higher energy branch and the − indicates the lower energy branch of the resonance condition $\omega_b = \pm(m\omega_d - \omega)$. *(a)* and *(b)* are for $f$= 2 mHz and *(c)* is for $f$=5 mHz. Solid lines denote 85° pitch angle case and dashed lines indicate 5° case.

plot for $O^+$ at $f$=5 mHz. No background convection or corotation electric field $E$ is included in Figure 5. The fact that there are two resonant energy ranges for each L value was first pointed out by *Southwood et al.* [1969]. Motivated by the ISEE-1 observations reported by *Cladis and Lennartsson* [1989], we have focused on the lower energy resonance here. Work on the generation mechanism for compressional Pc 5's [*Chen and Hasegawa,* 1988] has concentrated on the higher $H^+$ energy resonance, as has a recent study of the effect of Pc 5s on higher energy $H^+$ using gyroaveraged Hamiltonian test particle equations of motion [*Chan et al.*, 1989*a*].

All test particle runs in this study are for antisymmetric wave fields, consistent with compressional Pc 5 observations by *Takahashi et al.* [1985, 1987]. The symmetric wave has little effect on oxygen and hydrogen over the range of parameters surveyed. If a diagram similar to Figure 5 is plotted for the symmetric resonance condition, $2\omega_b = \pm(m\omega_d - \omega)$ or $m\omega_d = \omega$, larger m values are required for resonance than are typically observed.

Our expressions for $\delta W$, $\delta L$ and $\delta\mu_t$, which contain the same resonant denominators, have distinct phase relationships which can be compared with the test particle results. $\delta W$ and $\delta L$ are 180° out of phase, as one would expect if $\mu$ is well conserved. $\delta\mu_t$ and $\delta L$ are in phase, suggesting that the magnitude of $B$ dominates the leading term $W_\perp/B$ plotted in Figures 3c and 4c. The picture is further complicated by the non-negligible change in total kinetic energy associated with the time independent electric fields. In particular, we have seen examples where ions near resonance lose enough energy to follow equipotential streamlines out to the magnetopause. Particles which are turned around by a combination of energy loss to the Pc 5 and the convection electric field will be lost from the ring current.

## Conclusions

The mechanisms responsible for the reduction of enhanced storm time ring current fluxes may be different for hydrogen and oxygen. Examination of Figure 5 shows an almost continuous band of resonant energies from a few to 100 keV for $O^+$ with waves around a few mHz having m values ranging from 40-140. By contrast, there is a gap in resonant energy for hydrogen between 1-25 keV where resonance does not occur for observed m values.

Hydrogen on the low energy branch is below the cutoff energy for eastward drift vs. westward corotation for $K_p$=1.5 (see Cladis and Lennartsson, 1986). Since $H^+$ is resonant at higher energies than $O^+$ for the same parameters on the upper branch, the Pc 5 effect is typically smaller for $H^+$; that is to say, the energy loss or gain is a smaller percentage of the total particle energy. Thus, while Pc 5s may contribute to radial diffusion of the more energetic $H^+$ (Chan *et al.*, 1989), we do not expect Pc 5 interaction to cause a dramatic reduction in the $H^+$ flux of the type reported by Cladis and Lennartsson for $O^+$.

In the region where loss of oxygen ions is observed, ion cyclotron waves (ICWs) are observed as well, especially when cold helium ions are abundant in the region [*Mauk et al.*, 1981; *Young et al.*, 1981; *Roux et al.*, 1982]. In a hybrid particle simulation study where ions were treated as particles, electrons as a fluid and electromagnetic fields were solved for self-consistently [*Qian et al.*, 1989], we found that oxygen ions interact very weakly with ICWs. EMIC waves driven unstable by the ring current loss cone are more effective at causing pitch angle diffusion of ring current hydrogen than oxygen. In the plasmasphere helium is the dominant minor ion species, and the presence of a smaller amount of cold oxygen (e.g., $n_O=0.5\, n_H$, $n_{He}=0.1\, n_H$) does not significantly affect the EMIC wave mode structure, i.e., there is no significant amplitude around the oxygen gyrofrequency in our simulations. This explains the absence of significant pitch angle diffusion for ring current oxygen by interaction with EMIC waves, even for cases which were run with the above background cold oxygen included. The wave frequencies excited appear to be too high to cause oxygen pitch angle diffusion. Thus EMIC wave interaction may be more important for ring current hydrogen and Pc 5 interaction more important for oxygen (Hudson, *et al.*, 1990).

All test particle calculations shown in the present study were initialized with particles injected at 0900 LT in order to simulate the ISEE-1 measurements (Cladis and Lennartsson, 1986). More recently, we have begun a study with trajectory calculations initialized at dusk and ions drifting across the entire dayside where compressional Pc 5 micropulsations are observed. We find that in most cases oxygen in the ten to tens of keV energy range will lose enough energy due to the convection field to become resonant with Pc 5s in the observed frequency and wavenumber range. The oscillation in $W$ at $\Delta\omega$ then allows the ions to be swept away to the magnetopause by the convection electric field. This process appears to be insensitive to the precise initial conditions as long as the resonance condition is satisfied as particle energy is decreased by the convection electric field. The effectiveness of Pc 5s in removing oxygen is most apparent when they are allowed, as in nature, to drift across the dayside from dusk.

We have examined oxygen interaction with Pc 5s modeled as a single coherent mode, as is sometimes observed [*Takahashi et al.*, 1985]. More generally, there may be a spectrum of frequencies and azimuthal mode numbers present over a range of $L$ values, which would contribute to radial diffusion, as has been shown numerically by *Chan et al.* [1989*a*]. We have estimated a radial diffusion coefficient due to $e_\phi$ which is comparable to previous estimates from fluctuations in the convection electric field [*Cornwall*, 1968], $D \sim 2\times10^{-5}\, L^6\, R_E^2$/day. Estimates for the radial diffusion coefficient due to $b_\parallel$ are comparable [*Chan et al.*, 1989*b*]. We have only examined single particle interaction with a single wave mode in the present study, but plan to do multiple wave mode studies in the future.

The Pc 5 wave magnetic fields in the present model are assumed to vary simply as $1/L^{n+1}$, with $n$ a variable parameter. Better knowledge of the spatial dependence and mode structure of Pc 5s may become available with combined electric and magnetic field data from the CRRES satellite, launched into a highly elliptical orbit with 18° inclination and apogee at 6.3 $R_E$. We hope to improve our model as more information about latitudinal and radial wave mode structure, relative phases and amplitudes of the magnetic and electric field components becomes available. Meanwhile, it appears likely that drift-bounce resonant effects of compressional Pc 5s on low energy oxygen are sufficiently great when compared with low energy hydrogen to provide a test for the model.

# Appendix

Here we show the general structure of the perturbation field in dipole coordinates. Using the dipole coordinates $\alpha = 1/L$ (flux coordinate), $\phi$ (azimuthal angle) and $\delta$ as defined in (1), the relation $\mathbf{b} = \nabla \times A$ with $\mathbf{A} = A_\phi \hat{\phi}$ is written as

$$\begin{Bmatrix} b_\alpha \\ b_\delta \end{Bmatrix} = \frac{B}{R_E B_0} \frac{1}{r \cos\lambda} \begin{Bmatrix} \partial_\delta (A_\phi r \cos\lambda) \\ -h\, \partial_\alpha (A_\phi r \cos\lambda) \end{Bmatrix} \tag{A-1}$$

which becomes, with the help of the field line relation $r = R_E L \cos^2\lambda$,

$$\begin{Bmatrix} b_\alpha \\ b_\delta \end{Bmatrix} = \frac{B}{R_E^2 B_0} \left\{ L^{-1} \cos^{-3}\lambda \begin{bmatrix} \partial_\delta \\ \partial_\alpha \end{bmatrix} (R_E L (\cos^3\lambda) A_\phi) \right\} \tag{A-2}$$

We assume that the vector potential is separable in the coordinates $\alpha$, $\lambda$

$$A_\phi(\alpha, \lambda) = B_0 R_E\, \alpha^n\, \mathrm{H}(\lambda)\, e^{i(m\phi - \omega t)} \tag{A-3}$$

and use the relations (A-1) to obtain

$$\begin{Bmatrix} b_\alpha \\ b_\delta \end{Bmatrix} = \frac{B}{\alpha} \begin{Bmatrix} \cos^2\lambda \\ \sin 2\lambda \end{Bmatrix} (1+3\sin^2\lambda)^{-1} \; \partial_\lambda \left[\alpha^{n-1} (\cos^3\lambda)\, \mathrm{H}(\lambda)\right]_{\substack{\alpha \\ \delta}} e^{i(m\phi-\omega t)} \quad \text{(A-4)}$$

where the subscripts denote that the derivatives are to be taken at constant $\alpha$, $\delta$, respectively.

The perturbed electric field is obtained from $\nabla \times \mathbf{e} = -\frac{1}{c}\,\partial_t\, \mathbf{b}$, giving for either one of the two components of $\mathbf{b}$

$$e_\phi = \frac{i\omega}{c} R_E\, B_0\, \alpha^n\, \mathrm{H}(\lambda)\, e^{i(m\phi-\omega t)} \quad \text{(A-5)}$$

The integer $n$ should be determined by the experimental dependence of the fields as a function of L-shell. Due to the spatial differential relation between $e_\phi$ and $b_\alpha$, $b_\delta$, the magnetic fields have $B\,\alpha^{n-2} \sim \alpha^{n+1}$ vs. the $\alpha^n$ dependence of $e_\phi$. In the simulation study we took $n=0$ and $\mathrm{H}(\lambda) = \sin k\lambda$. For the case $n=0$, $e_\phi$ is independent of $L$.

## Acknowledgments

This work was supported by Air Force Contract F19628-87-K-0038, NASA grants NAGW-1652, NAGW-1626 and NAG5-1098 and NSF grant ATM-8918774. M. K. Hudson also acknowledges support from NSF grant DII 89-02856 while visiting Stanford University. Computations were performed on the Convex C1 at Dartmouth, and the SDSC Cray. The authors would like to thank R. Lysak, N. Otani and M. Temerin for helpful discussions, and to thank particularly A. Chan and X. Li for improving the Pc 5 wave field model and for computations and discussions subsequent to the Huntsville meeting.

## References


Birmingham, T. J., Pitch angle diffusion in the Jovian magnetodisc, *J. Geophys. Res.*, *89*, 2699, 1984.

Catto, P. J., W. M. Tang, and D. E. Baldwin, Generalized gyrokinetics, *Plasma Physics*, *23*, 639, 1981.

Chan, A. A., E. Duvall, and L. Chen, Nonlinear interaction of energetic ring current protons with magnetospheric hydromagnetic waves, *Trans. Am. Geophys. Union*, *70*, 1276, 1989*b*.

Chan, A. A., L. Chen, and R. B. White, Diffusion of energetic ring current protons by nonlinear resonance with magnetospheric hydromagnetic waves, *Geophys. Res. Lett.*, *16*, 1133, 1989*a*.

Chen, L., and A. Hasegawa, On magnetospheric hydromagnetic waves excited by energetic ring current particles, *J. Geophys. Res.*, *93*, 8763, 1988.

Cheng, C. Z., and C. S. Lin, Eigenmode analysis of compressional waves in the magnetosphere, *Geophys. Res. Lett.*, *14*, 884, 1987.

Cladis, J. B., and W. E. Francis, The polar ionosphere as a source of the storm time ring current, *J. Geophys. Res.*, *90*, 3465, 1985.

Cladis, J. B., and O. W. Lennartsson, On the loss of $O^+$ ions (< 7 keV/e) in the ring current during the recovery phase of a storm, in *Ion Acceleration in the Magnetosphere and Ionosphere*, Tom Chang, ed., AGU, Washington, D. C., 1986.

Cohen, R., G. Rowlands, and J. H. Foote, Nonadiabaticity in mirror machines, *Phys. Fluids*, *21*, 627, 1978.

Cornwall, J. M., Diffusion processes influenced by conjugate-point wave phenomena, *Radio Sci.*, *3*, 740, 1968.

Cummings, W. D., R. L. O'Sullivan, and P. J. Coleman, Jr., Standing Alfvén waves in the magnetosphere, *J. Geophys. Res.*, *74*, 778, 1969.

Hamlin, D. A., R. Karplus, R. C. Vik and K. M. Watson, Mirror and azimuthal drift frequencies for geomagnetically trapped particles, *J. Geophys. Res.*, *66*, 1, 1961.

Hasegawa, A., Drift mirror instability in the magnetosphere, *Phys. Fluids*, *12*, 2642, 1969.

Hastie, R. J., G. D. Hobbs, and J. B. Taylor, Non-adiabatic behavior of particles in inhomogeneous magnetic fields, in *Plasma Physics and Controlled Nuclear Fusion Research*, p. 389, International Atomic Energy Agency, Vienna, 1969.

Higbie, P. R., D. N. Baker, R. D. Zwickl, R. D. Belian, J. R. Asbridge, J. F. Fennell, B. Wilken, and C. W. Arthur, The global Pc 5 event of November 14-15, 1979, *J. Geophys. Res.*, *87*, 2337, 1982.

Hudson, M. K., S. Qian and I. Roth, Ion loss from the ring current and plasmaspheric heating in *Physics of Space Plasmas, SPI Conf. Proc. and rep. Series*, **9**, 97, 1990.

Hudson, M. K., A. A. Chan, X. Li and I. Roth, Ring current ion interaction with Pc 5 Micropulsations, in *Physics of Space Plasma, SPI Conf. Proc. and Rep. Series*, **10**, in press, 1991.

Kivelson, M. G., D. W. Southwood, Charged particle behavior in low-frequency geomagnetic pulsations, 4. Compressional waves, *J. Geophys. Res.*, *90*, 1486, 1985.

Mauk, B. H., C. E. McIlwain, and R. L. McPherron, Helijm cyclotron resonance within the earth's magnetosphere, *Geophys. Res. Lett.*, *8*, 103, 1981.

Maynard, N. C., and J. J. Chen, Isolated cold plasma regions: observations and their relation to possible production mechanisms, *J. Geophys. Res.*, *80*, 1009, 1975.

Nagano, H., and T. Araki, Long-duration Pc 5 pulsations observed by geostationary satellites, *Geophys. Res. Lett.*, *10*, 908, 1983.

Press, W. H., B. P. Flannery, S. A. Teukolsky, and W. T. Vetterling, *Numerical Recipes, The Art of Scientific Computing*, Cambridge University Press, 1988.

Qian S., M. K. Hudson, and I. Roth, Particle simulation of ion heating in the ring current, accepted by *J. Geophys. Res.*, **95**, 1001, 1990.

Roux, A., S. Perrault, J. L. Rauch, C. de Villedary, G. Kremser, A. Korth, and D. T. Young, Wave particle interactions near $\Omega_{He^+}$ observed onboard GEOS 1 and 2, 2. Generation of ion cyclotron waves and heating of $He^+$ ions, *Geophys. Res. Lett.*, *87*, 8174, 1982.

Schulz, M., and L. J. Lanzerotti, *Particle Diffusion in the Radiation Belts*, p. 16-18, Springer-Verlag, New York, 1974.

Southwood, D. J., The behavior of ULF waves and particles in the magnetosphere, *Planet. Space Sci.*, *21*, 53, 1973.

Southwood, D. J., A general approach to low-frequency instability in the ring current plasma, *J. Geophys. Res.*, *81*, 3340, 1976.

Southwood, D. J., and M. G. Kivelson, Charged particle behavior in low-frequency geomagnetic pulsations, 1, Transverse waves, *J. Geophys. Res.*, *86*, 5643, 1981.

Southwood, D. J., J. W. Dungey, and R. J. Etherington, Bounce resonant interaction between pulsations and trapped particles, *Planet. Space Sci.*, *17*, 349, 1969.

Takahashi, K., and P. R. Higbie, Antisymmetric standing wave structure associated with the compressional Pc 5 pulsation of November 14, 1979, *J. Geophys. Res.*, *91*, 11163, 1986.

Takahashi, K., P. R. Higbie, and D. N. Baker, Azimuthal propagation and frequency characteristic of compressional Pc 5 waves observed at geostationary orbit, *J. Geophys. Res.*, *90*, 1473, 1985.

Takahashi, K., J. F. Fennel, E. Amata, and P. R. Higbie, Field-aligned structure of the storm time Pc 5 wave of November 14–15, 1979, *J. Geophys. Res.*, *92*, 5857, 1987.

Takahashi, K., R. W. McEntire, A. T. Y. Lui, and T. A. Potemra, Ion flux oscillations associated with a radially polarized transverse Pc 5 magnetic pulsation, *J. Geophys. Res.*, *95*, 3717, 1990.

Volland, H., A semiempirical model of large-scale magnetospheric electric field, *J. Geophys. Res.*, *78*, 171, 1973.

Young, D. T., S. Perrault, A. Roux, C. de Villedary, R. Gendrin, A. Korth, G. Kremser, and D. Jones, Wave particle interactions near $\Omega_{He^+}$-rich plasma, *J. Geophys. Res.*, *86*, 6755, 1981.

# AN EARLY-STAGE PLASMASPHERE REFILLING MODEL BASED ON A KINETIC APPROACH WITH TRAPPING DUE TO ION HEATING AND PITCH-ANGLE SCATTERING

J. Lin, J. L. Horwitz, G. R. Wilson and C. W. Ho

Department of Physics and
Center for Space Plasma and Aeronomic Research
The University of Alabama in Huntsville
Huntsville, Alabama 35899

Abstract. We hypothesize that, in the early stage of replenishment of plasmaspheric flux tubes following magnetic storms, two important processes in the trapping and accumulation of plasma occur through stochastic wave-driven pitch-angle scattering and perpendicular heating of inflowing ionospheric particles. By treating the motion of particles as basically adiabatic but incorporating velocity-space diffusion associated with (1) pitch-angle scattering arising from ion interactions with left-hand circularly polarized electromagnetic waves and (2) perpendicular ion heating due to electromagnetic ion cyclotron waves, we calculate (a) bouncing ion trajectories, and (b) large-scale density evolution of 'polar wind' ionospheric ions flowing onto a closed L=4 flux tube. It is found that relatively modest power levels of wave-particle interactions lead to important effects on ion trajectories including decreasing mirror latitudes and sometimes significant and abrupt equatorial entrainment of these ions. Also, significant general accumulation of plasma all along the magnetic flux tube with up to 10 or more ions/cc in the equatorial regions after about 12 hours following initiation of refilling is observed.

## Introduction

Starting with the classic work by Banks et al. [1971], several groups have attempted to model the process of plasmasphere replenishment which occurs during the recovery phase of magnetospheric storms.

Modeling Magnetospheric Plasma Processes
Geophysical Monograph 62


In most treatments, an initially empty flux tube appropriate to the outer part of the plasmasphere (say L=4) is taken to be the starting configuration. Banks et al. [1971] were the first to treat the problem by using a hydrodynamic model involving supersonic polar wind streams from conjugate ionospheres, which resulted in the formation of a high density plasma slab at the magnetic equator.

Various authors, such as Grebowsky [1972], Khazanov et al. [1984] Singh and Torr [1988], and Rasmussen and Schunk [1988] have treated the problem of plasmasphere refilling with increasing sophistication. These last two used a multiple streaming formulation which allowed for the possibility of counter-streaming flows without necessarily forcing the formation of a 'collision' and a high-density plasma slab, as is the tendency with the single-stream approach. There have also been some kinetic investigations designed to elucidate some of the microphysical processes which are involved in the refilling process. For example, Schulz and Koons [1972] suggested that counter-streaming ion beams in the refilling flux tube would lead to wave-particle interactions which would pitch-angle scatter ions and lead to initial accumulation. Singh and Hwang [1988] examined the effects on 'average' ion motions and energy gains for ions injected onto closed field lines and subjected to perpendicular heating by ion cyclotron waves.

In this paper, we present a new, kinetic model for the early-stage or collisionless period of refilling, based on a hypothesis that the principal processes influencing the plasma accumulation are wave-particle interactions which pitch-angle scatter and perpendicularly heat incoming ions. We will show that our kinetic description provides a much different view and different physical insights on the important processes occurring during the early-stage refilling process than the hydrodynamic approaches previously used.

## Description of the Model

We consider a kinetic model in which individual ion trajectories as well as plasma density evolution along an outer plasmaspheric flux tube are calculated including the effects of pitch-angle scattering and energy diffusion due to wave-particle interactions. (For the initial study, the ambipolar field is not included in the model.) We will describe the theoretical model and the method of calculation of the effects of wave-particle interactions. In the following description, 'perpendicular'and 'parallel' refer to the components of the parameters perpendicular and parallel to the magnetic field line, respectively.

We track the motions of individual protons along an L=4 field line by integrating their equations of motion. The particles are subject to the effects of the magnetic mirror force, gravity, pitch-angle scattering, and perpendicular heating.

The calculation of the wave-particle interaction impulses is guided by the discussion given in Singh and Torr [1988], who indicated that both pitch-angle scattering and perpendicular heating should occur in the refilling region. Singh and Hwang [1987] and Singh and Torr [1988] suggested that perpendicular ion heating caused by electromagnetic ion cyclotron waves should play an important role in the energization and trapping of protons. For the actual calculation of the perpendicular energization, we use the following expression for the perpendicular energy diffusion (e.g.,Sturrock,1966; Cladis and Francis, 1985):

$$D_{\perp\perp} = \frac{q^2}{8m} \cdot W_\perp \cdot P_h \tag{1}$$

where $q$, $m$, $W_\perp$, are the charge, mass, perpendicular energy of the particles, respectively, and $P_h$ is the power spectral level in units of $V^2m^{-2}Hz^{-1}$.

We have used (1) to compute impulsive changes in the perpendicular energy in a given time step $\Delta t$ for the advance of the proton, using a Gaussian random generator with standard deviation:

$$\sigma_\perp^2 = 2 \cdot D_{\perp\perp} \cdot \Delta t \tag{2}$$

The generated energy increment is added to the previous perpendicular energy in the following manner:

$$W_{\perp n} = W_\perp + dW + 2\sqrt{W_\perp \cdot dW} \cdot cos\theta \tag{3}$$

where $W_{\perp n}$ is the new perpendicular energy calculated from the previous time (before time advance) but modified to incorporate conservation of the magnetic moment during the change in B. $dW$ is the generated energy increment. $\theta$ is an angle generated randomly between 0° and 180°. Based on previous work by Gendrin [1968], Singh and Torr [1988] suggested that pitch-angle diffusion of low-energy protons would result from their interaction with left-hand circularly polarized electromagnetic waves which could be generated from hot ring current particles. Gendrin [1968] and Singh and Torr [1988] obtained the following expression for the pitch-angle scattering coefficient:

$$D_{\alpha\alpha} = 1.6\times10^{-3} \cdot P_\alpha \cdot \left(\frac{v_a}{v_\parallel}\right)^{2/3} \left[1 + \left(\frac{v_a}{v_\parallel}\right)^{2/3} \cdot cos^2\alpha\right]^2 \tag{4}$$

where $\alpha$ is the pitch-angle, $P_\alpha$ is the power spectral density of the left-hand polarized electromagnetic ion cyclotron waves in units of $\gamma^2 Hz^{-1}$, $v_a$ is the Alfven velocity and $v_\parallel$ is the parallel velocity of the particles. Similar to the perpendicular heating described above, we generate impulsive changes in pitch-angle using a Gaussian random generator with standard deviation:

$$\sigma_\alpha^2 = 2 \cdot D_{\alpha\alpha} \cdot \Delta t \tag{5}$$

Having generated a pitch-angle change using (4) and (5), we compute a new pitch-angle and therefore the new parallel and perpendicular velocities for actual advance of the proton velocity.

For the simulation of the polar wind stream injection and consequent density evolution, we inject ensembles of simulation particles at altitude of 1800 km in the northern and southern topside ionospheres.

In this paper, the injection of simulation ions is turned on at time t=0 and maintained as constant and equal streams of ions from both topside ionospheres. The form of the injected distribution is taken to be the upgoing half of a drifting bi-Maxwellian distribution:

$$\begin{aligned} f(v_\parallel, v_\perp) &= n_o \frac{(m/2k\pi)^{3/2}}{\sqrt{T_\parallel} T_\perp} \cdot exp\left[-\frac{m(v_\parallel - u_{so})^2}{2kT_\parallel} - \frac{mv_\perp^2}{2kT_\perp}\right] \quad , v_\parallel \text{ upward} \\ f(v_\parallel, v_\perp) &= 0 \quad , v_\parallel \text{ downward} \end{aligned} \tag{6}$$

The parameters $n_o$, $T_\perp$, $T_\parallel$, and $u_{so}$, are taken to be 50 ions/cc, 3000° K, 3000° K, 0 m/sec, respectively, in the examples in this paper. Note, however, that the up-going drift velocity for this distribution with these parameters is 1.9 km/sec.

We calculate the density as a function of time and distance along the field line at the L=4 flux tube simply by computing the number of simulation ions in a particular spatial cell and converting to density. The techniques for ion distribution injection at the boundary and calculation of the density evolution are basically the same as those used by Wilson et al. [1990].

## Results

### (a) Characteristic ion trajectories for early-stage refilling

In this section, we describe example trajectories and their properties for bouncing protons interacting with waves either responsible for perpendicular heating or pitch-angle scattering or both. Figure 1 illustrates the time variation in latitude. In each case, a proton is injected at latitude λ=55°, with initial energy E=0.25 eV and pitch-angle α=82°. As shown in Figure 1a, in which no wave-particle interactions are involved, the ion travels from northern hemisphere into southern hemisphere, without any bouncing before escaping at λ=-55°. Ions which reach one hemisphere at the injection latitude with downward velocity are considered to be lost to the atmosphere and further calculation of these trajectories are suppressed. In Figure 1b, a case in which power spectral level density for perpendicular heating is held constant along the field line at $10^{-14}V^2m^{-2}Hz^{-1}$ and no pitch-angle scattering was included, now shows bouncing at the southern hemisphere. In Figure 1c, no perpendicular heating is involved, but the pitch-angle scattering

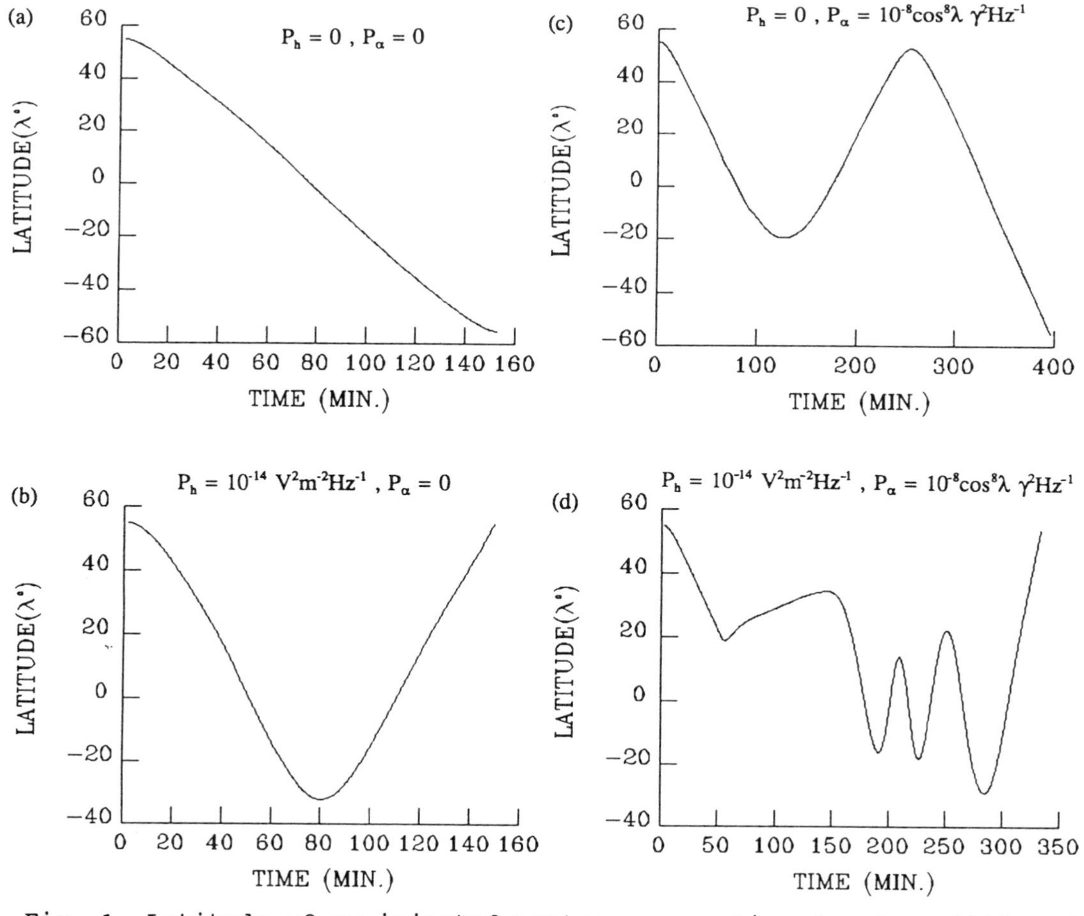


Fig. 1. Latitude of an injected proton versus time for four different cases of wave-particle interactions.

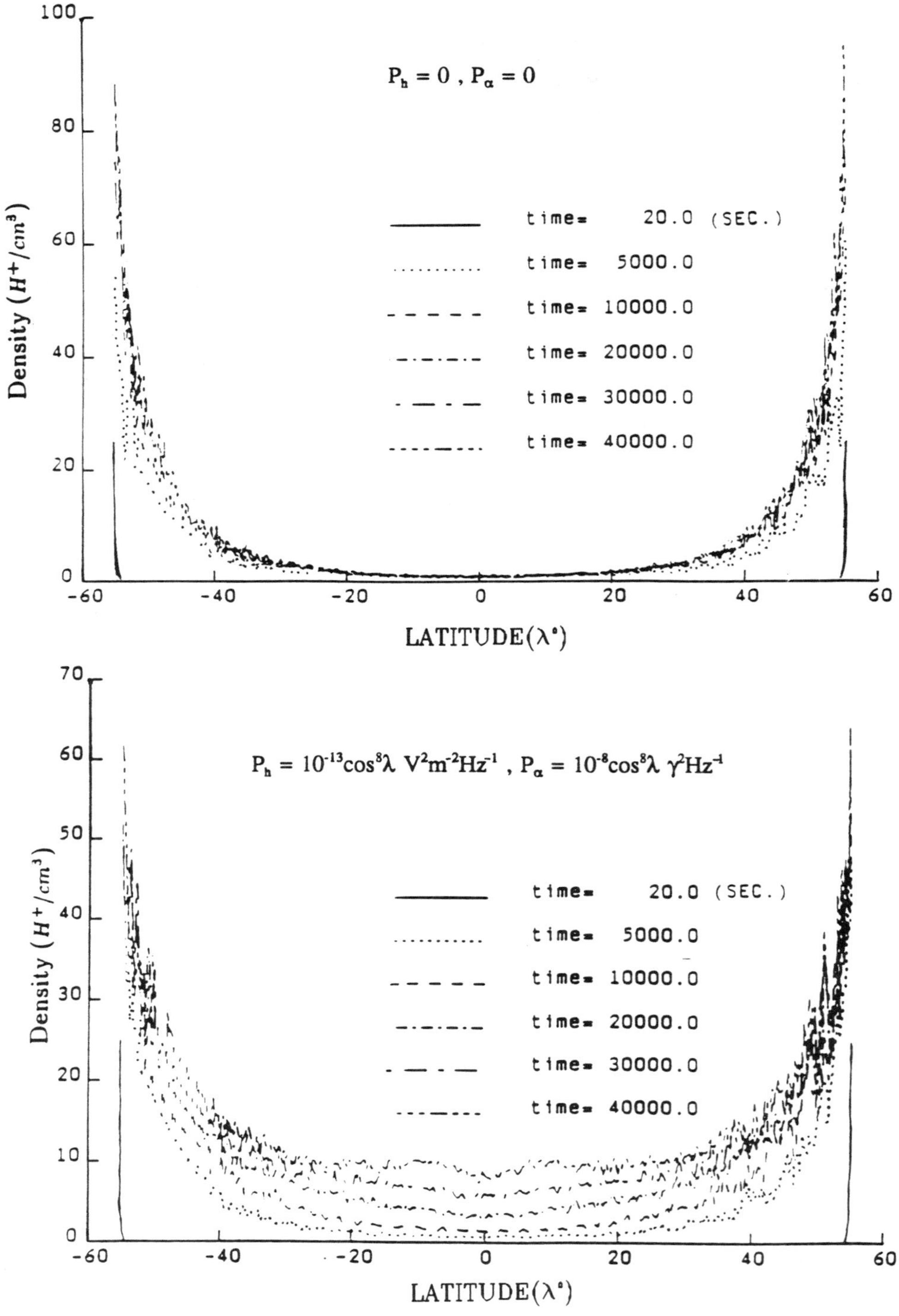


Fig. 2. The time evolution of the ion density profiles contrasted between the case when only adiabatic motion is involved and the case when perpendicular heating and pitch-angle sacttering are involved.

is turned on, with a variation of $10^{-8}\cos^8\lambda$ $\gamma^2 Hz^{-1}$, making the effect strongly peaked about the equator. By itself the only effect of this pitch-angle scattering is that the ion subsequently bounces once at both the southern hemisphere and northern hemisphere before exiting at the southern hemisphere. Finally, the results of both energy heating and pitch-angle scattering are illustrated in Figure 1d. Again, for this case, the ion bounces around the equatorial region, with a mirror latitude of 20° to -30°. Note that the heating and pitch-angle scattering effects are stochastic in nature, and it is not possible to predict the details of a given ion trajectory. However, in general, the effects of increasing the perpendicular heating are to increase the tendency to bounce more frequently and to mirror at lower latitudes. For ions with initially small equatorial pitch angles, the pitch-angle scattering will, on average, tend to produce similar effects. And, again in general, including both effects tends to be synergistic in producing lengthened intervals of equatorial entrainment and bouncing.

(b) Evolution of density profiles during refilling

Figure 2 contrasts the temporal evolution of the latitudinal profiles of the densities for (lower panel) when perpendicular heating and pitch-angle scattering are involved, and both effects are concentrated at the magnetic equator, and (upper panel) when neither effect is on and only adiabatic motion of the ions is involved. The adiabatic motion-only case involves no trapping and accumulation, and basically reaches equilibrium in one characteristic ion bounce period. This equilibrium involves very low densities. The effects of the heating and pitch-angle scattering, however, induce enhanced accumulation and after about twelve hours the equatorial ion densities exceed 10 ions/cc. In Figure 3, we calculate the total number of ions in an L=4 flux tube (1 $cm^2$ in area at the base altitude of 1800 km) as a function of time for four different effect combinations: Curve (a) involves no perpendicular heating or pitch-angle scattering; Curve (b) has equatorially-concentrated pitch-angle scattering but no perpendicular heating; Curve (c) includes equatorially-concentrated perpendicular heating but no pitch-angle scattering; and Curve (d) has both equatorially-concentrated perpendicular heating and pitch-angle scattering. We see that, in that order, the total number of ions in the flux tube accumulates to higher levels and takes longer to reach an equilibrium state. Indeed, at the time of 40,000 seconds, the case with both perpendicular heating and pitch-angle scattering has not yet reached its asymptotic state.

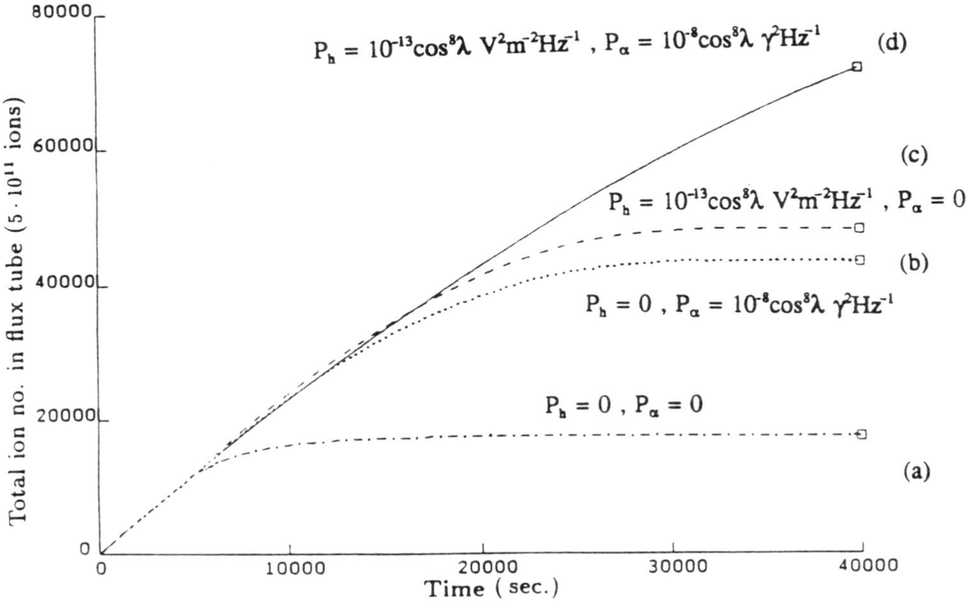


Fig. 3. The time variation of the total number of ions in a flux tube for the four cases of wave-particle interactions used.

## Conclusion

In this report, we have used illustrative example cases to indicate some of the expected properties of the early stages of plasmasphere refilling from a kinetic perspective when wave-particle interactions might be involved. Because of its kinetic treatment of the ions the model is able to describe the complex and random bouncing trajectories of individual particles and large-scale density evolution along the flux tube. Though not discussed in this paper, it is also able to demonstrate ion velocity effects (such as velocity dispersion and double stream) which cannot be incorporated into a hydrodynamic model.

Acknowledgments. This work was supported under NASA grants NAGW-2128 and ATM 8911799/NSF to the university of Alabama in Huntsville.

## References

Banks, P. M., A. F. Nagy and W. I. Axford, Dynamical behavior of thermal protons in the mid-latitude ionosphere and magnetosphere, Planet. Space Sci., 19, 1053-1067, 1971.

Cladis, J. B. and W. E. Francis, The polar ionosphere as a source of the storm time ring current, J. Geophys. Res., 3465, 90, 1985.

Gendrin, R., Pitch angle diffusion of low energy protons due to gyroresonant interactions with hydrodynamic waves, J. Atmos. Terr. Phys., 30, 1313, 1968.

Grebowsky, J. M., Model development of supersonic trough wind with shocks, Planet. Space Sci., 20, 1923, 1972.

Khazanov, G. V., M. A. Kuen, Yu. V. Konikov and I. M. Sidorov, Simulation of ionosphere-plasmasphere coupling taking into account ion inertia and temperature anisotropy, Planet. Space Sci., 32, 585, 1984.

Rasmussen, C. E. and R. W. Schunk, Multistream hydrodynamic modeling of interhemispheric plasma flow, J. geophys. Res., 93, 14557, 1988.

Schulz, M. and H. C. Koons, Thermalization of colliding ion streams beyond the plasmapause, J. Geophys. Res., 77, 248-254, 1972.

Singh, N. and K. S. Hwang, Perpendicular ion heating effects on the refilling of the outer plasmasphere, J. Geophys. Res., 92, 13531, 1987.

Singh, N. and D. G. Torr, Review of microscopic plasma processes occurring during refilling of the plasmasphere, Adv. Space Res., 8, 5, 1988.

Sturrock, P.A., Stochastic accelerations, Phys. Rev., 141, 186, 1966.

Wilson, G. R., C. W. Ho, J. L. Horwitz, N. Singh and T. E. Moore, A new kinetic model for time-dependent polar plasma outflow: Initial results, Geophys. Res. Lett., 17, 263, 1990.

# MODELLING OF PLASMASPHERIC FLOWS WITH AN EQUATORIAL HEAT SOURCE FOR ELECTRONS

S. M. Guiter and T. I. Gombosi

Space Physics Research Laboratory, University of Michigan, Ann Arbor, Michigan 48109-2143

Abstract. A time-dependent three fluid ($H^+$ and $O^+$ ions and electrons) one-stream interhemispheric plasma flow model is used to investigate plasmaspheric refilling. In the model, the coupled time-dependent hydrodynamic equations (continuity, momentum and energy) of a two-ion quasineutral, currentless plasma are solved for a closed geomagnetic field line. For the present set of calculations, an L = 2 field line was used. Energy deposition from the magnetosphere was simulated by including an equatorial heat source for electrons. A steady-state solution was found and used as the initial condition in a subsequent simulation of the effect of a density depletion. The density depletion was modeled by reducing the densities by an arbitrary factor above 2500 km altitude, while keeping the velocities and temperatures unchanged. Shock structures develop which move up the field line, meeting and reflecting at the equator. The reflected shocks are absorbed by the dense neutral atmosphere when they reach the model flux tube boundaries. After the upwelling streams collide, refilling occurs from the equator downward for two to three hours; thereafter, there is steady refilling from the ionospheres upward.

## 1. Introduction

An important unresolved question in magnetospheric physics is that of refilling of plasmaspheric flux tubes following density depletions caused by magnetic storms [Park, 1970]. There have been questions as to whether the refilling starts from the equator downward or from the ionosphere upward [Banks et al., 1971]. To address this question, various models for interhemispheric flows on magnetic flux tubes have been used.

The models can be divided into one-stream and two-stream models. One-stream models have been developed by Moffet and Murphy [1973], Young et al. [1980], Khazanov et al. [1984], and Singh et al. [1986b]. Singh [1988] and Rasmussen and Schunk [1988] developed two-stream models. The first two of these models assume diffusive equilibrium. Khazanov et al. [1984] neglect electron inertia. The last three of these models assume that the electron number density is related to the electric potential by a Boltzmann relation, which implicitly assumes diffusive equilibrium for electrons. One-stream models do not distinguish the streams coming from the two hemispheres, whereas two-stream models do. Thus, in the one-stream models, a shock wave system might be formed near the equator. In two-stream models the streams can interpenetrate and shocks do not automatically form. However, the interpenetrating streams can excite waves which can pitch-angle scatter the ions, leading to ion trapping. Thus, an important question is whether the collisionless interaction of the streams will result in shocks or ion trapping at the equator.

Banks et al. [1971] postulated that after flux tube depletion, the resulting flows would be supersonic and that the collision of these streams would result in a pair of electrostatic shocks. Schulz and Koons [1972] developed a two-step mechanism for ion trapping: first ions are trapped due to pitch-angle scattering as a result of wave-particle interactions; when the density becomes sufficiently large, Coulomb collisions are effective in thermalizing the plasma. Schulz and Koons [1972] also suggested that shock formation was unlikely, because the interaction of the individual plasma streams with the background plasma will not excite waves, which would dissipate the energy of the ion streams. They did not consider the interaction of the plasma streams themselves. More recent simulations by Singh et al. [1986a,b] showed that in a collisionless plasma, shocks would develop if $T_e>3T_i$. Since $T_e \sim T_i$ in the conjugate ionospheres, this would require preferential electron heating. Another potential mechanism for ion trapping is perpendicular ion heating.

The purpose of this paper is to present results which have been found using a time-dependent, one-stream hydrodynamic model with three species. It is an adaptation, for closed field lines, of a model for polar wind flows developed by Gombosi et al. [1985]. This model takes into account the effects of ionization, charge exchange, recombination, collisions, heat conduction, and allows for external heat sources; it includes $O^+$ and $H^+$ ions and electrons. No diffusive equilibrium assumptions are used and it includes an energy equation for each species, which means that shocks can be self-consistently modeled. In an earlier version of the model [Guiter and Gombosi, 1990] energy deposition was simulated by assuming a downward electron heat flux at the equator; now this is done using an equatorially confined heat source. The results found in the present case very closely match those found with the earlier version. A flux tube with L = 2 was used because then the density at the equator is large enough so that Coulomb collisions are effective in thermalizing and coupling the streams from the conjugate hemispheres. However, due to the one-stream nature of the model, shocks must form when the streams meet at the equator and so this work is not a proper test of the ion trapping mechanism proposed by Schulz and Koons [1972]. The boundaries of the flux tube are at 200 km altitude, which is low enough so that shocks are not reflected from the boundaries. Steady-state results were found and used as initial conditions in a subsequent simulation of the effect of a 5-fold flux tube density depletion above 2500 km altitude. Although L = 2 flux tubes are not usually depleted, this study should be useful for clarifying the stages of the refilling process at higher L values.

## 2. Model

The model includes the time-dependent coupled continuity, momentum, and energy equations for $O^+$ and $H^+$ ions and the energy equation for electrons. As a result of the strong geomagnetic field only motions along magnetic field lines are important; field, line curvature effects are neglected. Quasineutrality is assumed and no field-aligned electric currents are allowed. With these constraints, the governing equations become:

Modeling Magnetospheric Plasma Processes
Geophysical Monograph 62

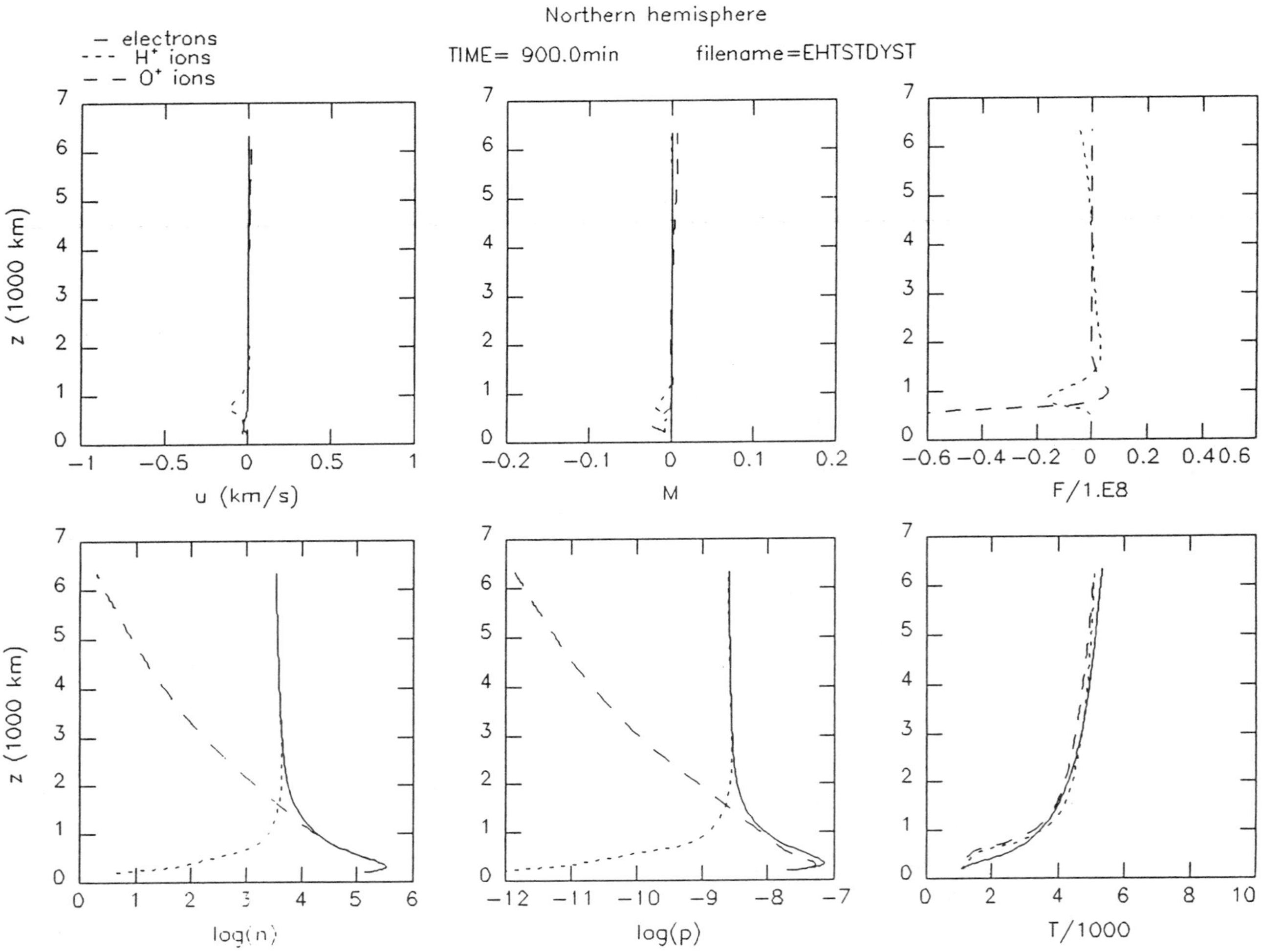


Fig. 1. Steady state solution for the northern (a) and southern (b) hemispheres. The six panels display flow velocity, hydrodynamic Mach number, particle flux (normalized to a reference altitude of 1000 km), number density, pressure, and temperature profiles between 200 km and 6500 km altitude.

$$\frac{\partial}{\partial t}\left(A\rho_i\right) + \frac{\partial}{\partial r}\left(A\rho_i u_i\right) = AS_i$$

$$\frac{\partial}{\partial t}\left(A\rho_i u_i\right) + \frac{\partial}{\partial r}\left(A\rho_i u_i^2\right) + A\frac{\partial p_i}{\partial r} = $$
$$= A\rho_i\left(\frac{e}{m_i}E_{\parallel} - g\right) + A\frac{\delta M_i}{\delta t} + Au_i S_i$$

$$\frac{\partial}{\partial t}\left(\frac{1}{2}A\rho_i u_i^2 + \frac{1}{\gamma_i - 1}Ap_i\right) + $$
$$+ \frac{\partial}{\partial r}\left(\frac{1}{2}A\rho_i u_i^3 + \frac{\gamma_i}{\gamma_i - 1}Au_i p_i\right) = $$
$$= A\frac{\delta E_i}{\delta t} + AQ_i + \frac{\partial}{\partial r}\left(A\kappa_i\frac{\partial T_i}{\partial r}\right) + $$
$$+ A\rho_i u_i\left(\frac{e}{m_i}E_{\parallel} - g\right) + Au_i\frac{\delta M_i}{\delta t} + \frac{1}{2}Au_i^2 S_i$$

$$\rho_e\left(\frac{\partial T_e}{\partial t}\right) = \frac{\left(\gamma_e - 1\right)m_e}{kA}\frac{\partial}{\partial r}\left(A\kappa_e\frac{\partial T_e}{\partial r}\right) - $$
$$- \rho_e u_e\frac{\partial T_e}{\partial r} - T_e\left[S_e + \frac{\left(\gamma_e - 1\right)}{A}\rho_e\frac{\partial\left(Au_e\right)}{\partial r}\right] + $$
$$+ \frac{m_e\left(\gamma_e - 1\right)}{k}\left[Q_e + \frac{\delta E_e}{\delta t}\right]$$

$$E_{\parallel} = -\frac{1}{en_e}\left[\begin{array}{l}\dfrac{\partial}{\partial r}\left(p_e + u_e^2\rho_e - \sum\limits_{ions}\dfrac{m_e}{m_i}\left(p_i + u_i^2\rho_i\right)\right) - \\ - \dfrac{\delta M_e}{\delta t} + \sum\limits_{ions}\dfrac{m_e}{m_i}\left(\dfrac{\delta M_i}{\delta t}\right) - \left(u_e - u_i\right)S_i\end{array}\right] - $$
$$- \frac{1}{en_e}\frac{A'}{A}\left(u_e^2\rho_e - \sum_{ions}\frac{m_e}{m_i}u_i^2\rho_i\right)$$

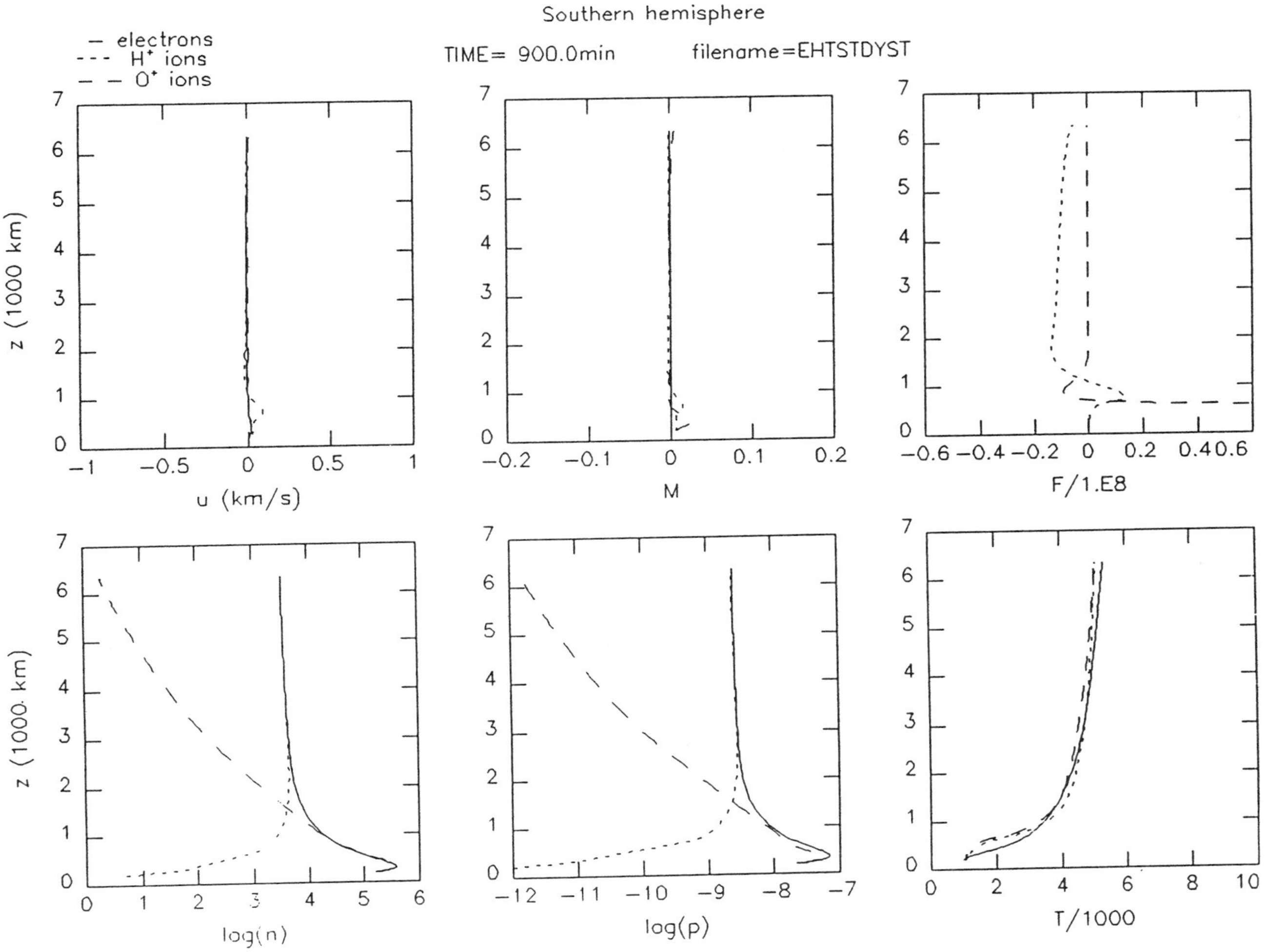


Fig. 1. (continued)

In these equations t is time, r is distance along magnetic field lines, B is the magnetic field strength, $A \sim B^{-1}$ is the cross-sectional area of a flux tube, m is the particle mass, ρ is the mass density, n is the number density, u is the field-aligned flow velocity, p is the pressure, T is the temperature, e is the electron charge, k is Boltzmann's constant, γ is the specific heat ratio, κ is the heat conductivity, S is the net mass production rate, $E_{\|}$ is the polarization electric field, g is the component of the gravitational acceleration parallel to the magnetic field line, δM/δt is the momentum exchange rate, δE/δt is the energy exchange rate, and Q is the external heating rate. The subscript i refers to either of the ions, whereas the subscript e refers to the electrons.

In this model the collision terms are assumed to have the following form:

$$\frac{\delta M_e}{\delta t} = \sum_t n_s m_s v_{st} \left(u_t - u_s\right)$$

$$\frac{\delta E_s}{\delta t} = \sum_t \frac{n_s m_s v_{st}}{m_{s+m_t}} \left[3k \left(T_{t-T_s}\right) + m_t \left(u_t - u_s\right)^2\right]$$

In these equations the summations run over all species, and the neutral velocity is assumed to be zero; $\nu_{st}$ is the momentum transfer collision frequency of species s with species t. The velocity-dependent correction factors [cf. Schunk, 1977; Burgers, 1969; Tanenbaum, 1967] have been approximated by one. This approximation is not expected to influence the results significantly because at the altitudes used in the model the relative drifts and momentum transfer collision frequencies are small.

The neutral atmosphere model used was MSIS-86 [Hedin, 1987]; this includes $N_2$, $O_2$, O, and H. It is important to note that this model is not symmetric about the magnetic equator. Also, in the present set of calculations the neutral atmosphere was not allowed to vary with time. The $O^+$ ions are produced by photoionization, whereas the $H^+$ ions are created by charge transfer only. $O^+$ is removed by chemical reactions with $N_2$ and $O_2$; $H^+$ is removed by charge transfer with O. The ion-ion, ion-neutral, and ion-electron collision frequencies and the heat conductivities used in this model were taken from Raitt et al. [1975].

In the model, an L=2 flux tube connects two external reservoirs, each at an altitude of 200 km; these reservoirs represent photochemically controlled regions of the ionosphere. The ions in the stationary reservoirs

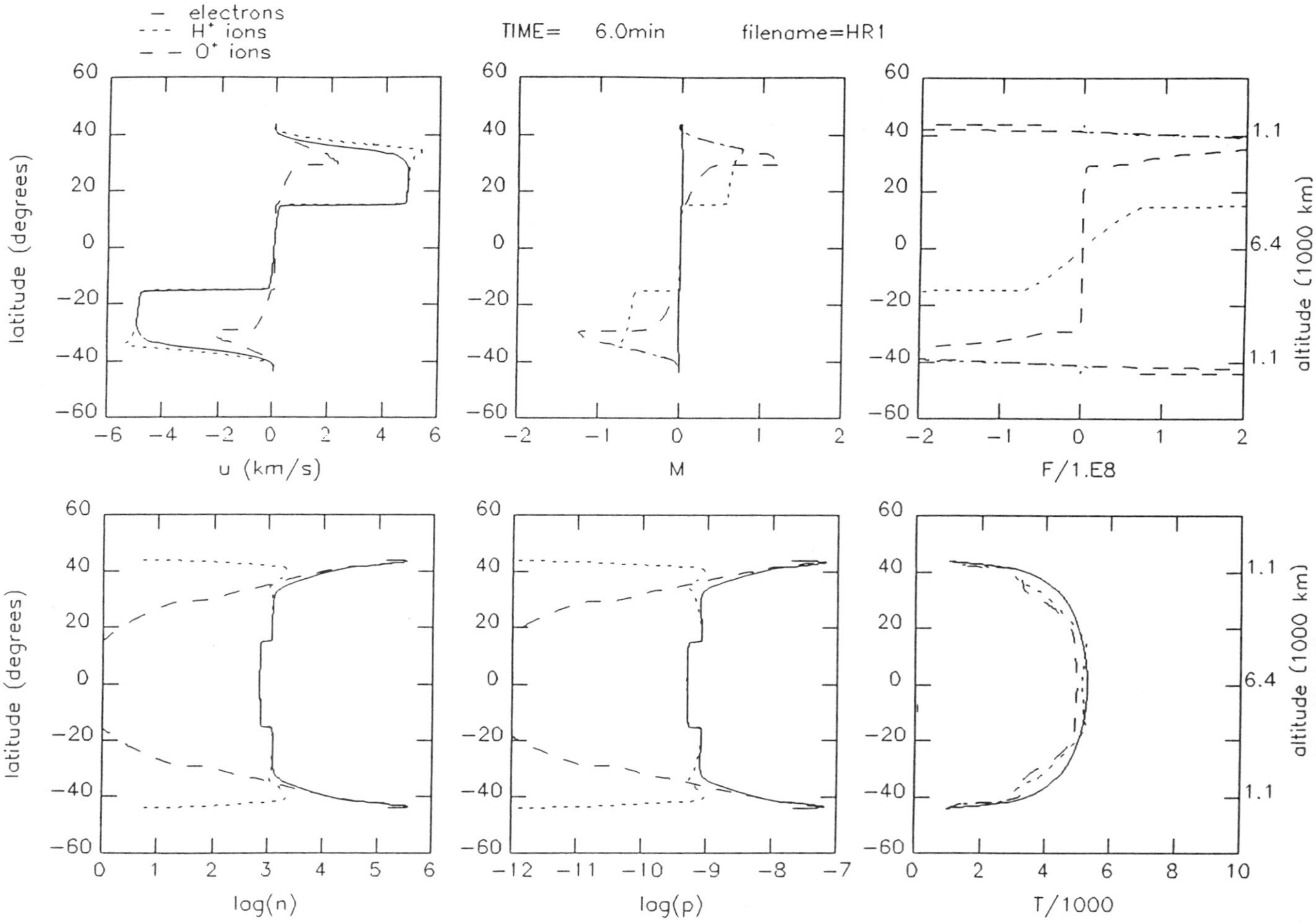


Fig. 2. Solution for t = 6 minutes in the 5-fold density depletion case. The six panels display flow velocity, hydrodynamic Mach number, particle flux (normalized to a reference altitude of 1000 km), number density, pressure, and temperature profiles between -40 and +40 degrees magnetic latitude.

are assumed to be in chemical and thermal equilibrium with the neutral atmosphere, while the electron temperature in the reservoirs is assumed to be 1000 K. Since the reservoirs are at 200 km altitude, the neutral species densities are high enough so that downward propagating shocks are absorbed at the boundaries rather than being reflected from them. An electron heat source of $6.3\times10^{-12}$ erg cm$^{-3}$ s$^{-1}$ confined to within ten degrees of the magnetic equator is used to simulate energy deposition from the magnetosphere. For this work the model was run assuming equinox conditions during solar minimum (Sept. 22, 1988) at noon local time and no external ion heating.

The coupled time-dependent partial differential equations are solved using a combined Godunov scheme/Crank-Nicholson method with dimensional splitting.

## 3. Summary of Results and Discussion

### 3.1 Steady State

Figures 1a and 1b show the steady-state altitude profiles for the northern and southern hemispheres. The six panels in each figure display flow velocity, hydrodynamic Mach number, particle flux, number density, pressure and temperature between 200 km and 6500 km altitude. It should be noted that positive flows are upward in the northern hemisphere but downward in the southern hemisphere. An important result is counter-streaming of $O^+$ and $H^+$ ions with $O^+$ moving up, between about 700 km and 1000 km altitude in the southern hemisphere and about 700 km and 1200 km altitude in the northern hemisphere; this is in agreement with the results of Young et al. [1980]. The magnitudes of the counterstreaming fluxes are not identical; usually the $H^+$ flux is greater than the $O^+$ flux. The counterstreaming is a result of the $O^+$ moving up to replace $O^+$ that is lost in charge exchange reactions with H at higher altitudes, while the $H^+$ that is created there moves down. Counterstreaming in steady-state was predicted by Young et al. [1979], with the sum of the fluxes being zero for symmetric flux tube conditions but nonzero for nonsymmetric conditions.

The results are not symmetric about the magnetic equator; this is due to the asymmetry of the neutral atmosphere model. As an example, the $O^+$ flux at 1000 km is about twice as large in the northern hemisphere than in the southern hemisphere.

Below about 700 km, the fluxes of both ions are downward, but now the $O^+$ flux is much greater than the $H^+$ flux which is practically zero.

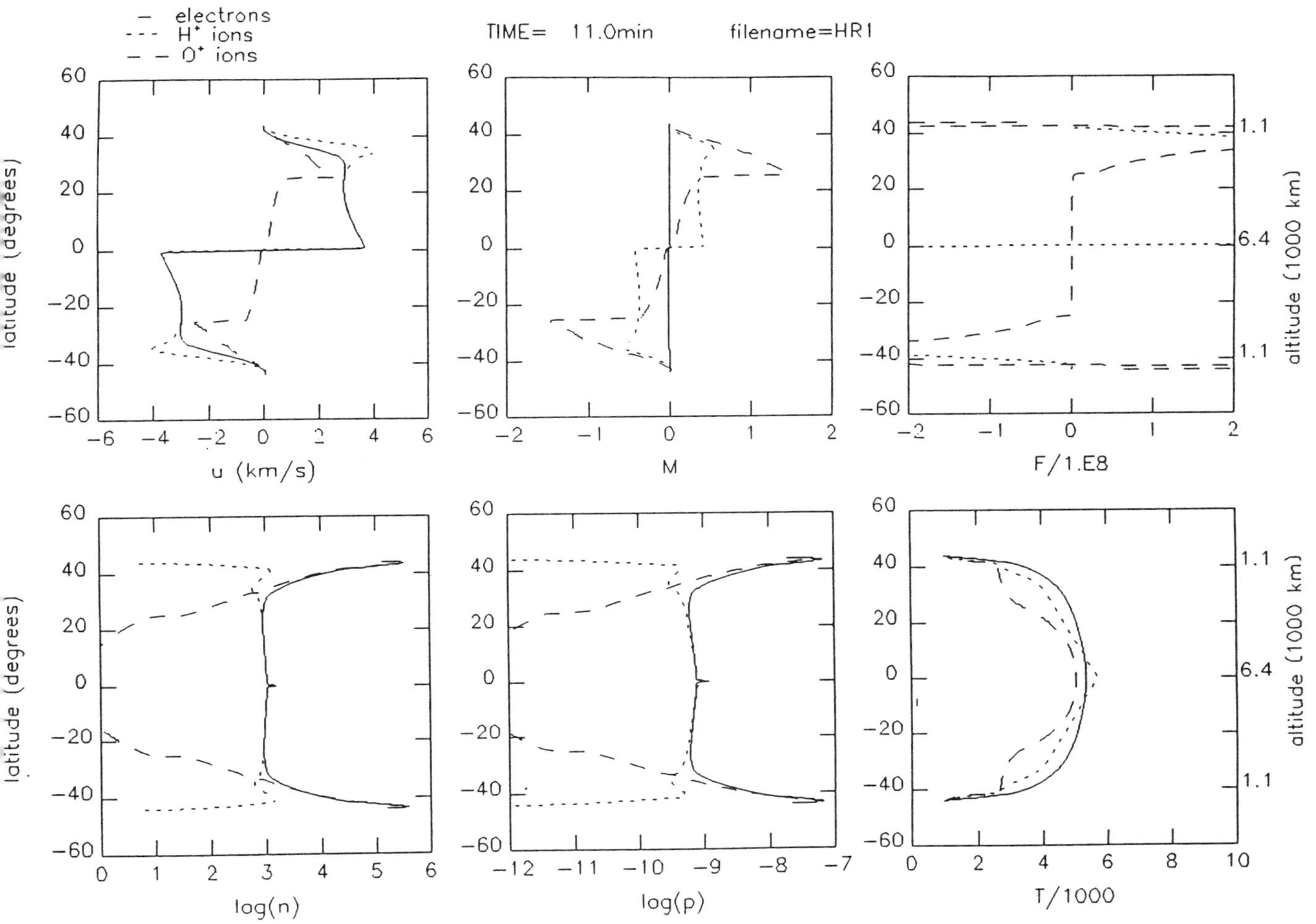


Fig. 3. Solution for t = 11 minutes in the 5-fold density depletion case. The six panels display flow velocity, hydrodynamic Mach number, particle flux (normalized to a reference altitude of 1000 km), number density, pressure, and temperature profiles between -40 and +40 degrees magnetic latitude.

The $O^+$ flows down to replace $O^+$ which is lost in reactions with $O_2$ and $N_2$ at lower altitudes.

The temperatures of the ions and electrons are comparable for most latitudes. $T_e$ is greater than $T(O^+)$ for all points on the flux tube; $T(H^+)$ is greater than $T_e$ for altitudes between 1000 km and 2500 km. In that altitude region the hydrogen ions are heated as a result of frictional interactions with the $O^+$ ions and the neutral species.

### 3.2 Density Depletions

Next, the effect of a 5-fold density depletion was modeled. In order to simulate such conditions, the steady-state densities for all species were reduced by a factor of five above 2500 km altitude; however, the temperatures and velocities were not changed. The density depletion was also assumed to be instantaneous. This type of initial profile was used in order to simulate refilling after a strong magnetic storm; because of it, shock structures clearly develop. These structures move upward because the ion and electron gases expand into the low density regions. The model was allowed to run for three hours of simulated time.

For this case, the result was similar to a blast wave. The blast wave is the result of a strong discontinuity in the density of the medium; the disturbance moves into the low density region with a speed which exceeds the local sound speed of the upstream medium. Figures 2 through 4 show results at t = 6, 11, and 14 minutes for this case. It should be noted that positive latitudes refer to the northern hemisphere. Shocks develop which move upward, as shown by the velocity profiles of Figure 2; this figure also shows that the $H^+$ shocks are outrunning the $O^+$ shocks. This occurs because the $H^+$ ions are sixteen times lighter. The propagation speed of the $H^+$ shock is about 12.5 km/s; the propagation speed of the $O^+$ shock is initially 1.8 km/s but by 30 minutes this has increased to 2.8 km/s. The $O^+$ shock speed increases because the $O^+$ velocity increases behind the shock, which implies that the shock velocity must increase because of mass flux conservation relationships. The $H^+$ shocks are supersonic but the $O^+$ shocks are subsonic, with respect to the ion-acoustic speed (~ 11.7 km/s). The $H^+$ flow velocities approach the sonic velocity at 1 minute and

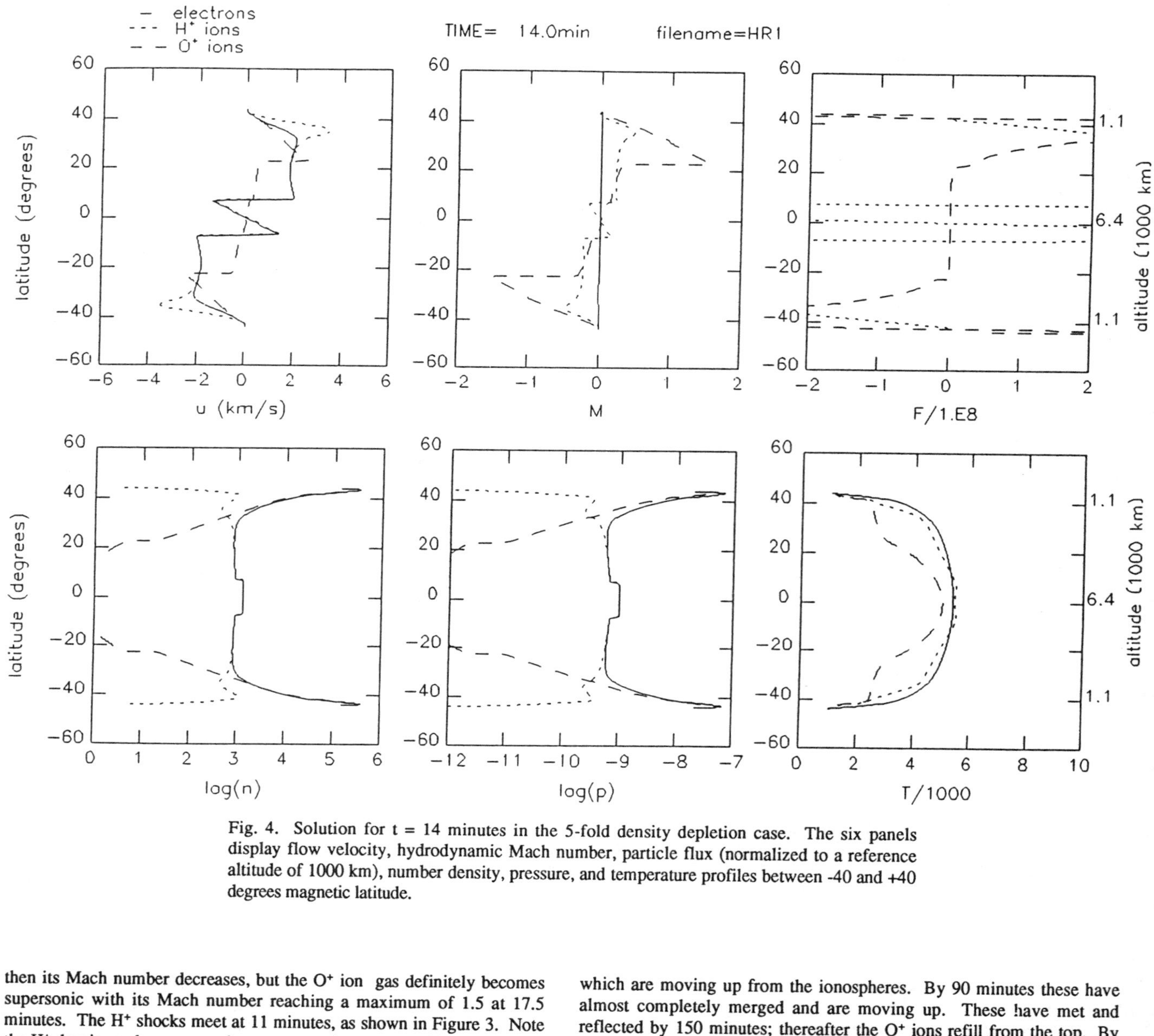


Fig. 4. Solution for t = 14 minutes in the 5-fold density depletion case. The six panels display flow velocity, hydrodynamic Mach number, particle flux (normalized to a reference altitude of 1000 km), number density, pressure, and temperature profiles between -40 and +40 degrees magnetic latitude.

then its Mach number decreases, but the $O^+$ ion gas definitely becomes supersonic with its Mach number reaching a maximum of 1.5 at 17.5 minutes. The $H^+$ shocks meet at 11 minutes, as shown in Figure 3. Note the $H^+$ density and pressure enhancement at the equator. Also note the $H^+$ temperature increase near the equator. This figure also shows that the $O^+$ temperature decreases behind the $O^+$ shocks, due to adiabatic cooling. After this, the shocks reflect and the $H^+$ ions start refilling from the top, as can be seen in Figure 4. The $O^+$ maximum flow velocity decreases after 17.5 minutes due to increased collisions with the refilling hydrogen ions.

Figures 5 and 6 show results at t = 45, and 60 minutes in the 5-fold density depletion case. The $O^+$ shocks meet at 45 minutes as shown in Figure 5. This figure also shows a strong $O^+$ density and pressure enhancement at the equator and an equatorial $O^+$ temperature increase. The flux profiles in this figure also show counterstreaming of $O^+$ and $H^+$ ions between about 1100 km and 4900 km altitude. The $O^+$ shocks reflect and after this the $O^+$ ions start refilling from the top; this is shown in Figure 6. These reflected shocks move down and merge with reverse shocks which are moving up from the ionospheres. By 90 minutes these have almost completely merged and are moving up. These have met and reflected by 150 minutes; thereafter the $O^+$ ions refill from the top. By 180 minutes the shocks have almost decayed and there is steady refilling; there are upward $H^+$ fluxes in both hemispheres. These upward fluxes are expected since the equatorial $H^+$ density is still lower than the steady-state value.

When the upwelling ion streams for a particular ion species collide the temperature of that ion increases sharply at the equator. This is due to heating caused by the collision of the upwelling streams. In addition, at that time the ion flow velocity at the equator becomes very small due to the one-stream nature of the model and, hence, the density must increase.

## 4. Conclusion

A time-dependent, one-stream hydrodynamic model for plasmaspheric flows has been used to investigate plasmaspheric refilling. The model

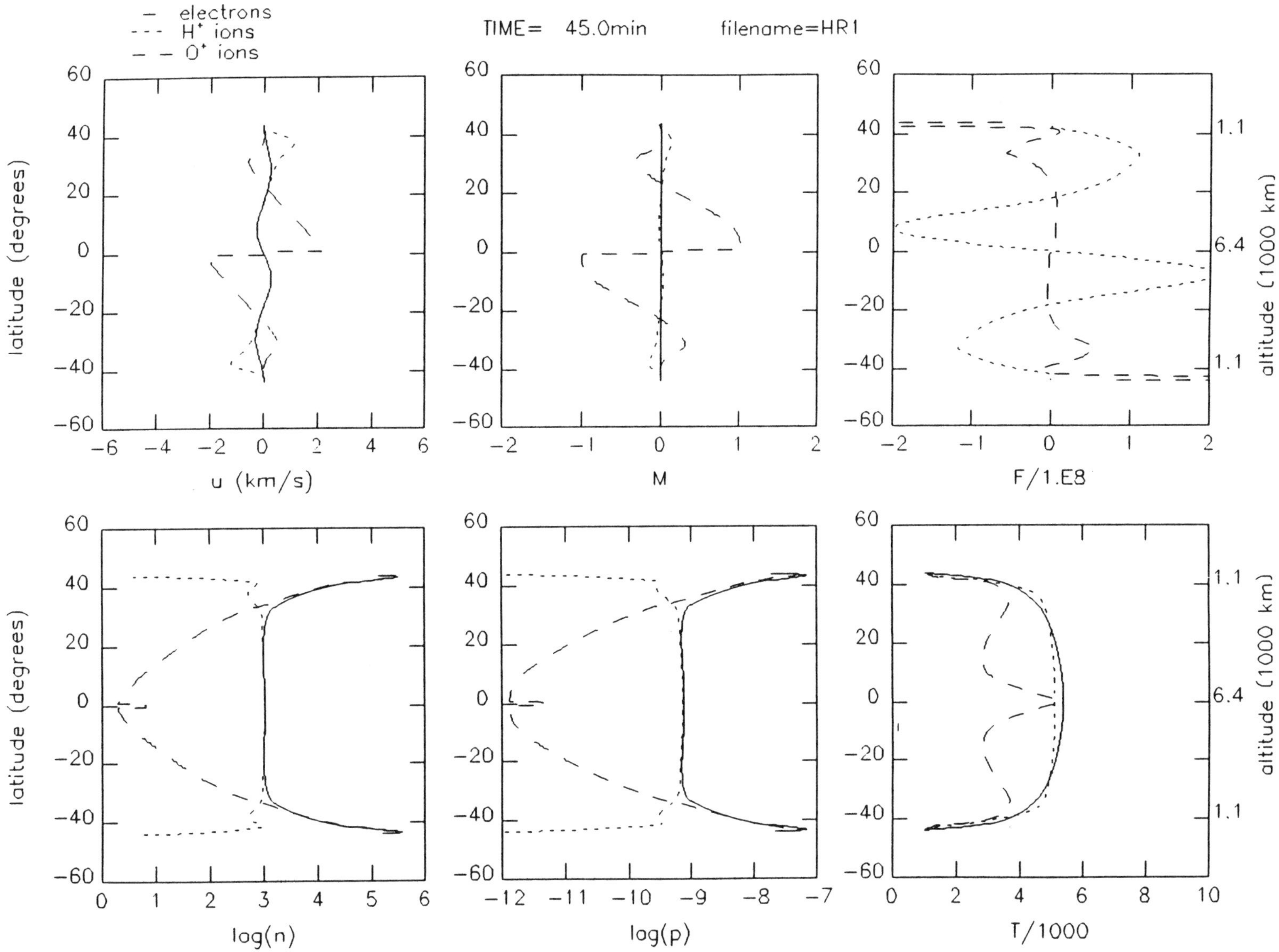


Fig. 5. Solution for t = 45 minutes in the 5-fold density depletion case. The six panels display flow velocity, hydrodynamic Mach number, particle flux (normalized to a reference altitude of 1000 km), number density, pressure, and temperature profiles between -40 and +40 degrees magnetic latitude.

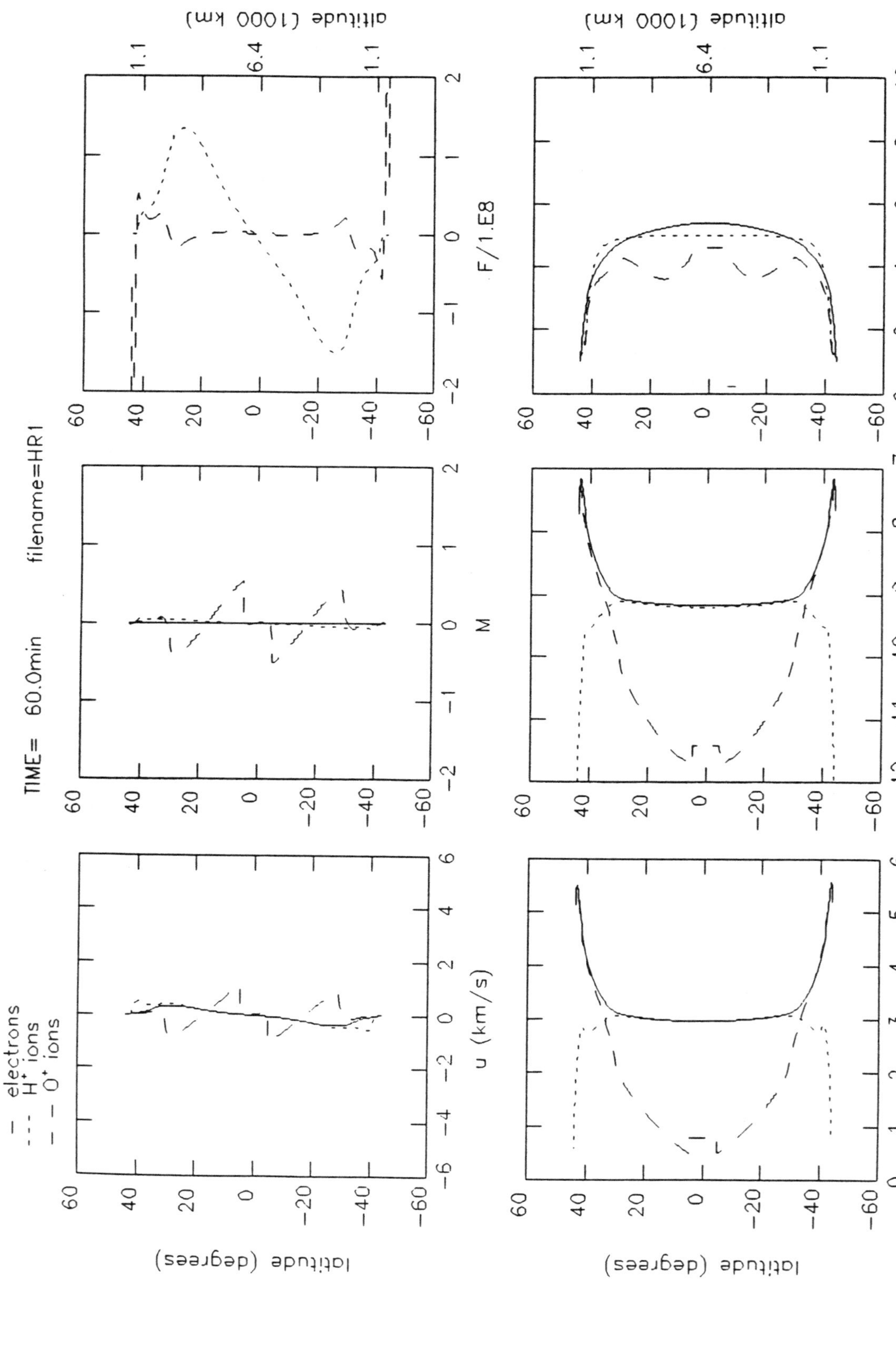


Fig. 6. Solution for t = 60 minutes in the 5-fold density depletion case. The six panels display flow velocity, hydrodynamic Mach number, particle flux (normalized to a reference altitude of 1000 km), number density, pressure, and temperature profiles between -40 and +40 degrees magnetic latitude.

includes $O^+$ and $H^+$ ions and electrons. Energy deposition from the magnetosphere was simulated by including an electron heat source confined to within ten degrees of the magnetic equator. A steady-state solution was found and used in a simulation of the effect of a 5-fold density depletion. Since ion streams from conjugate hemispheres are not distinguished in the model a shock pair must form at the equator when the upwelling ion streams collide; however, since an L = 2 flux tube was used for this work the equatorial density should be high enough for Coulomb collisions to be effective in thermalizing the streams. The results found matched those found using an earlier version of the model in which a downward electron heat flux at the equator was used to simulate magnetospheric energy deposition.

Acknowledgements. The authors are indebted to Drs. T. E. Moore, A. F. Nagy and C. E. Rasmussen for illuminating discussions. This work was supported by NSF grant ATM-8908183 and by NASA grant NAGW-1619. One of us (SMG) was supported by NASA Marshall Space Flight Center through the graduate student training grant NGT-50368. Acknowledgement is also made to the National Center for Atmospheric Research, sponsored by the National Science Foundation, for the computing time used in this research.

References

Banks, P. M., A. F. Nagy, and W. I. Axford, Dynamical behavior of thermal protons in the mid-latitude ionosphere and magnetosphere, *Planet. Space Sci., 19*, 1053, 1971.

Burgers, J. M., *Flow Equations for Composite Gases*, Academic, New York, 1969.

Gombosi, T. I., T. E. Cravens, and A. F. Nagy, A time-dependent theoretical model of the polar wind: Preliminary results, *Geophys. Rev. Lett., 12*, 167, 1985.

Guiter, S. M., and T. I. Gombosi, The role of high speed plasma flows in plasmaspheric refilling, *J. Geophys. Res., 95*, 10427, 1990.

Hedin, A. E., MSIS-86 thermospheric model, *J. Geophys. Res., 92*, 4649, 1987.

Khazanov, G. V., M. A. Kuen, Y. V. Konikov, and I. M. Sidorov, Simulation of ionosphere-plasmasphere coupling taking into account ion inertia and temperature anisotropy, *Planet. Space Sci., 32*, 585, 1984.

Moffet, R. J., and J. A. Murphy, Coupling between the F-region and protonosphere: Numerical solution of the time-dependent equations, *Planet. Space Sci., 21*, 43, 1973.

Park, C. G., Whistler observations of the interchange of ionization between the ionosphere and the protonosphere, *J. Geophys. Res., 75*, 4249, 1970.

Raitt, W. J., R. W. Schunk, and P. M. Banks, A comparison of the temperature and density structure in the high and low speed thermal proton flows, *Planet. Space Sci., 23*, 1103, 1975.

Rasmussen, C. E., and R. W. Schunk, Multistream hydrodynamic modeling of interhemispheric plasma flow, *J. Geophys. Res., 93*, 14557, 1988.

Schulz, M., and H. C. Koons, Thermalization of colliding ion streams beyond the plasmapause, *J. Geophys. Res., 77*, 248, 1972.

Schunk, R. W., Mathematical structure of transport equations for multispecies flows, *Rev. Geophys. Space Phys., 15*, 429, 1977.

Singh, N., Refilling of a plasmaspheric flux tube - Microscopic plasma processes, in Modelling Magnetospheric Plasma, ed. by T. E. Moore and J. H. Waite, Jr., *Geophys. Monograph 44*, American Geophysical Union, Washington, D. C., p. 87, 1988.

Singh, N., H. Thiemann, and R. W. Schunk, Studies on counterstreaming plasma expansion, *Physica Scripta, 33*, 355, 1986a.

Singh, N., R. W. Schunk, and H. Thiemann, Temporal features of the refilling of a plasmaspheric flux tube, *J. Geophys. Res., 91*, 13433, 1986b.

Tanenbaum, B. S., *Plasma Physics*, McGraw-Hill, New York, 1967.

Young, E. R., D. G. Torr, and P. G. Richards, Counterstreaming of $O^+$ and $H^+$ ions in the plasmasphere, *Geophys. Res. Lett., 6*, 925, 1979.

Young, E. R., D. G. Torr, P. Richards, and A. F. Nagy, A computer simulation of the midlatitude plasmasphere and ionosphere, *Planet. Space Sci., 28*, 881, 1980.

# $O^+$, $H^+$, and $He^+$ Densities from the 200-1600 km altitude ionosphere at Arecibo: A comparison of theory and measurement

P. G. Richards

*Computer Science Department and Center for Space Plasma and Aeronomic Research, University of Alabama in Huntsville*

D. G. Torr

*Physics Department and Center for Space Plasma and Aeronomic Research, University of Alabama in Huntsville*

K. L. Miller

*Physics Department and Center for Atmospheric and Space Sciences, Utah State University*

*Abstract.* We have compared measured $O^+$, $H^+$, and $He^+$ densities over the altitude range 200-1600 km at Arecibo on February 10, 1972 with densities calculated by our comprehensive ionosphere-plasmasphere model. There is excellent agreement between measurement and model for all three ions, indicating that the theory of the major ion chemistry appears to be well understood. Comparison was also made between the model and measured electron and ion temperatures. There was good agreement between the model and the data for the ion temperature and at most altitudes for the electron temperature during the daytime. However, in the region of the peak electron density, the measured daytime temperature was much larger than the model temperature. The model electron temperatures did agree very well with the measured temperatures for similar thermospheric conditions in February 1977. At night, the measured electron and ion temperatures were higher than the modeled temperatures, which were close to the neutral temperature, indicating a larger plasmaspheric heat reservoir than predicted by the model.

## 1. Introduction

The purpose of this paper is to determine how well our current understanding of the physics and chemistry of the ionosphere and thermosphere can account for the measured behavior of the topside ionospheric electron and ion density and temperature structure at Arecibo (L=1.4) on February 10, 1972. Hagen and Hsu [1974] provide ion density and temperature profiles at 14:36 and 22:30 local times on February 10, 1972. This appears to be the only published paper giving altitude profiles of the light ions at a single location. From a modeling point of view, Arecibo is a good location for this study because the plasmaspheric flux tube is expected to be in a quasi-equilibrium state most of the time. Thus, the topside densities are not expected to be affected by the stage of flux tube refilling after magnetic storms which is important for higher latitudes and is difficult to quantify. In this paper, we find excellent agreement between the measured densities and those calculated by our comprehensive field line interhemispheric plasma (FLIP) model. The temperatures measured on February 10, 1972 are significantly higher than the model temperatures which are in good agreement with measured temperatures on February 14, 1977.

## 2. Model

The FLIP model, which has been developed over a period of more than ten years, has been described previously by Richards and Torr [1988] and more recently by Torr et al. [1990]. The $He^+$ chemical and physical processes have been discussed by Newberry et al. [1989]. The main component of this one-dimensional model calculates the plasma densities and temperatures along entire magnetic flux tubes from 80 km in the northern hemisphere through the plasmasphere to 80 km in the southern hemisphere. The model, which is called the field line interhemispheric plasma (FLIP) model, uses a tilted dipole approximation to the Earth's magnetic field. The equations solved are the continuity and momentum equations for $O^+$, $H^+$, and $He^+$, as formulated for the topside ionosphere by St. Maurice and Schunk [1976]. Collisions between ions and neutrals have been included in order to extend the equations into the $E$ and $F$ regions.

The electron and ion temperatures are obtained by solving the energy equations Schunk and Nagy, [1978]. Electron heating due to photoelectrons is provided by a solution of the two-stream photoelectron flux equations using the method of Nagy and Banks [1970]. The solutions have been extended to encompass the entire field line on the same spatial grid as the ion continuity and momentum equations. With the latest cross sections and solar EUV fluxes as inputs, Richards and Torr [1984] have demonstrated that the model photoelectron fluxes are in good agreement with the measured fluxes of Lee et al. [1980].

The FLIP model calculates the vibrational distribution of $N_2$ in order to take into account the strong dependence of the $O^+ + N_2 \rightarrow NO^+ + N$ reaction rate on the degree of vibrational excitation of $N_2$. It has been shown previously by Richards and Torr [1986] that this effect may be important at solar maximum in summer but is usually not important at other times under magnetically quiet conditions. Vibrationally excited $N_2$ has a long lifetime and is produced efficiently

Modeling Magnetospheric Plasma Processes
Geophysical Monograph 62

by photoelectron and thermal electron excitation and also by a number of chemical reactions but is quenched efficiently at low altitudes by atomic oxygen.

## 3. MODEL PARAMETERS

In order to simulate the ionosphere, the FLIP model requires three key inputs: the solar EUV flux, the neutral atmosphere, and the meridional component of the neutral wind. These three inputs are discussed below.

The solar EUV flux is related to the level of solar activity, which was moderate during the early part of 1972 as indicated by the 10.7 cm solar flux. The $F10.7$ index for February 10, 1972 was 116. In the FLIP model we use the solar EUV fluxes in the 37 wavelength intervals proposed by Torr et al. [1979]. We linearly scale the EUV fluxes in each wavelength bin independently with F10.7 in the following way. For F107=71 we use the standard F74113 fluxes except we have doubled the fluxes in the wavelength range below 250 Å for reasons discussed by Richards and Torr [1984,88]. For other values of F10.7 we do a simple linear scaling of this modified F74113 spectrum for each of the 37 wavelength bins. This is achieved by using the standard 1976 reference spectrum (SC#21REFW, F10.7=68) and the measured fluxes for day 78348 F10.7=206). Under this scaling the 304 Å solar EUV flux intensity was a factor of 1.3 higher on February 10, 1972 (F10.6=116) than the intensity on day 74113.

The mass spectrometer and incoherent scatter (MSIS) model Hedin, [1987] supplies the neutral densities and temperatures used in these calculations. The magnetic activity was low during the period under study here (Ap=4).

The meridional neutral wind affects the ion densities above about 200 km by affecting the rate at which the $O^+$ ions diffuse downward. The most obvious effects are on the $F2$ peak density ($NmF2$) and the height at which the peak density occurs ($hmF2$). Normally the wind is poleward during the day, enhancing the downward diffusion to a region of increased loss, thereby reducing the peak density ($NmF2$) and lowering the height of the layer. At night the wind becomes equatorward, inhibiting the downward diffusion, raising the height of the layer, and helping to maintain the nighttime $F2$ layer density. Richards and Torr [1986] and Miller et al. [1986] demonstrated that this relationship between the meridional neutral wind and $hmF2$ could be employed to estimate neutral winds from ground-based measurements of $hmF2$. One uncertainty in these winds is the magnitude of the $O^+$ – O collision frequency. Burnside et al. [1987] have presented evidence that the value being used in most models should be increased by a factor of 1.7. This significantly reduces the magnitude of the meridional winds derived from $hmF2$. In these calculations we have used the smaller collision frequency. Another uncertainty is the possibility that the wind derived from $hmF2$ includes the effects of electric fields. At midlatitudes, it is not possible to distinguish between the effects of electric fields and neutral winds purely from local measurements of $hmF2$.

While the collision frequency and the electric fields are important issues where the magnitude of the neutral wind is concerned, they are irrelevant to the ionospheric modeling in this paper which is primarily concerned with plasma densities and temperatures. In this case, it is most important to be able to reproduce $hmF2$ and the mechanisms that produce the variations are not important. Reproducing the correct $hmF2$ ensures that the chemical production and loss rates are appropriate at the peak height.

Because of the need to match the observed $hmF2$, the neutral wind input is well constrained for a particular neutral atmosphere and is not a free parameter.

## 4. DATA

The data used in this comparison were obtained with the radar beam directed in the zenith at Arecibo which is located at (18 N, 67 W). Thus the radar beam cuts through a range of L-values from 1.35 at 200 km to 1.6 at 1500 km. Full details on the measurements are provided in the papers by Hagen and Hsu [1974] and Emery et al., [1981]. The model was run for L=1.5 which represents an altitude of about 800 km in the radar beam. Thus the model parameters that are presented as altitude profiles actually cover a latitude range from 23 degrees at 200 km to 14 degrees at 1500 km along the field line. There are two possible ways that the variation in latitude (L) could affect the comparison. First, the difference in latitude could produce differences in photoproduction rates through the solar zenith angle and neutral density latitudinal variations. However, the actual variation in the photoproduction is small because the main $F_2$ photoproduction region is between 200 and 500 km where the latitude changes by only 3 degrees. Second, the L value could affect the nighttime densities and temperatures because it determines the size of the plasmaspheric particle and energy reservoir. This is not a problem because the difference in L-value is small.

Unfortunately, the full diurnal variation of $NmF2$ and $hmF2$ are not available for February 10, 1972. Hagen and Hsu [1974] provide ion density and temperature profiles only at 14:36 and 22:30 local times. We have used the diurnal $hmF2$ variation for February 14-15, 1977 Emery et al. [1981] to deduce the winds and obtain the correct $hmF2$ in the model. The 1977 layer height at 14:36 and 22:30 is very close to the observed values for February 10, 1972. The F10.7 value of 87 on February 14, 1977 was comparable to that on February 10, 1972 (F10.7=116).

## 5. DENSITY RESULTS

Figure 1 shows the wind (dot-dash line) derived from the observed $hmF2$ (dashed line) along with the calculated $hmF2$ (solid line). The diurnal wind variation is generally poleward during the day and equatorward at night but there are some interesting features, notably, a large poleward wind just prior to sunrise and an equatorward incursion between 12:00 and 14:00 hours. The solid line is the $hmF2$ variation obtained from the FLIP model when the derived wind is used. Ideally, the solid line and the dashed line would be identical but the wind calculation makes the assumption that $hmF2$ responds immediately to changes in the neutral wind. In fact, the $hmF2$ curves in Figure 1 indicate that there is an approximate 30 minute time delay when the wind speed changes rapidly. This indicates that more accurate winds could be obtained from $hmF2$ if this phase difference

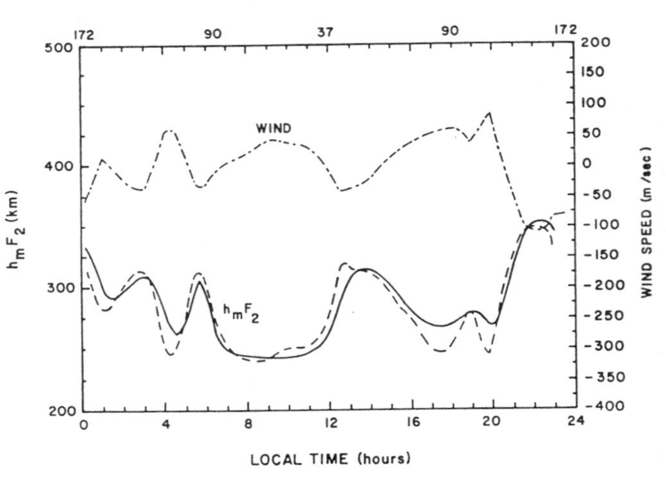


Fig. 1. The diurnal variation of $hmF2$ and the neutral wind derived from $hmF2$ for February 10, 1972. The calculated $hmF2$ using the derived winds in the FLIP model is given by the solid line and the measured $hmF2$ by the dashed line. Note the 30 minute phase difference between the calculated and measured $hmF2$, the poleward surge in the wind just before sunrise, and the equatorward wind incursion beginning just before noon local time.

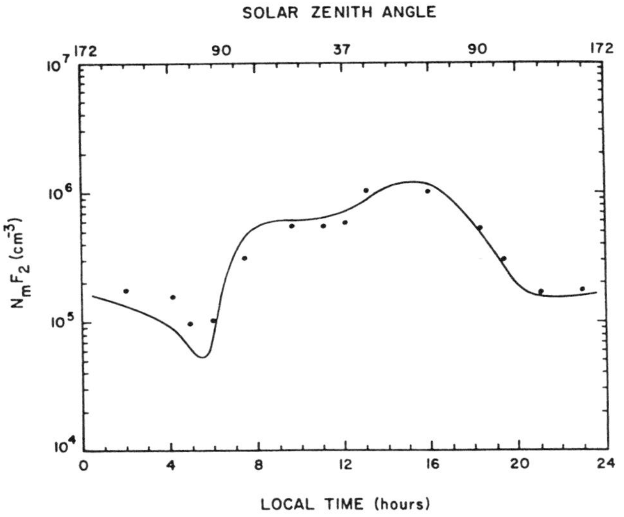


Fig. 2. The variation of the modeled peak electron density for February, 12 1972 (solid line) and the measured peak electron density (dots) for February 14, 1977.

were taken into account. However, a half hour phase shift is not a serious problem for modeling the ion densities.

The calculated diurnal variation of $NmF2$ is compared with the measured variation in Figure 2. The agreement is excellent but it must be kept in mind that the measured height and density variations shown apply to February 1977 (F10.7=87), a period of slightly lower solar activity than February 10, 1972 (F10.7=116). We have also done the calculations for 1977 and the $NmF2$ variation agrees very well with that measured in 1977. The small increase in the production rate

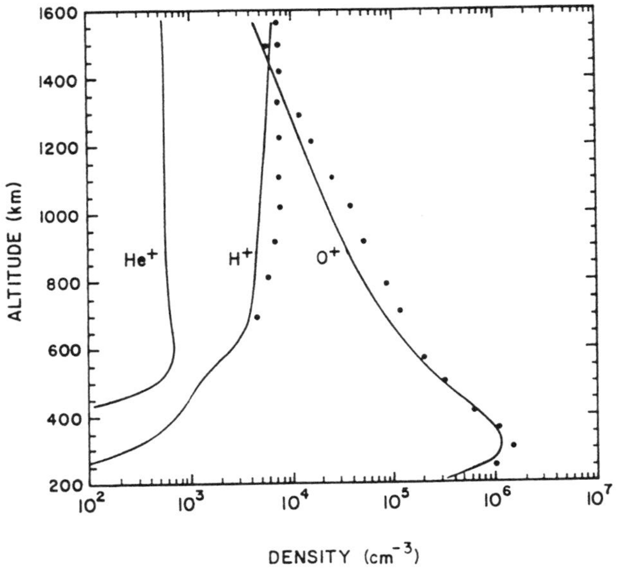


Fig. 3. Altitude profiles of the densities of the three major topside ions at 14:36 local time on February 10, 1972. The theoretical densities from the FLIP model are given by the solid lines while the measured densities are indicated by the dots. The $He^+$ density was too low to measure.

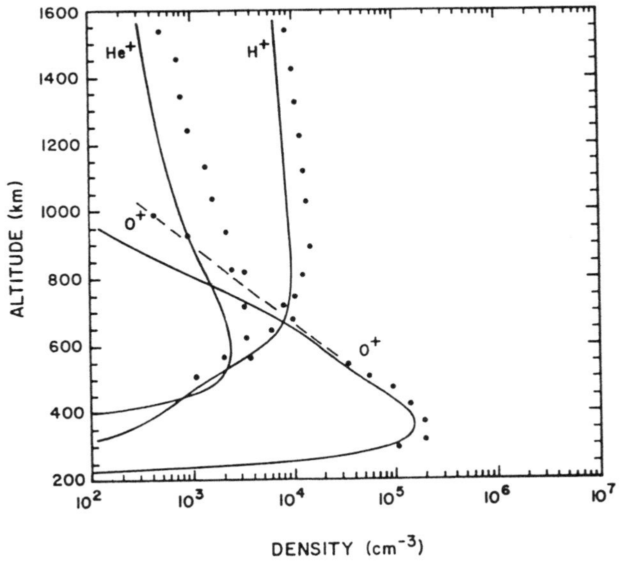


Fig. 4. Altitude profiles of the densities of the three major topside ions at 22:30 local time on February 10, 1972. The theoretical densities from the FLIP model are given by the solid lines while the measured densities are indicated by the dots. The dashed line was drawn to identify the dots associated with the measured $O^+$ densities.

by solar EUV photons from 1972 to 1977 is offset by the increased loss rate from increased molecular neutral densities at $hmF2$. The neutral wind variation is also similar in 1977 and 1972 but the winds are more equatorward in 1977.

Figure 3 shows that the agreement between the calculated and measured $O^+$ and $H^+$ densities is very good at 14:36 hours local time. The modeled $He^+$ density is also shown but the measured densities are not available because they are too low. The modeled $He^+$ density is only about 10% of the $H^+$ density and about 1% of the total density under these conditions.

The nighttime densities are shown in Figure 4. Once again the agreement is very good although there is a factor of two difference between the modeled and measured $He^+$ densities above 600 km. Although the $He^+$ to $H^+$ density ratio is comparable to the daytime ratio, $He^+$ makes up a larger fraction of the total density due to the large decrease in the $O^+$ density. In the region where $H^+$ becomes dominant, there are significant differences in the measured and calculated scale heights of $O^+$. This may be a result of differences between the measured and calculated temperatures or simply to difficulty of measuring the low concentrations of $O^+$.

## 6. TEMPERATURE RESULTS

The electron and ion temperatures at 14:36 and 22:30 for 1972 are shown in Figure 5. At night, both the model and data indicate that the electron and ion temperatures are equal. During the day, the model electron temperature is higher than the ion temperature at all altitudes. The ion temperature is close to the neutral temperature below 400 km altitude but increases with altitude as the cooling to the neutral gases decreases until by 800 km it approaches the electron temperature. At these high altitudes the radar ion and electron temperatures are the same.

It has been shown previously by Newberry et al. [1989] and by Horwitz et al. [1990] that normal coulomb collisions of photoelectrons traversing the plasmasphere produce insufficient heating to raise the plasma temperatures to their observed values. The additional heating may result from pitch angle scattering of a fraction of the photoelectrons as they traverse the plasmasphere, to ring current heating, or to

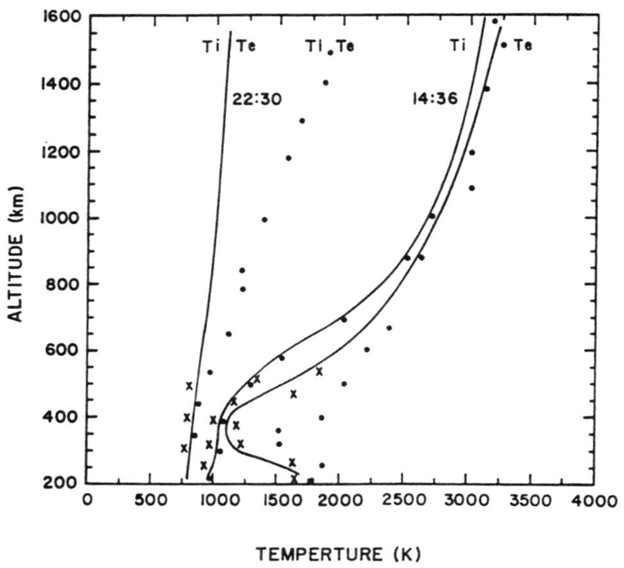


Fig. 5. Measured and calculated electron temperatures for both 14:36 and 22:30 local times. The model electron temperatures are represented by the solid lines while the dots represent the measured temperatures on February 12, 1972. The crosses indicate the measured temperatures on February 14, 1977. At night, both the model and data indicate that the electron and ion temperatures are equal. During the day, the electron temperature is higher than the ion temperature which is close to the neutral temperature at low altitudes and close to the electron temperature above 800 km where the measured ion and electron temperatures are indistinguishable. Note the good agreement between the calculated and measured ion temperatures in the daytime.

some other unidentified source. The local time variation of the data presented by Newberry et al. [1989] suggested that at L=2 the additional heat source is related to the solar zenith angle and therefore most likely due to pitch angle trapping of photoelectrons. To achieve the agreement between the measured and modeled daytime temperatures in this study, a 20% trapping of photoelectrons was applied.

At 14:36, the measured and calculated ion temperatures are in excellent agreement at all altitudes but the measured electron temperature for 1972 is significantly higher than both the 1977 measured temperatures and the calculated temperatures. Although the difference between the measured and calculated temperatures is only 400° K near 400 km, it can readily be shown that a large increase in electron heating rate would be needed to bring the temperatures into agreement. We can assume that the heat flux is small at 400 km and the electron temperature should be close to its local equilibrium value owing to the high electron-ion cooling rate. In the $F$ region, the main sink for electron energy is Coulomb collisions with the ions. The cooling rate is proportional to $N_e^2\ (Te\text{-}Ti)$ where $N_e$ is the electron density and $Te$ and $Ti$ are the electron and ion temperatures. Given the same $N_e$, the factor of 4 electron-ion temperature difference that is evident in Figure 5 implies a factor of 4 difference in electron heating rate. This is difficult to understand when it is noted that the temperatures agree reasonably well near 200 km and above 600 km. Also, the 1977 measured temperatures agree well with the calculated temperatures. We would not expect a factor of 4 increase in heating rate when the 10.7 cm radio flux increases from 87 to 116.

At 22:30, the 1972 measured temperatures are much larger than the calculated temperatures above 600 km. The model predicts a rápid draining of the plasmaspheric heat content after conjugate sunset at 19 hours local time; the temperatures rapidly reverting back to the neutral temperature. However, the measured temperatures indicate that there is still significant heat being conducted from the plasmasphere as late as 22:30. Our 22:30 temperatures are very close to the 2:42 temperatures published by Hagen and Hsu [1974]. The 1977 measurements do not go to a high enough altitude to indicate if the measured temperatures were different on that day.

## 7. Conclusions

The comparisons presented in this paper suggest that the basic physics and chemistry of the F region and topside ionosphere over Arecibo is well understood for quiet conditions at solar minimum. There is good agreement between calculated and measured densities of all three of the major ions in this region ($O^+$, $H^+$, and $He^+$). Horwitz et al. [1990] have also found similar good agreement between the FLIP model topside ion densities and satellite measurements at a range of latitudes at solar maximum. The advantage of the radar is that it produces full altitude profiles whereas the satellite altitude coverage is severely limited.

There are still some problems in the thermal structure. The model reproduces the F region temperatures observed in 1977 but not the much higher temperatures observed in 1972. The measured temperatures in 1972 imply much greater local heating rates and a greater plasmaspheric heat reservoir than either the model or the 1977 measurements. The difference between the measured temperatures is particularly difficult to understand when the solar and magnetic activity, and the electron density on the two days are so similar. The accurate modeling of the ionosphere in this case was possible because of the availability of $hmF2$ from a similar day for deriving appropriate neutral winds.

*Acknowledgments.* This work was supported by NSF grants ATM-8713693, ATM-8716036, ATM-8907808, and ATM-8714461; and NASA grants NAGW-922, and NAGW-996 at The University of Alabama in Huntsville. Support for K. L. Miller was provided by NSF grant ATM-8715367. This research was performed on the Space Plasma Interactive Data and Simulation Laboratory (SPIDASL) system at UAH, under the auspices of the Center for Space Plasma and Aeronomic Research. SPIDASL, a clustered VAX computer system, was developed under an EPSCoR (Experimental Program to Stimulate Competitive Research) grant to UAH, with significant contributions from the National Science Foundation, the State of Alabama, and The University of Alabama in Huntsville.

## References

Burnside, R. G., C. A. Tepley, V.B. Wickwar, The $O^+$-O collision cross-section: Can it be inferred from aeronomical measurements? *Annales Geophysicae, 5A*, (6), 343-350, 1987.

Emery, B. A., R. M. Harper, S. Ganguly, J. C. G. Walker, and G. A. Giles, Arecibo Ionospheric Data, August 1974-May 1977, NAIC 163, *National Astronomy and Ionosphere Center*, May 1981.

Hagen, Jon B., and Peter You-Sen Hsu, The structure of the protonosphere above Arecibo, *J. Geophys. Res., 79*, 4269-4275, 1974.

Hedin, A. E., MSIS-86 thermosphere model, *J. Geophys. Res., 92*, 4649, 1987.

Horwitz, J. L., R. H. Comfort, P. G. Richards, M. O. Chandler, C. R. Chappell, P. Anderson, W. B. Hanson, and L. H. Brace, Plasmasphere-Ionosphere coupling II: Ion composition measurements at plasmaspheric and ionospheric altitudes and comparison with modeling results, *J. Geophys. Res., 95*, 7949-7960, 1990.

Lee, J. S., J. P. Doering, T. A. Potemra, and L. H. Brace, Measurements of the ambient photoelectron spectrum from Atmosphere Explorer, I, AE-E measurements below 300 km during solar minimum conditions, *Planet. Space Sci., 28*, 947, 1980.

Miller, K. L., D. G. Torr, and P. G. Richards, Meridional winds in the thermosphere derived from measurement of *F2*–layer height, *J. Geophys. Res.*, *91,* 4531, 1986.

Nagy, A. F., and P. M. Banks, Photoelectron fluxes in the ionosphere, *J. Geophys. Res.*, *75*, 6260, 1970.

Newberry, I. T., R. H. Comfort, and P. G. Richards, Thermal $He^+$ in the plasmasphere: comparison of observations with numerical calculations, *J. Geophys. Res.*, *94*, 15265-15276, 1989.

Richards, P. G., and D. G. Torr, An investigation of the consistency of the ionospheric measurements of the photoelectron flux and solar EUV flux, *J. Geophys. Res.*, *89*, 5625, 1984.

Richards, P. G., and D. G. Torr, A method for extracting meridional winds from ionosonde measurements by using ionospheric models, in Proceedings of the *Thermosphere Dynamics Workshop* II held at Beltsville/Calverton, Maryland, October 3–5, 1984, edited by H. G. Mayr and N. J. Miller, NASA conference publication 2398, 369, 1986.

Richards, P. G., and D. G. Torr, A factor of 2 reduction in the theoretical *F2* peak electron density due to enhanced vibrational excitation of $N_2$ in summer at solar maximum, *J. Geophys. Res.*, *91*, 11331, 1986.

Richards, P. G., and D. G. Torr, Ratio of photoelectron to EUV ionization rates for aeronomic studies *J. Geophys. Res.*, *93*, 4060, 1988.

Schunk, R. W., and A. F. Nagy, Electron temperatures in the *F* region of the ionosphere: Theory and observations, *Rev. Geophys.*, *16*, 355, 1978.

St. Maurice, J.-P., and R. W. Schunk, Diffusion and heat flow equations for the mid-latitude topside ionosphere, *Planet. Space Sci.*, *25*, 907, 1976.

Torr, M. R., D. G. Torr, R. A. Ong, and H. E. Hinteregger, Ionization frequencies for major thermospheric constituents as a function of solar cycle 21, *Geophys. Res. Lett.*, *6*, 771, 1979.

Torr, M. R., D. G. Torr, P. G. Richards, and S. P. Yung, Mid low latitude model of thermospheric emissions 1. $O^+(^2P)$ 7320 Å and $N_2(^2P)$ 3371 Å, *J. Geophys. Res.*, in press, 1990.

# A STUDY OF THE STATISTICAL BEHAVIOR OF ION TEMPERATURES FROM DE 1 / RIMS

P. D. Craven+, R. H. Comfort#, D. L. Gallagher+, and R. West*

Abstract. Retarding Ion Mass Spectrometer (RIMS) observations made from the Dynamics Explorer 1 satellite are averaged over one minute intervals and used to derive over 20,000 measurements of hydrogen ion temperatures. Temperatures are sorted by L-shell, local time, magnetic (Kp) and solar activity (F10.7). Average temperatures increase with L-shell. On lower L-shells, L < 3, the average temperatures increase with F10.7 but are generally less than 0.8 eV. No dependence on geomagnetic activity is found for L-shells less than 2. On higher L-shells, the average temperature increases with Kp, in some cases reaching a maximum of about 25 eV, but is usually less than 15 eV. In the lower L-shells, diurnal effects are small, less than 0.15 eV, for low F10.7 and less than 0.1 eV for high F10.7. On the higher L-shells the temperature appears to be independent of day-night conditions and of F10.7 flux. The behavior of the average temperature is indicative of both an ionospheric and an outer plasmaspheric heat source for plasmaspheric ions.

## Introduction

This investigation examines the statistical behavior of the temperature of magnetospheric hydrogen ions as a function of L-shell, local time, and solar and geomagnetic activity. The purpose of this study is to determine the qualitative behavior of the temperature with these physical parameters as a precursor to developing an empirical model of $H^+$ temperature in the plasmasphere. Although our main interest is in the plasmaspheric ions, we examine the behavior of the ion temperature on L-shells from 1 to 6. An empirical model of the ion temperatures can be used to study the average effects of different heating mechanisms through its representation of the average distribution of energy throughout the magnetosphere. Such a model can be used as a check on heating studies such as that of Chandler et al. [1988] or as a base with which to compare the results from computer models such as the Field Line Interhemispheric Plasma (FLIP) model of Richards and Torr [1985]. Ion temperature is a reflection of the efficiencies of heat sources and sinks (eg. solar flux, photoelectrons, ring current, particle precipitation, density of neutrals, etc.) that are active in the magnetosphere and ionosphere, so that, in effect, a study of the statistical behavior of ion temperature is also a study of the heat sources and sinks and of heat transport.

+ NASA/Marshall Space Flight Center, Space Science Laboratory, Huntsville, AL, 35812

# Center for Space and Aeronomy Research, University of Alabama in Huntsville, Huntsville, AL, 35809

* Boeing Computer Support Services, Inc., Boeing Corporation, Huntsville, AL, 35814

Modeling Magnetospheric Plasma Processes
Geophysical Monograph 62


## Analysis Procedure

The data base is composed of over 20,000 hydrogen ion temperatures derived from measurements made by the Retarding Ion Mass Spectrometer (RIMS) on the Dynamics Explorer 1 (DE 1) satellite. Each temperature is derived from a one minute average of the retarding potential analyzer (RPA) data from RIMS. The RIMS was described in detail by Chappell et al. [1981]. Temperatures are calculated using the thin sheath model of Comfort et al. [1982], following a procedure described by Comfort et al. [1985]. In this case, however, the procedure is applied to RPA data from the RIMS -Z end head which views perpendicular to the orbit plane. Since DE 1 is in a polar orbit, this head also views perpendicular to the Earth's magnetic field. No temperatures derived from the radial head of RIMS are included in this study.

In applying this procedure to the end head, we have assumed that the expression used by Comfort et al. [1985] can be applied by setting the flow velocity into the detector to zero, ie. by setting the Mach number in their equation 1 to zero. A rigorous study involving numerical integration of the flux integral [Whipple et al. , 1974, equation (19)] suggests that this procedure will result in differences between temperatures derived from the end and radial heads of less than 10%. Comparisons of temperatures derived from the radial head with those from the end head for times near the beginning of the mission show that the -Z end head temperatures follow those from the radial head but at a lower value. This behavior is particularly apparent at high Mach numbers (low altitudes). On evening passes, differences between the temperatures derived from the -Z end head and those derived from the radial head are a maximum of about 50 per cent; differences are less on the morning passes in this early part of the mission.

Only temperatures derived for $H^+$ are used since $H^+$ is the dominant ion, typically comprising 70% to 85% of the plasmaspheric number density. The temperature of the next most common ion, $He^+$, is generally within 10% of the $H^+$ temperature [Comfort et al., 1988]. We used 1000 Km to represent the top of the ionosphere and set this as a lower altitude limit. No temperatures derived from measurements made below 1000 Km are included in the study.

Temperatures are taken from measurements made in 1981, 1982, and 1983. As described by Chappell et al. [1981], RIMS has the capability for biasing the aperture plate relative to spacecraft ground. However, when a potential bias is applied to the aperture plate, the interpretation of the data becomes more complicated. Because of this, no temperatures found during a period when the aperture was being biased are used in this study.

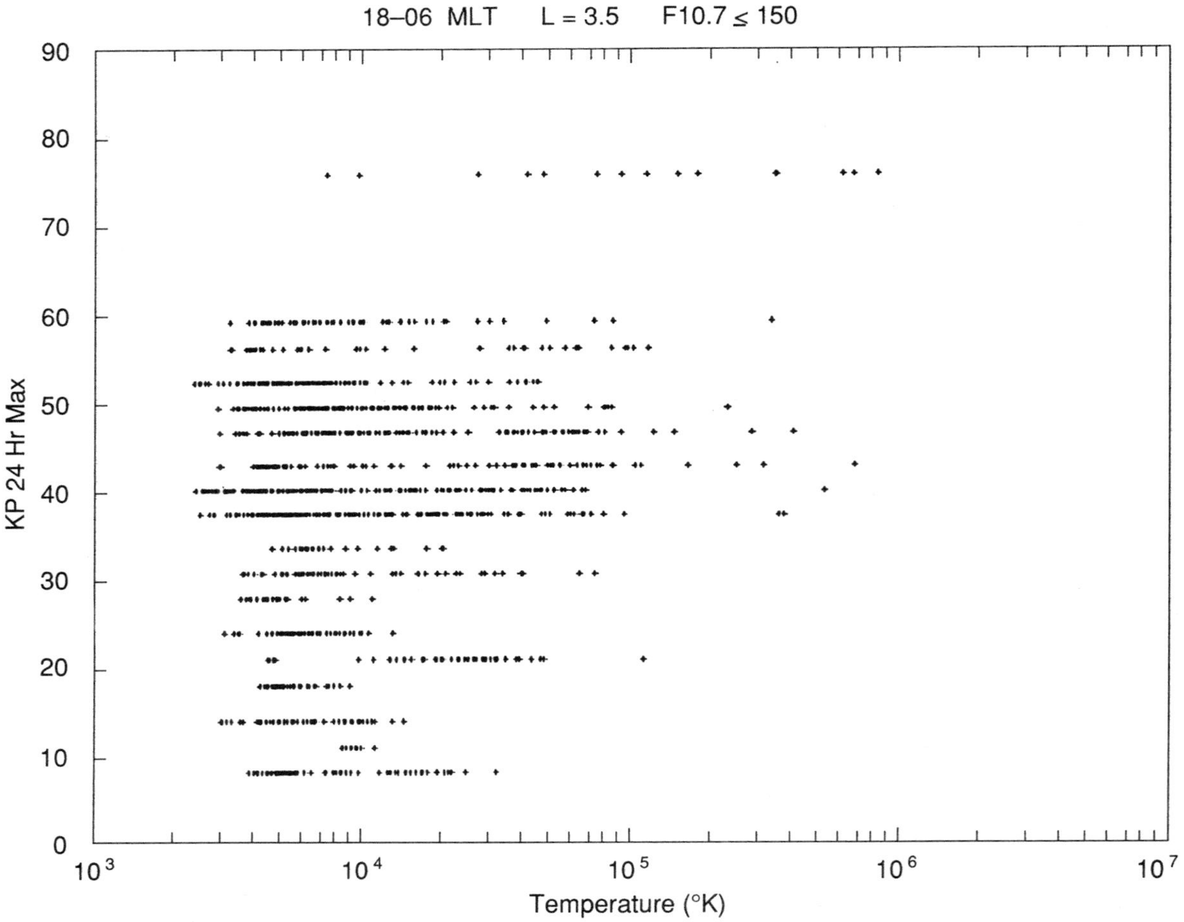


Fig. 1. Typical scatter plot of ion temperature as a function of Kp. This plot is for the L = 3.5 range, nightside magnetic local time and low solar activity (F10.7 ≤ 150).

Each derived temperature value is associated with a local time, L-shell, solar activity level (F10.7) and geomagnetic activity level (Kp). These are obvious parameters with which to correlate ion temperature; F10.7 and local time for solar effects, Kp and L-shell for ring current, particle precipitation and heat conduction effects. Local times are divided into two 12 hour bins, one for dayside (from 0600 through noon to 1800 hours) and one for nightside (from 1800 through midnight to 0600). Solar activity is divided into two ranges using a 30 day average of the solar F10.7 flux; low activity includes F10.7 ranges to 150 ($10^{-22}$ w/m$^2$/hz), and high includes ranges above 150. Geomagnetic activity is measured using the maximum 3 hour Kp level in the 24 hours preceding the time of the temperature measurement. This index for geomagnetic activity was used in recognition of the fact that ion temperature responds to changing magnetic conditions, but not necessarily to that activity represented by the Kp coincident with the time of the measurement. The L-shells are grouped into bins of one unit, ie. an L-shell designation of 2.5 includes all L-shells from 2.0 up to but not including 3.0.

In grouping together all temperatures occurring on an L-shell, we have assumed that the temperature is constant along that L-shell. Previous unpublished RIMS results have shown that the field aligned gradient is an order of magnitude less than the transverse gradient. Data presented by Comfort et al. [1985] suggest that the difference in temperature between high and low altitude along the same field line is less than 0.5 eV. Therefore, any scatter in the temperatures resulting from this assumption is small in comparison to other effects. For this reason, the assumption of a constant temperature along a field line should give reasonably good first order results.

Diurnal effects are probably not independent of solar activity effects and because of this, the day night groupings are kept distinct in this study. We assume that the solar effects and the geomagnetic effects are independent of one another.

## Typical Data

Figure 1 shows a typical scatter plot of ion temperature versus Kp for nightside, L=3.5, and low solar activity. Figure 2 shows the ion temperatures as a function of the 30 day average F10.7 for nightside and L=1.5. No distinction is made in Figure 2 for the

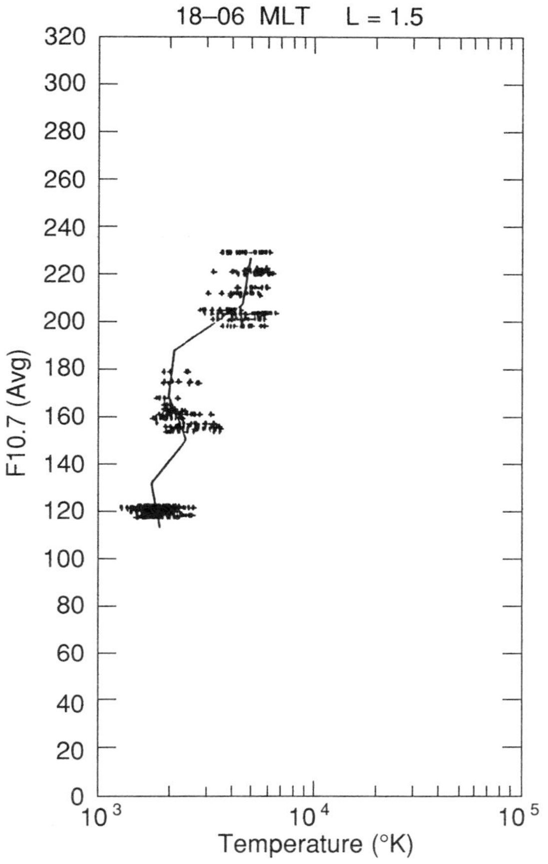


Fig. 2. Scatter plot of ion temperature as a function of F10.7. All Kp are included in the data which is for the L = 1.5 range and nightside magnetic local times.

geomagnetic activity, all the encountered levels are included in the plot. The wide range of temperatures (about 3 orders of magnitude) seen in Figure 1 is typical of the temperatures encountered at almost all levels of either solar or geomagnetic activity on L-shells of 3.5 and above. Because of this large scatter in the measured temperatures, the temperature was converted to electron volts (eV). Figure 3 shows for L-shells of 3.5, 4.5, and 5.5, the occurrence frequencies of the temperatures sorted into 1 eV bins for the two F10.7 groups and daytime. The total number of samples on which the histogram is based is also given in the figure. There is no grouping on Kp for these histograms. Although not shown, on the lower L-shells, L < 3.5, temperatures of 1eV or less have occurrence frequencies of greater than 90% for both F10.7 groups. On the higher L-shells, the occurrence frequency for the higher temperatures increases (see Figure 3) while that for temperatures less than 1 eV significantly decreases. The L=3.5 range appears to be a transition region in that the low temperature still dominates but higher temperatures (out to 50 eV) start to appear. Occurrence frequencies for the temperatures above 10 eV are slightly higher for F10.7 > 150 than they are for F10.7 < 150.

Figure 4 shows nightside temperatures in the same format used in Figure 3. The nightside temperature behavior is similar to the dayside in that for L < 3.5 the temperatures are 1 eV or less for both F10.7 groups 90% of the time. There is no obvious temperature variation with F10.7, at least to the 1 eV resolution of this data set. Nightside temperatures for L > 3.5 are lower than those during the daytime as evidenced by the fact that there is a larger fraction of the data on a given field line with temperatures of 1 eV or less (compare Figures 3 and 4). For both nightside F10.7 groups, there is a significant peak occurrence frequency for L = 5.5 ($5 \leq L < 6$) in the 2 eV bin that does not appear in the dayside temperatures.

In order to investigate the behavior of the low temperatures, the temperatures of 10 eV or less were sorted into 0.1 eV bins. The temperature histograms (not shown at this resolution) for the nightside L = 5.5 range exhibit a peak in temperature between about 0.7 and 2.0 eV while the histograms of the dayside temperatures tend to be less peaked and more evenly distributed over the full range from about 0.3 to 10.0 eV. In addition, in this L range, the nightside temperatures of 0.1 to 0.3 eV occur much less often than on the dayside. With this resolution (0.1 eV), there is a very sharp drop to near zero in the occurrence frequency of temperatures less than the modal temperature (the temperature with the greatest frequency of occurrence), the drop is sharper for F10.7 < 150 than for F10.7 > 150.

Changes in temperature due to solar activity are more likely to be seen on low L-shells where geomagnetic effects do not penetrate. Using the 1 eV resolution data, it appears that geomagnetic effects are least on L-shells less than 3 and do not penetrate at all, in this data, to L-shells less than 2. The temperatures on these L-shells are less than 5 eV with over 90% of the samples having a temperature of less than 1 eV. Using this observation, only temperatures of 2 eV or less are used to study the solar effects. Figure 5 shows the temperatures averaged over the 0-2 eV range, for the two F10.7 groups for both dayside and nightside. No sorting on Kp has been done for this data, all Kp are included. On L-shells greater than 3, there is no consistent difference in the data between temperatures for high and low F10.7 or for day or nightside conditions, and the temperature tends to increase with L-shell independent of solar or diurnal effects. For L-shells less than three, temperatures for low F10.7 conditions are less than those for high F10.7 conditions by about 0.2 eV on the nightside and about 0.3 eV on the dayside. Nightside temperatures for low F10.7 levels are less than those for the dayside by about 0.1 eV. However, diurnal differences for high F10.7 are much smaller, about 0.05 eV.

Since the ion temperature changes rapidly at dawn and at dusk, the local time of the measurements is important when considering diurnal variations in the temperature. Figure 6 shows a histogram of the temperatures as a function of magnetic local time. The local time is divided into 24 one hour bins. This figure shows that the measurements are concentrated around 0800-1100 and 1900-2200 magnetic local time. Peak dayside temperatures should occur in the 0800-1100 local time sector. Temperatures in the 1900-2200 local time sector should be near the nightside minimums so that the diurnal variation presented above should be representative of the change to be expected.

The dependence of the temperatures on geomagnetic activity (Kp) is shown in Figure 7 in which the average temperature on each L-shell range is plotted as a function of Kp for both dayside and nightside and high and low F10.7. Recall that the Kp value used in this study is the largest value in the 24 hours preceding the observation. The data set used for this figure includes the full range (0-50 eV) and has a 1 eV resolution. The increase of the average tempera-

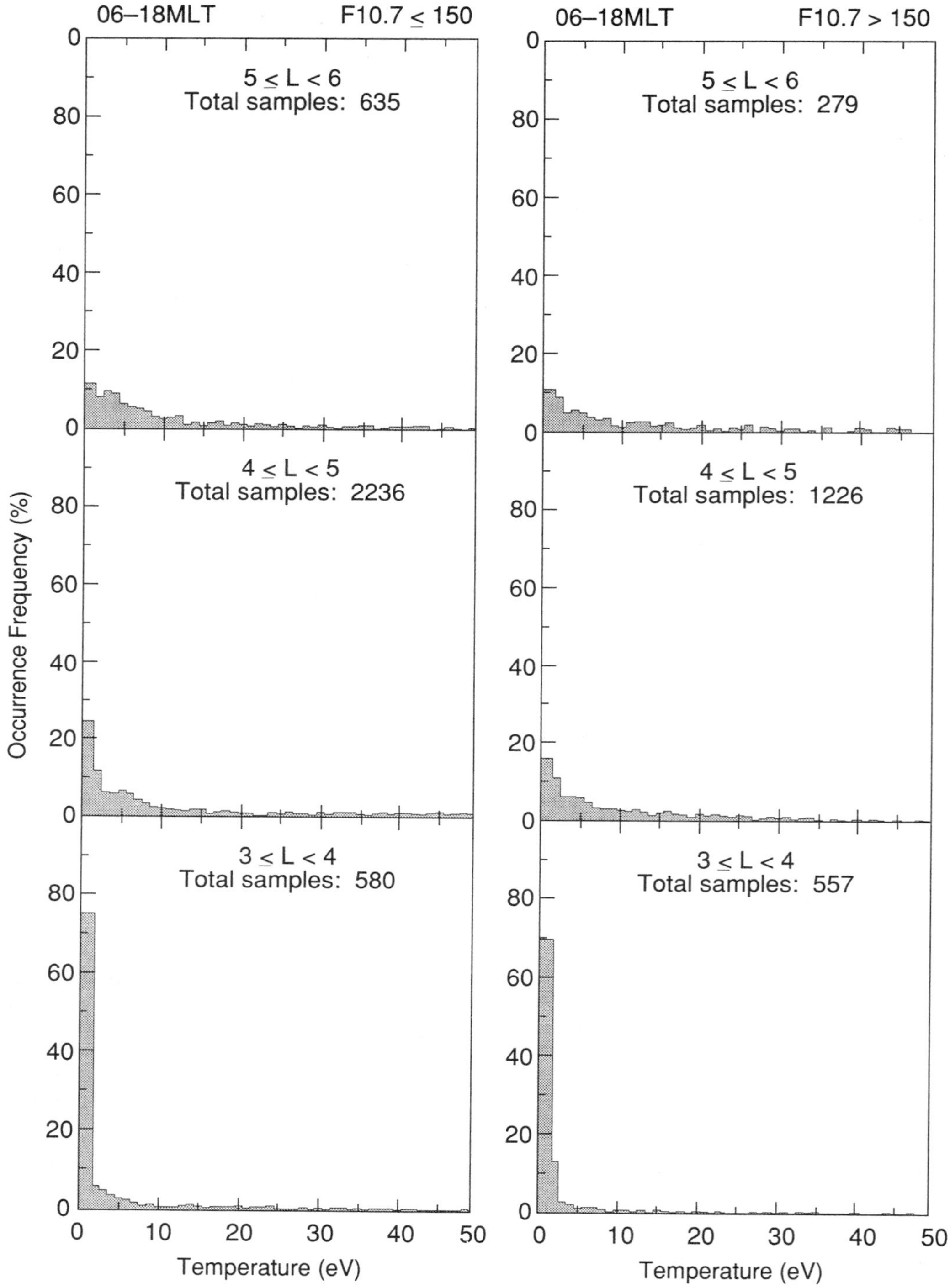


Fig. 3. Occurrence frequencies for dayside ion temperatures. The temperature resolution is 1.0 eV and the full 0 - 50 eV range of RIMS is shown. All encountered Kp levels are included in the plot. Low F10.7 on the left, high on the right. No values are shown for L < 3.5 because 90% of the temperatures are less than 1 eV.

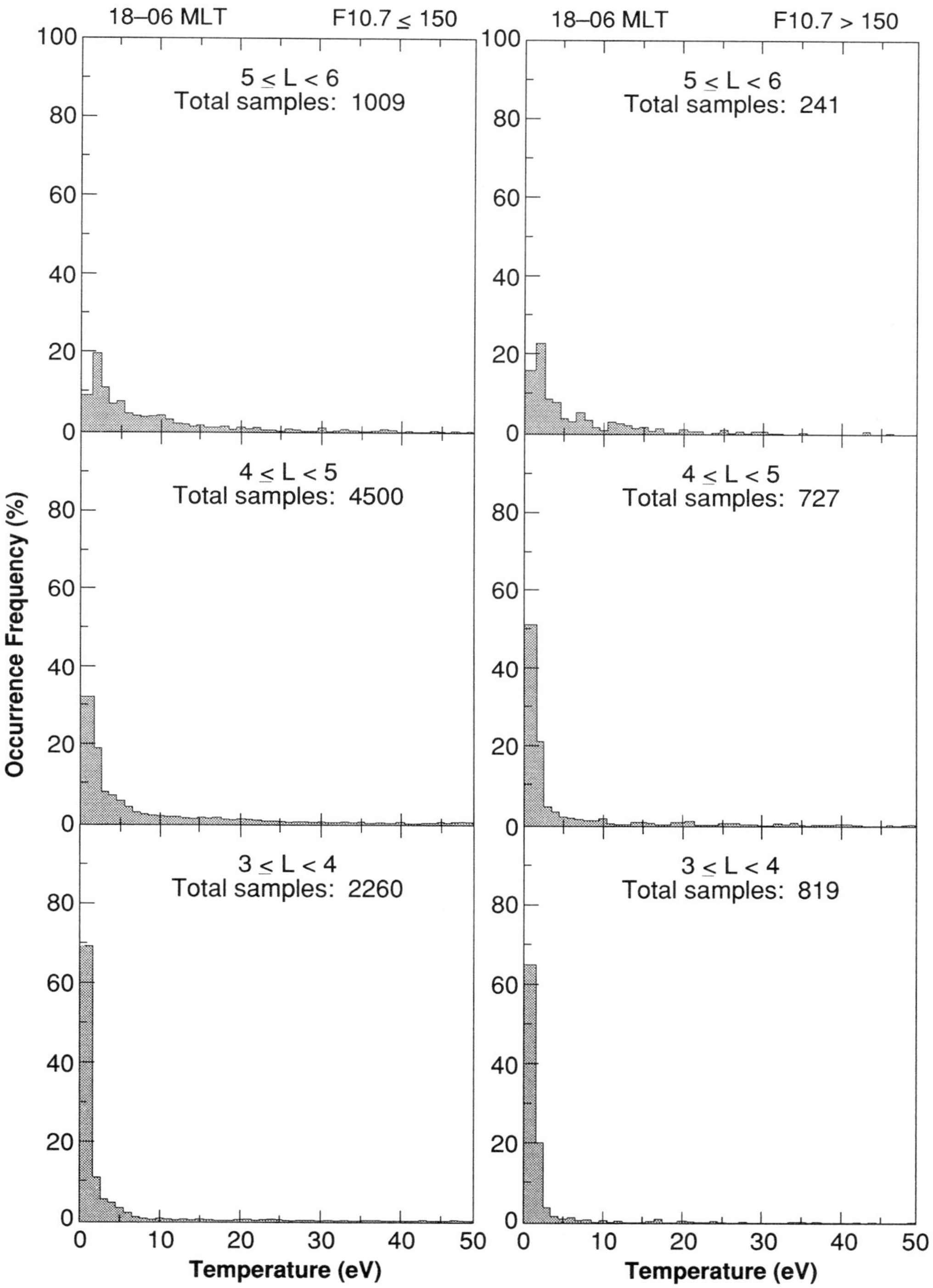


Fig. 4. Same as for Fig. 3 but for nightside local times.

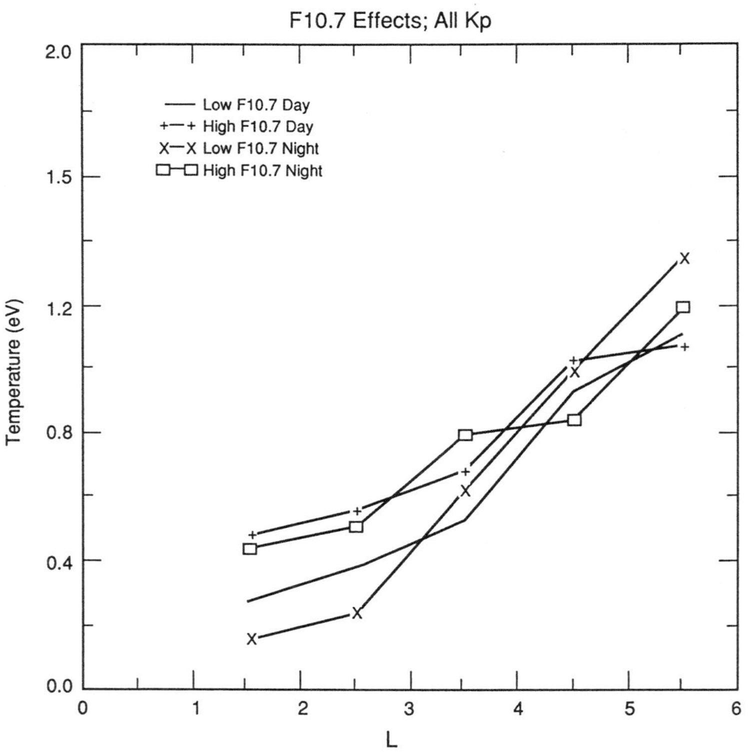


Fig. 5. The average temperature as a function of L-shell for both nightside and dayside, high and low solar activity. All Kp are included in the plot. The temperature resolution of the data set is 0.1 eV and the average is taken from 0 - 2 eV.

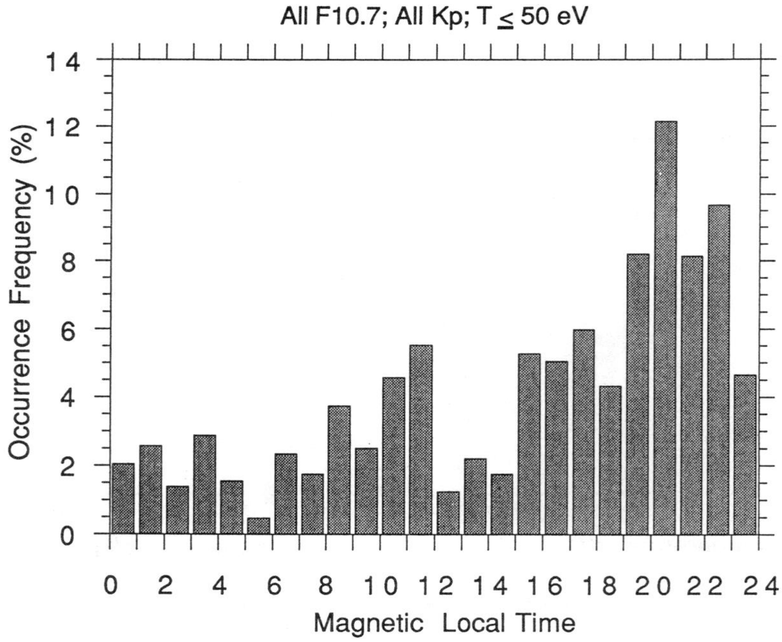


Fig. 6. A histogram of the temperatures as a function of magnetic local time. All temperatures from 0 to 50 eV are included.

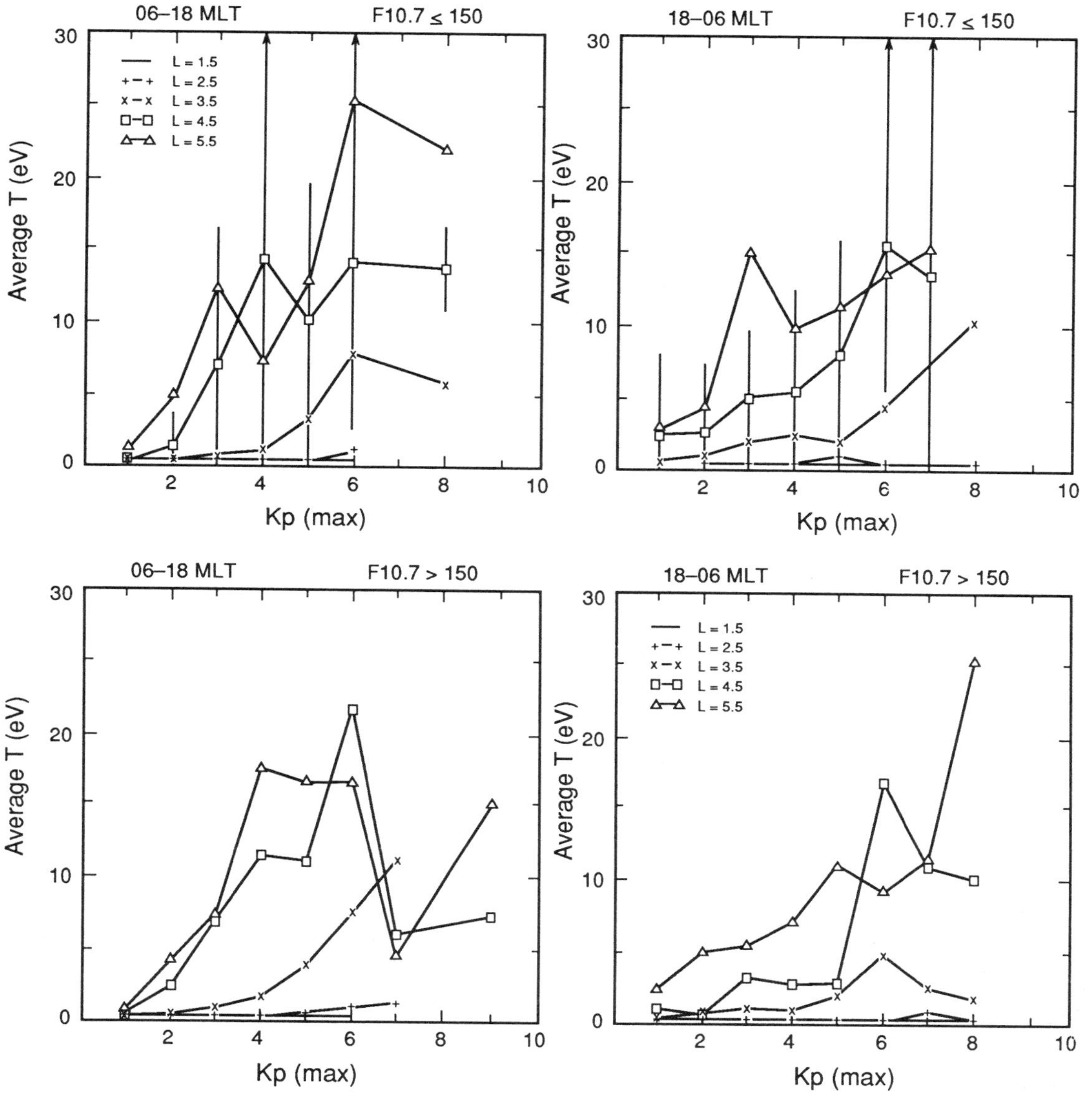


Fig. 7. Temperature as a function of Kp for each L-shell. Dayside temperatures are on the left, nightside on the right. Low F10.7 in the top panels and high F10.7 in the bottom panels. Temperature resolution is 1 eV. The error bars are drawn on the L = 4.5 curve in the low F10.7 panels.

ture with L-shell is apparent in this figure. There is no apparent temperature dependence on Kp for L ranges less than 3.5. For Kp less than about 6, the magnitude of the temperature increase with Kp gets larger with increasing L-shell. There is no apparent correlation in this data between temperature and Kp for L > 3.5 and Kp greater than about 6. The error bars shown in Figure 6 are for L = 4.5 and represent one standard deviation from the mean, except that they were not extended below 0 eV or above the upper limit on the plot. The variability in this data typically increases with L-shell and Kp as reflected in the error bars shown in the figure.

It should be noted that the observations do not include the full range of Kp values, 0 to 9, at each L-shell. Figure 8 shows the number of samples at each encountered Kp level for all the data. No samples are available for Kp=0, and there is a sharp drop in the number of observations for Kp greater than 6. Average temperatures are well behaved for the lower Kp values (Kp < 3) in that the scatter about the mean is small (<1 eV). For the higher values of Kp and L, the scatter in the temperatures is large so that the average temperatures are not as meaningful as those for lower Kp and L.

## Discussion

Several statistical trends can be seen in this data relating the average temperature to F10.7, L-shell, Kp, and diurnal effects. In

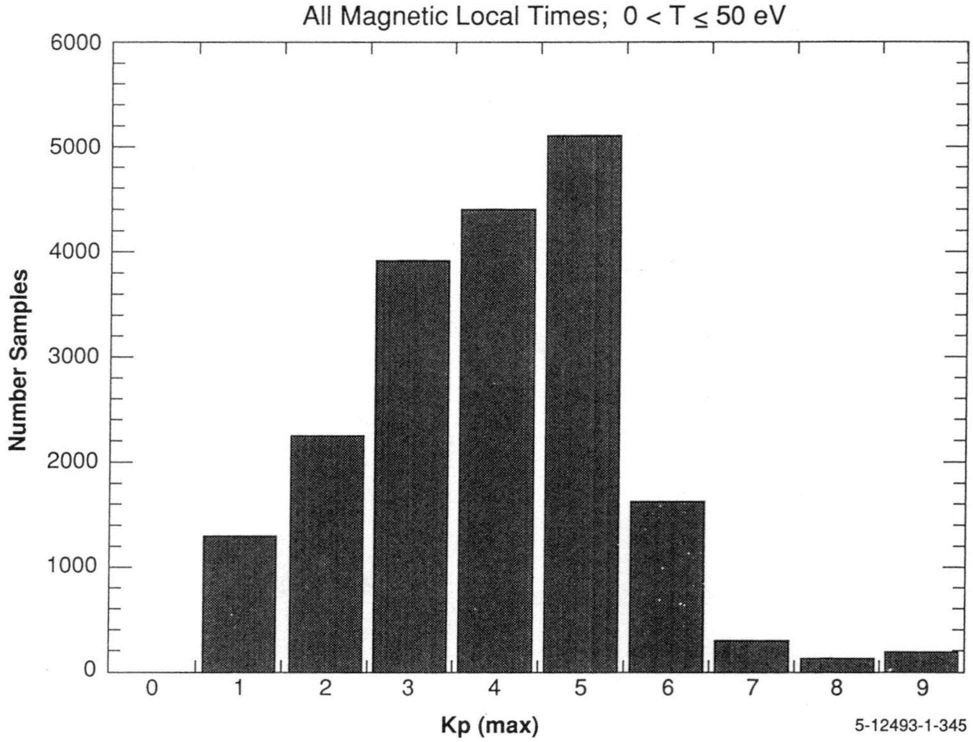


Fig. 8. The number of temperature samples at each Kp level. All F10.7 and local times are included in the data.

general, there is considerable variation in the average temperatures measured on all L-shells (see Figure 7). These variations are greatest in the regions in which heat sources connected with geomagnetic activity are dominant. The variations are probably due to the fact that no provisions have been made for taking into account the length of time the heating had been taking place. Also, heating mechanisms which relate to Kp but do not vary directly with it ( eg. wave-particle interactions, stochastic heating, etc.) are not treated in a quantitative manner. However, the statistical behavior of the average temperature should indicate general qualitative effects that will help in developing an empirical model.

We have used two temperature ranges and resolutions to study the ion temperature behavior. The full 0-50 eV range , 1 eV resolution has been used to study the relatively large effects associated with geomagnetic activity. The smaller 0-10 eV range, 0.1 eV resolution, has been used to study the smaller effects due to solar activity and diurnal changes.

Two characteristics of the observed temperature behavior are independent of the temperature range and resolution used. First, the average temperature increases with L value. Second, for $L > 2.5$, the average temperature on a given L-shell increases with Kp for Kp above some threshold value. For L's of 2.5 and below, the temperature appears to be independent of Kp (Note, however, there is no data for $Kp > 6$). For $L > 2.5$, the level of Kp at which the temperature becomes dependent on Kp decreases with increasing L-shell. This effect is consistent with a plasmapause location for the heating region that moves inward with increasing geomagnetic activity. In this data set, there appears to be a limit to the inward movement in the L-shell range of 2.5 ($2 \leq L < 3$).

Neither solar activity nor diurnal effects are apparent in the average temperature calculated using the full 50 eV range and the 1.0 eV resolution. Diurnal effects are noticeable in the histograms, but not in the behavior of the average temperature taken over the full 50 eV range. Solar activity effects are not apparent in either the histograms or in the behavior of the average temperature. The

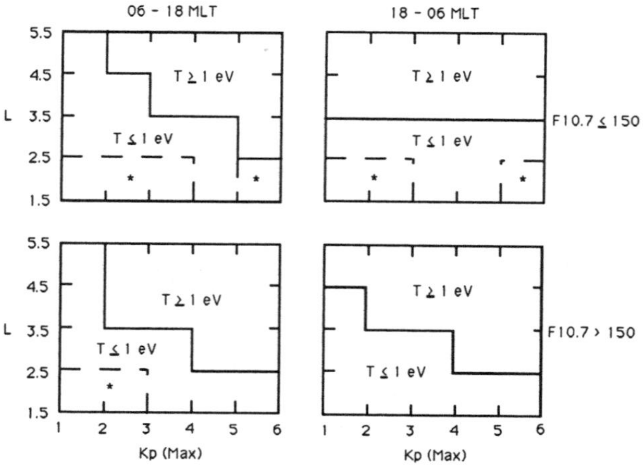


Fig. 9. The boundary between regions in which the temperature is 1 eV or less and regions in which temperatures greater than 1 eV are encountered.

effects of solar activity and day/night conditions are most apparent in temperatures less than 2 eV. For L < 4, the majority of the measured temperatures are less than 1 eV for all conditions (see Figure 3 and 4).

Temperatures of 1 eV or less are measured everywhere. Temperatures greater than 1 eV are encountered on the higher L-shells; but the range of L-shells on which the temperatures are above 1 eV, varies with Kp and F10.7. The relation of Kp and L to temperatures greater than 1 eV can be seen by drawing the boundary between regions in L- Kp space in which the measured temperatures are 1 eV or less and those in which the temperatures are 1 eV or more. This boundary is shown in Figure 9 for day and night magnetic local times for both low and high solar activity. For dayside magnetic local times, the boundary moves to higher L-shells for lower Kp. For both F10.7 ranges, there is a region in which the dayside temperature is less than 1 eV for all conditions; this is on all L-shells, but for the lowest Kp. For nightside magnetic local times, the L-shell of 2.5 appears to be the boundary for all Kp values when F10.7 < 150. For high solar activity and nightside local times, the higher temperatures move into the region of low Kp and high L-shells in which the temperature is 1 eV or less on the dayside. The regions of L-Kp space marked with an asterisk in the figure are ones in which no temperatures were measured.

The distribution of temperatures in altitude is shown in Figure 10. In this figure, the occurrence frequency for the temperatures less than 1 eV is shown as a function of altitude in the top panel, the occurrence frequency for temperatures between 1 and 50 eV is shown in the bottom panel. In both cases, the occurrence frequency is calculated using the number of samples in the total data set. One can see from this figure that the 1 eV occurrence frequencies are within a factor of 2 everywhere except at apogee where the satellite spends relatively more time. The higher temperatures have greater occurrence frequencies at the higher altitudes, reflecting the increase of temperature with L-shell. There is also an increase in the altitude sampling toward the higher altitudes near apogee reflecting the fact that the satellite spends relatively more time at apogee than at other altitudes and possibly also an increase in the relative amount of time RIMS was operated at the higher altitudes in the later years.

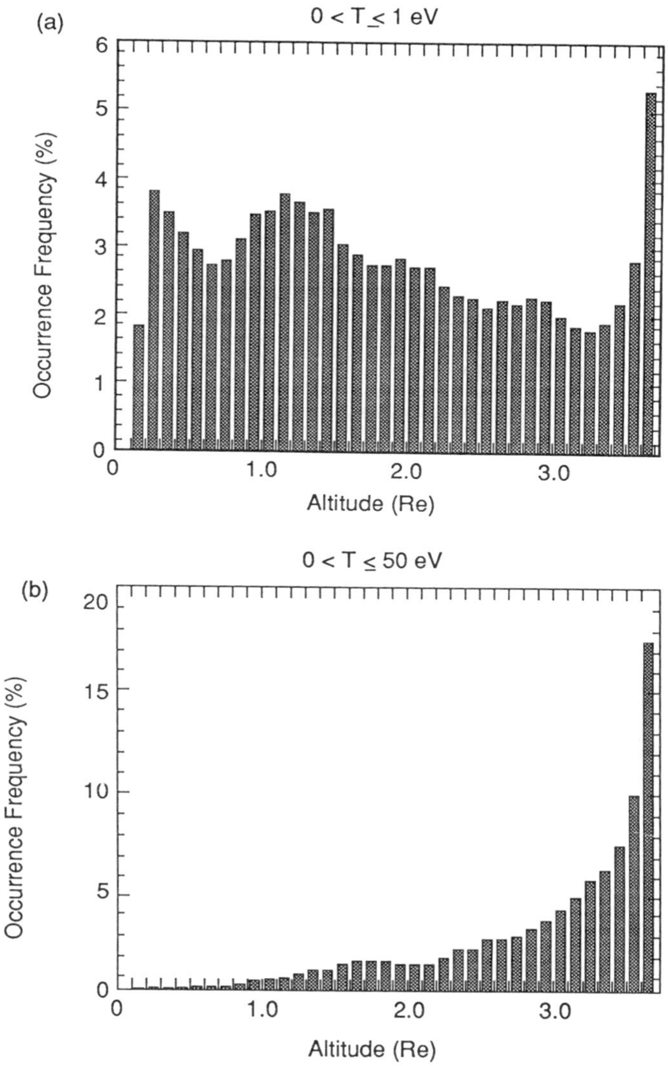


Fig. 10. A histogram of the occurrence frequencies of temperatures as a function of altitude. The altitude is in units of Earth radii. Occurrence frequencies are shown for T ≤ 1eV in the top panel, and for 1 < T ≤ 50 eV in the bottom panel.

## Summary

The behavior of the RIMS temperature data indicates that solar effects do not heat the $H^+$ to temperatures greater than about 1 eV; in fact it appears that in the plasmasphere (L < 4.5), the solar effects are limited to around 0.5 eV. Diurnal changes in the average temperature are smaller than the changes due to solar effects. In the plasmasphere where the foot of the field lines rotate in and out of the sun, heat gained on the dayside is lost on the nightside. From the fact that the difference between day and night temperatures for high F10.7 is less than for low F10.7, it appears that during high solar activity, the conditions and processes change so that thermal balance is achieved on the nightside at a higher temperature than for low solar activity. Why the dayside and nightside temperatures are so close together for high solar activity needs further investigation. One possibility is that there is a nightside source of heat that becomes more prominent during high solar activity.

Temperatures greater than 1 eV appear to be strongly related to geomagnetic influences as represented by Kp. The occurrence of ion temperatures above 1 eV depends on both the L-shell and Kp. The higher L-shells show a dependence on Kp above some threshold value; the threshold Kp going to smaller values as L goes to larger values. This behavior is obviously related to the level of geomagnetic activity, with the effects of the higher levels of activity able to penetrate further into the plasmasphere. In this data set, there is a limit to the depth of penetration (L=2.5), but this limit must be taken with caution because there is little data for levels above Kp=6.

The spread of the temperatures on L-shells greater than 3 suggests a dynamic situation involving heating, cooling, mixing, and transport processes, possible all operating simultaneously, a situation that cannot be adequately treated in a statistical study of ion temperatures. The situation for both the dayside and nightside on L-shells less than 3 is less dynamic, as indicated by the smaller spread in the temperature for all F10.7 levels. On L-shells between 1 and 2 (L=1.5), there is an apparent transition in the spread of the dayside temperatures. The transition occurs at a F10.7 level of about 150. Above this level, the temperatures are spread over the range from 1500 °K to 10,000 °K. Below this level, the temperature has a small spread (approximately 1600 °K) about 3500 °K on the dayside. The spread and shift of the nightside temperatures can be seen in Figure 2. The temperatures for the lower F10.7 levels is consistent with photoelectrons as a heat source. Heat is transferred in collisions between ions and thermal electrons, which are in turn heated by ionospheric photoelectrons. The fact that effects of solar activity are seen only at energies near an eV is probably indicative of the limit of the heating that can be attributed to this source.

The spread in temperatures on the higher L-shells and the fact that the temperature is apparently independent of F10.7, would seem to indicate that another more efficient heat source for the ions is operating in this region. The Kp dependence of the ion temperatures on the higher L-shells is a strong indicator that the heat source is related to the ring current or activity near the ring current, ie. wave particle interactions.

The lower temperatures on the nightside for the lower L-shells, probably reflect the loss of the ionospheric photoelectrons from the sunlit conjugate ionosphere. If the smaller diurnal change obtained for high solar activity is actually representative of the diurnal variation, it would indicate a change in the process of heat exchange on the nightside between high and low solar activity. One possibility is the presence of a nightside heat source that is stronger during high solar activity. However, this possibility and others need further study to determine the cause of the difference.

In terms of modeling the ion temperature, three regions must be considered. One is the inner plasmasphere (L<3.5) in which solar and diurnal effects dominate the heating. The second is a transition region (L=3.5) in which both solar and geomagnetic source effects operate and in which temperatures can reach the ten's of eV range. The third is the outer plasmasphere (L>3.5) in which the dominate heat source is related to geomagnetic effects and in which a wide range of temperatures can be encountered. In both this region and the transition region, there is a need for a more detailed study using other parameters (eg. Dst, AE, source location, time history, etc.) in order to more accurately quantify the behavior of the temperatures. The temperature behavior shown in this study gives a first order behavior that any global model must replicate.

Acknowledgments. The authors wish to thank C. R. Chappell, the RIMS Principal Investigator, for making the data set available to us. We are also grateful to the referees for their suggestions and comments. In addition, we thank the editors, G. Wilson and M. Chandler for their patience during the preparation of the manuscript for this report. RHC was supported in part for this work under NASA grant NAG8-134 with the University of Alabama in Huntsville.

## References

Chandler, M. O., J. U. Kozyra, J. L. Horwitz, R. H. Comfort, and L. H. Brace, Modeling of the Thermal Plasma in the Outer Plasmasphere - A Magnetospheric Heat Source, in Modeling Magnetospheric Plasma (ed.'s T. M. Moore and J. H. Waite, Jr.), AGU Geophysical Monograph Series, 44, 101, 1988.

Chappell, C. R., S. A. Fields, C. R. Baugher, J. H. Hoffmann, W. B. Hanson, W. W. Wright, H. D. Hammack, G. R. Carignan, and A. F. Nagy, The Retarding Ion Mass Spectrometer on Dynamics Explorer-A, Space Sci. Instrum., 5, 477, 1981.

Comfort, R. H., C. R. Baugher, and C. R. Chappell, Use of the Thin Sheath Approximation for Obtaining Ion Temperatures from the ISEE 1 Limited Aperture RPA, J. Geophys. Res., 87, 5109, 1982.

Comfort, R. H., J. H. Waite, Jr., and C. R. Chappell, Thermal Ion Temperatures from the Retarding Ion Mass Spectrometer on DE 1, J. Geophys. Res., 90, 3475, 1985.

Comfort, R. H., I. T. Newberry, and C. R. Chappell, Preliminary Statistical Survey of Plasmaspheric Ion Properties from Observations by DE 1 / RIMS, in Modeling Magnetospheric Plasma (ed.'s T. M. Moore and J. H. Waite, Jr.), AGU Geophysical Monograph Series, 44, 107, 1988.

Richards, P. G. and D. G. Torr, Seasonal, Diurnal, and Solar Cyclical Variations of the Limiting $H^+$ Flux in the Earth's Topside Ionosphere, J. Geophys. Res., 90, 5261, 1985.

Whipple, E. C., J. M. Warnock, and R. H. Winkler, Effect of Satellite Potential on Direct Ion Density Measurements through the Plasmapause, J. Geophys. Res., 79, 179, 1974.